Cognitive Science Perspectives on Verb Representation and Processing

Roberto G. de Almeida • Christina Manouilidou
Editors

Cognitive Science Perspectives on Verb Representation and Processing

Springer

Editors
Roberto G. de Almeida
Department of Psychology
Concordia University
Montreal
Québec
Canada

Christina Manouilidou
Department of Philology
University of Patras
Rio Patras
Greece

ISBN 978-3-319-10111-8 ISBN 978-3-319-10112-5 (eBook)
DOI 10.1007/978-3-319-10112-5
Springer Cham Heidelberg New York Dordrecht London

Library of Congress Control Number: 2014952280

Printed on acid-free paper

Springer is part of Springer Science+Business Media (www.springer.com)

Preface

The present volume is a snapshot of current research on verb representation and processing, embodying the rich interdisciplinary perspective of cognitive science. It brings together different linguistic perspectives on verbs as well as contributions from psycholinguistics, neurolinguistics, and cognitive neuroscience, showcasing the diversity of theoretical approaches and types of empirical evidence bearing on the nature of verb structure and meaning. Together with linguistic theory and its empirical (cross-linguistic, distributional) evidence, we present data stemming from behavioral experiments, cases of aphasia, brain imaging, and language acquisition. We think empirical evidence gathered from different methods is crucial for supporting (or refuting) linguistic postulates and, more broadly, for understanding the architecture of language and its interfaces with other cognitive domains.

The volume originates from the conference *Verb Concepts: Cognitive Science Perspectives on Verb Representation and Processing* we organized at Concordia University in Montreal, in 2008. We take this opportunity to thank conference participants for their presentations and contributions to discussions, and also the Psycholinguistics and Cognition Lab crew, at Concordia, for their help before, during and after the conference. Although the inspiration for the volume was that conference, the current selection of papers does not constitute its proceedings. Guest speakers and other presenters were invited to submit chapters related to their presentations but not limited to them, and to provide a wider scope of their research, keeping an interdisciplinary readership in mind. We are certainly very grateful to the authors for their contributions, their rewritings, and specially their patience with the many delays in the making of the volume. We want to believe that the time elapsed between initial submission and final writing was important for strengthening hypotheses, evaluating data, and bringing new insights to the fore.

Chapters were selected by the editors and anonymously reviewed by numerous colleagues representing various fields (from theoretical linguistics to cognitive neuroscience). We are grateful, in particular, to the following for their work reviewing the chapters: Catherine Anderson (McMaster University), Alan Bale (Concordia University), Evelyn Ferstl (University of Freiburg), Alan Garnham (University of Sussex), Roberta Golinkoff (University of Delaware), Dana Isac (Concordia University), Nina Kazanina (University of Bristol), Katalin Kiss (Hungarian Academy of Sciences),

Marcus Vinicius Lunguinho (Universidade de Brasília), Asifa Majid (Max Planck Institute for Psycholinguistics, Nijmegen), Martha McGinnis (University of Victoria), Andrew Nevins (University College London), Despoina Papadopoulou (University of Thessaloniki), Pirita Pyykkonen (Norwegian University of Science and Technology), Anna Roussou (University of Patras), Stavroula Stavrakaki (University of Thessaloniki), Teresa Cristina Wachowicz (Universidade Federal do Paraná), and Brendan Weekes (University of Hong Kong).

Work on this book would not have been possible without the support from the Social Sciences and Humanities Research Council of Canada (grant 646-2008-060), Concordia's Faculty of Arts and Science and Department of Psychology.

Finally, we thank Springer, our publisher, and in particular Welmoed Spahr and Hana Nagdimov for their editorial support (and extended deadlines). We are also thankful to Springer's anonymous reviewers for valuable comments and also for suggesting that the volume could well serve as a text for a course on verbs from the interdisciplinary perspective of cognitive science. We certainly agree with them and had this in mind during the preparation of the volume.

Montreal and Patras
July 2014

Roberto G. de Almeida and Christina Manouilidou

Contents

Contributors

Sudha Arunachalam Department of Speech, Language, and Hearing Sciences, Boston University, Boston, MA, USA

Roelien Bastiaanse Center for Language and Cognition Groningen (CLCG), University of Groningen, Groningen, The Netherlands

William Croft Department of Linguistics, University of New Mexico, Albuquerque, NM, USA

Roberto G. de Almeida Department of Psychology, Concordia University, Montreal, QC, Canada

Brendan S. Gillon Department of Linguistics, McGill University, Montreal, Quebec, Canada

J. Gonzalez-Castillo Section on Functional Imaging Methods, Laboratory of Brain and Cognition, National Institute of Mental Health, Bethesda, MD, USA

E. Matthew Husband St. Hugh's College, University of Oxford, Oxford, UK

David Kemmerer Department of Speech, Language, and Hearing Sciences, Purdue University, West Lafayette, IN, USA

Department of Psychological Sciences, Purdue University, West Lafayette, IN, USA

Jean-Pierre Koenig Department of Linguistics and Center for Cognitive Science, University at Buffalo, Buffalo, NY, USA

Vicky T. Lai Department of Psychology, University of South Carolina and Max Planck Institute for Psycholinguistics, Columbia, SC, USA

E. Malaia University of Texas at Arlington, Arlington, TX, USA

Christina Manouilidou Department of Philology, University of Patras, Rio Patras, Greece

Alexandra Marquis École d'orthophonie et d'audiologie, Faculté de médecine, Université de Montréal, Montréal, QC, Canada

Gail Mauner Department of Psychology, Center for Cognitive Science, University at Buffalo, The State University of New York, Buffalo, NY, USA

Bhuvana Narasimhan Department of Linguistics, University of Colorado Boulder, Boulder, CO, USA

Paul M. Pietroski Department of Linguistics, University of Maryland, College Park, MD, USA

Artem Platonov Donders Institute, Radboud University Nijmegen, Nijmegen, The Netherlands

European Master's of Clinical Linguistics, Universities of Groningen, Groningen, The Netherlands

Douglas Roland Graduate School of Arts and Sciences, The University of Tokyo, Tokyo, Japan

Rushen Shi Département de Psychologie, Faculté des Sciences Humaines, Université du Québec à Montréal, Montréal, QC, Canada

Linnaea Stockall Department of Linguistics, Queen Mary University of London, London, UK

T. M. Talavage School of Electrical and Computer Engineering, Purdue University, West Lafayette, IN, USA

C. Weber-Fox Department of Speech, Language, and Hearing Sciences, Purdue University, West Lafayette, IN, USA

R. B. Wilbur Department of Speech, Language, and Hearing Sciences, Purdue University, West Lafayette, IN, USA

Linguistic Program, School of Interdisciplinary Studies, Purdue Universtiy, West Lafayette, IN, USA

Hongoak Yun Department of Psychology and Center for Cognitive Science, University at Buffalo, Koyang-si, Kyunggi-do, South Korea

Editor Biography

Dr. Roberto G. de Almeida is an Associate Professor in the Department of Psychology at Concordia University, in Montreal, Canada. He received his PhD in Psychology (Cognitive Science) from Rutgers University.

Dr. Christina Manouilidou is an Assistant Professor in the Department of Philology at the University of Patras, Greece. She received her PhD in Linguistics from the University of Ottawa, Canada.

Part I
Foundations

Chapter 1
The Study of Verbs in Cognitive Science

Roberto G. de Almeida and Christina Manouilidou

Verbs are the key elements of syntactic and semantic (or conceptual) representations of events and states.[1] Rarely this assertion requires much more elaboration in linguistic circles. This is so because virtually every linguist assumes that verbs—more than any other grammatical category—carry core semantic properties of the events and states that sentences describe, and also license a myriad of information about the nature of the syntactic arguments that are constitutive of grammatical sentences. Besides its importance in linguistics, the nature of events and the very notion of predicate-argument structure, crucial to understanding verb meaning, have long been prominent topics in analytic philosophy (e.g., Davidson 1967). Thus, while linguistics and philosophy have generally taken verbs to play a central role in the representation of linguistic meaning and in the conceptualization of events, other

Research for this article was supported in part by a grant from the Natural Sciences and Engineering Research Council of Canada (NSERC) to RGdA. We are grateful to Lila Gleitman and Merrill Garrett for their comments to an earlier version of this chapter.

[1] Just to be clear on what we mean by "semantic" and "conceptual": We take verb meanings and word meanings in general to be encoded in the mind/brain as *concepts*, i.e., mental particulars bearing content. Thus, the verb *drink* is a lexicalization of a particular event, which is encoded (or represented) as a concept. The concept, just like the lexical item itself, refers to any drinking event. We will use "semantic" and "conceptual" interchangeably although in some theoretical contexts—viz., linguistics—it might be more appropriate to use "semantic" to refer to the content and structure of token items. See Sect. 1.2, for further discussion.

R. G. de Almeida (✉)
Department of Psychology, Concordia University,
4171 Sherbrooke Street West, Montreal, QC H4B 1R6, Canada
e-mail: roberto.dealmeida@concordia.ca

C. Manouilidou
Department of Philology, University of Patras,
Rio Patras 26504, Greece
e-mail: chmanouilidou@upatras.gr

R. G. de Almeida, C. Manouilidou (eds.), *Cognitive Science Perspectives on Verb Representation and Processing*, DOI 10.1007/978-3-319-10112-5_1

cognitive science branches have not given the same attention to verb representation. The psychological literature on word meanings (or concepts), for instance, has for a long time focused primarily on concepts that are lexicalized by nouns (e.g., Smith and Medin (1981) and Murphy (2002) for reviews) with relatively few studies investigating how concepts lexicalized by verbs are represented (e.g., Gentner 1975; Kintsch 1974; Fodor et al. 1980; Miller and Johnson-Laird 1976). Until recently, a similar state of affairs could be found in the cognitive neurosciences. The vast majority of studies bearing on the so-called category-specific semantic deficits in patients with brain damage or disease have focused on dissociations within semantic categories labeled by nouns (see, e.g., Tyler and Moss 2001; Humphreys and Forde 2001). Comparatively few studies focusing on patterns of dissociation within categories labeled by verbs have been reported (e.g., Breedin et al. 1998; Grossman et al. 1996; Kemmerer et al. 2001).

Recent years, however, have brought us numerous aphasia and neuroimaging studies focusing on verb meaning and structure (e.g., Bastiaanse and van Zonnevelt 2005; Faroqi-Shah and Thompson 2010; Kemmerer 2006; Kemmerer and Eggleston 2010; Meltzer-Asscher et al. 2013; Manouilidou et al. 2009; Thompson et al. 2007, 2010).[2]This relatively recent surge of studies on the nature of verb representations in cognitive neuroscience and related fields can be attributed in part to the increasing cross talk between theoretical linguistics (e.g., Jackendoff 1990, 2002; Levin and Rappaport Hovav 2005) and the experimental wings of cognitive science, propelled in particular by the refinement of empirical hypotheses and methods. Despite recent progress, there are still great divides both between disciplines and within linguistics. We aim to contribute to narrowing these gaps with the present chapter as well as with the present volume—with contributions on verb meaning stemming from diverse linguistic-theoretical camps, philosophy, psycholinguistics, neurolinguistics, and cognitive neuroscience more broadly.[3]

In this chapter, we provide a general introduction to the domain of inquiry of verb representation and processing. We briefly discuss theories and experimental studies within particular areas of investigation—most notably, argument structure, thematic roles, and semantic/conceptual structure—aiming to show how different types of evidence (theoretical, cross-linguistic, experimental) support or refute linguistic-theoretical constructs and advance our knowledge of how events and states are conceptualized. Although these areas of investigation have been prolific in linguistics

[2] This brief survey is certainly nonexhaustive and leaves out a long tradition of psycholinguistic work on verb *argument structure* and *thematic roles* (see, e.g., Fodor et al. 1968, for an early account, and Sects. 1.3 and 1.4). For reasons of space, the present chapter does not discuss a vast literature on how verbs are acquired—i.e., on the origins of the link between token verbs and events and states (see, e.g., Gleitman and Gleitman 1992; Gleitman et al. 2005; see also Chaps. 12, 13). Our concern here is that qua mental particulars bearing content far less attention has been given to verbs in areas such as the psychology of concepts and categorization.

[3] We do not mean to legislate on disciplinary boundaries. We use traditional labels for these disciplines simply for convenience (see Sect. 1.1). For us, as for many others, linguistics, psycholinguistics, and related fields, are part of cognitive science, for what really matters is the explanatory adequacy of any given theory and its empirical evidence.

and psycholinguistics, we imposed some constraints on our discussion: (i) First, we keep the discussion on different views on argument structure and lexical semantics to a minimum. Most current views stem from syntactic and semantic theories from earlier periods of generative grammar (e.g., Jackendoff 1983; Katz 1972; McCawley 1972). We believe that the distinctions between various more current verb representation theories (e.g., Croft 2012; Dowty 1991; Levin and Rappaport Hovav 2005; Goldberg 1995; Jackendoff 1990; Pietroski 2005; Pustejovsky 1995) boil down to the type of semantic information that are constituents of verb meaning in terms of more elementary predicates, and how the constituents of semantic representation are mapped onto the arguments of a verb and event structure. We acknowledge that there are several important distinctions between these approaches, but one issue stands out: whether one conceives verb meaning to be atomic (e.g., Fodor 1998) or whether it is molecular (or *decompositional*; e.g., Jackendoff 1990; Levin and Rappaport Hovav 2005). This distinction is key to understanding how verb meaning and argument realization are attained and how sentences and their constituents are processed. We focus on this issue in Sect. 1.5. (ii) Second, we also attempt to limit our discussion of experimental results to those studies which we believe aim at elucidating the nature of verb *representation,* beyond their contribution to our understanding of language *processing*. We will see that this particular constraint is important vis-à-vis our view of how evidence for linguistic constructs are gathered. To advance a bit our discussion: We deem all methods—from native-speakers' intuitions to neuroimaging—as equally relevant in uncovering the nature of mental representations, for we take all relevant data (and in particular intuitions) to be *psychological* data. (iii) And finally, the most obvious constraint: Since our discussion cannot possibly cover all relevant topics (see chapters in the present volume and their references for a rather comprehensive coverage), we are compelled to focus on only a few representative issues within the domains we chose—ranging from argument structure to semantic/conceptual templates. With regard to this last constraint, but for reasons of familiarity, we bring to fore sample research from our own labs, whenever appropriate. Our goal is to provide the reader—perhaps most importantly the uninitiated on verb matters—enough background to venture into the readings collected here and beyond.

This chapter is organized as follows. We first establish the domain of investigation—verbs—as the object of different disciplines and methodologies, beyond traditional linguistic theorizing. Thus, we start off by laying out a few methodological points with regard to how we see the study of verbs and linguistic objects more generally from an interdisciplinary perspective characteristic of cognitive science. Then, we situate the study of verb representation and processing within cognitive science by providing a few criteria on what might be relevant for understanding verb content and structure. Following these two sections, we present brief discussions of three content areas that constitute basic types of representation bearing on verb meaning. We move "up the ladder" from purely structural aspects of verb representation to questions of semantic composition (and predicate *decomposition*). We conclude by providing a short overview of the accompanying chapters.

1.1 Verb Representation and Psychological Evidence

It has long been the tradition in linguistic theorizing to rely on a sort of purity of method, with native-speaker intuitions driving much of the analytic endeavor together with cross-linguistic and distributional sources of evidence (i.e., based on the "behavior" of certain constructions or the validity of certain postulates within a language and between languages). In principle, this grants a certain degree of autonomy for linguistics in the investigation of the principles underlying verb structure and meaning. Although we must recognize that to a large extent linguistic theory is predicated on native-speaker intuitions, cross-linguistic data, and distributional arguments, investigating how linguistic principles are represented in the mind/brain, how they emerge during language acquisition, how they have evolved in the species, and how they are used in utterance contexts are within the scope of many constituent cognitive sciences. As Chomsky (1986, p. 37) once put it, "there is no way of delimiting the kinds of evidence that might, in principle, prove relevant."

While in principle the empirical evaluation of linguistic postulates is subject to contributions from many different methods and data-gathering procedures, it is not clear the extent to which psycholinguistic and other experimental data are taken to be the basis of theoretical advances on linguistic *representations*.[4] This is so in part because psycholinguistics (as well as related experimental disciplines) has been mainly concerned with *processing* mechanisms, appropriately motivated by the idea that a theory of grammar characterizes the knowledge employed in language processing. But, the relatively low impact that experimental data has had on linguistic theory is also due to the long-held practice of distinguishing roles for linguistics and psycholinguistics, possibly on the assumption that core principles of the language faculty in the "narrow sense" (Hauser et al. 2002) might be beyond the reach of processing data. But it is not possible to know in advance which principles constitute the core computational domain of the language faculty or how much of the linguistic characterization of the grammar constitutes the knowledge used in processing mechanisms. Besides, native-speakers' intuitions—to mention one key method of gathering data on grammaticality—are psychological data, as Fodor et al. (1975) observed, and purity of method entails that *all* psychological data be taken into account in evaluating linguistic representations, including data brought about by psycholinguistic and cognitive–neuroscience methods (see de Almeida 2006). Moreover, if understanding (or producing) a sentence relies on recovering its representations, the

[4] We can think of a few linguistic theories whose goals are to be somehow compliant with language-processing constraints. Lexical-functional grammar (LFG; see Kaplan and Bresnan 1982), for instance, came out committed to "psychologically plausible processing mechanisms" (pp. 173–174). A similar commitment was made by head-driven phrase structure grammar (HPSG; see Pollard and Sag 1994) and, more recently, by Culicover and Jackendoff (2005). This of course is not equivalent to taking experimental evidence into account in postulating linguistic principles. In fact, as Pollard and Sag say, "[w]hereas it is reasonable to expect that further research into human language processing will produce specific results that inform the minute details of future linguistic theories, we do not yet know how to bring these considerations to bear." (p. 13).

processes involved in comprehension and production ought to be taken as informative on the very nature of the representations involved.[5] It should be clear by now that psychology—or nonbehavioristic psychology—is about the characterization of the internal mechanisms underlying observable behavior, on a par with linguistic competence.

Although this methodological discussion concerns mainly issues on the nature of empirical evidence for hypothetical core linguistic principles, it is also related to how we ought to investigate the faculty of language and its interfaces more broadly. Much of the present chapter—and volume—is concerned with issues taken to be at the interface between these hypothetical core principles of linguistic structure (e.g., argument structure) and systems of *interpretation*. It is perhaps at this interface where contributions from diverse cognitive science disciplines show greater promise. Experimental studies employing response-time measures (e.g., lexical decision, eye tracking), the recording of electrical currents (e.g., ERPs, MEG), patterns of blood flow (e.g., functional magnetic resonance imaging, fMRI), as well as on- and off-line studies with impaired populations (e.g., Alzheimer's patients, people with aphasia) have brought forth great insights not only on the nature of the mechanisms involved in processing but also on the nature of *representations*—even when these insights are not explicitly incorporated in linguistic theory. It is our view, however, that bridging the gap between experimental data and linguistic theorizing is also the mission of those conducting experimental work.[6]

It is with the aim of bridging this gap that we think that guiding assumptions common to the practice of theoretical linguistics and the experimental wings of cognitive science are required. Linguistic—and more broadly, cognitive—theorizing requires generalizations at different but often interacting levels. Let us assume (with Pylyshyn 1984) that these levels are the *biological,* the *semantic,* and the *symbolic*.[7] The *biological* level aims at explaining regularities in behavior—including here covert behavior such as linguistic processes—by appealing to the vocabulary of biology (or the neurosciences). Anatomical or neurophysiological descriptions of particular processes

[5] We are not implying that empirical data should necessarily determine theory change: data cannot be the sole basis of such change. Without being exegetic in our philosophy of science, we expect this to be a common guiding assumption (see, e.g., Laudan et al. 1986). What we are saying is that experimental data should be taken seriously in advancing theories on representations, if we are to rely on psychological evidence.

[6] As important as it is to provide support for linguistic claims, experimental data play an important role in refuting those claims, thus motivating theory change. There is by now a handful of such cases in psycholinguistics. See, for instance, experimental studies on the reality of empty categories—which has been a point of contention between different syntactic theories (e.g., Bever and McElree 1988). For a more recent case, see experimental studies and theoretical debates on the nature of so-called semantic coercion (e.g., de Almeida 2004; de Almeida and Dwivedi 2008; de Almeida and Riven 2012; Pylkkänen and McElree 2006). And as we show in Sect. 1.5, psycholinguistic evidence for and against verb-semantic decomposition lingers within reach of lexical-semantic theories.

[7] These are hardly new because perhaps most nonlinguists in cognitive science are keen on describing processes and representations at all these different levels. The case of vision—a hypothetical faculty akin to language—is paradigmatic (see, e.g., Pylyshyn 2004).

answer to a specific set of questions about particular manifestation of a cognitive system—the patterns of physical implementation, including here the neurological correlates of particular processes or knowledge. The *semantic* level of explanation accounts for certain regularities in behavior by appealing to what one knows about the world or to the content of representations. We can include here overt judgments of grammaticality or intuitions about the semantic content of linguistic expressions. And finally, the *symbolic* level is where explanations appeal to the symbols and rules that underlie certain types of behaviors, primarily the rule-governed behaviors such as parsing and rule-driven aspects of semantic interpretation. The reason why this level is important—perhaps the most important for the purposes of understanding the fixed linguistic and cognitive capacities—is that it provides a common ground for establishing the regularities in linguistic representations (the nature of the symbols, their combinatorial or syntactic properties) and how these symbols are put to use.

While these levels of description often involve their own vocabularies, theoretical postulates, and empirical predictions, in actual practice understanding a given domain of knowledge—say, the nature of verb representations—requires descriptions at all levels (see also Marr (1982) on this point). Also, in actual practice it is difficult to determine whether or not explanations at one level are independent of or immune to explanations at other levels. For instance, explaining the neuronal correlates of different verb classes also requires appealing to the symbolic level—which is the characterization of the rules and representations underlying the classes and their linguistic behavior. We expect that much of the theoretical and empirical investigations on verb representations and processes aim at characterizing the fixed properties of verb representations and at providing what Pylyshyn (1984) called "strong equivalence" of cognitive processes: That the rules we postulate are instantiated as rules in the system, perhaps realized as physical patterns corresponding to actual rule-following computations. Finding out the strong equivalence of particular cognitive/linguistic processes and the representations that these processes involve requires a concerted effort within cognitive science.

1.2 Verb Content and Structure

We now turn to more specific issues characterizing our cognitive science perspective on verb representation and processing. This section sets the stage for a review of some specific controversies on the nature of verb representation and processing, presented below. Our goal here is to briefly discuss criteria for sorting out linguistic and nonlinguistic aspects of verb meaning.

We take as the standard view—perhaps common to all theoretical approaches to verb representation—that verbs qua linguistic entities are lexicalizations of "happenings" (Levin and Rappaport Hovav 2005). This means that verbs are morphologically simplex or complex lexical items whose referents are events or states in the world.[8] We also assume that it is standard to take verb meanings to be encodings—i.e., *representations* in the mind/brain—of such happenings. This is much of what we take to be uncontroversial, for what exactly verbs pick out of these events or states, how they interact with other linguistic constituents, and how they are mentally represented and neurologically implemented are matters of great divide in the literature.[9]

Beyond their linguistic life, verb meanings—the representations of happenings—are *concepts* and thus might be represented "outside" the linguistic system proper, on the assumption that a line can be drawn between linguistic and nonlinguistic representations and processes. That is to say, whatever properties of the world (events/states) a verb picks out, these properties are available to other cognitive systems, perhaps at a central system common to different perceptual and cognitive domains. To wit, the verb *drink* picks out (or refers to) drinking events, whether this event is perceived linguistically (e.g., during sentence comprehension), visually (e.g., perceiving someone drinking), or is part of one's action (e.g., drinking). Moreover, because the very idea of *drinking* is a concept, it is available to thoughts and other higher-cognitive processes (reasoning, planning actions, making decisions, etc.). Concepts that are labeled by verbs, thus, are not necessarily linguistic entities, as they share with other concepts (the likes of DOG, LOVE, and BACHELOR) the property of being mental particulars that are constituents of thoughts and other cognitive processes. This in fact goes for any word meaning: Entertaining a thought entails entertaining a complex expression in the language of thought whose constituents are concepts (call it "the language of thought" hypothesis).[10]

As it is the case with any token lexical item, however, besides being a label for a particular concept, a verb also encodes *linguistic* information proper. This bears on the combinatorial properties of the verb (e.g., arguments and their hierarchy) as well as perhaps thematic-role information. We think that this assumption is somewhat controversial because not all theories acknowledge that verbs encode arguments, or that arguments bear thematic roles. These issues are discussed in Sects. 1.3–1.5. Leaving aside these controversies for now, following Grimshaw (1993), we can say

---

[8] We use "reference" in a broad sense to include events and states whether they are observable or not—i.e., within and beyond the "perceptual circle" to use Fodor and Pylyshyn's (2014) term. Thus, while *to drink* refers to an observable event, *to think* does not. In both cases, verbs are lexicalizations of the meanings of such happenings.

[9] In reality, not even the idea that verb meanings are mental representations (or neurologically encodings) of "happenings" is absolutely uncontroversial—for one might assume that there are no mental/linguistic representations but only behaviors to talk about (e.g., Quine 1960), or that word meanings are not encoded in the mind/brain (e.g., Putnam 1970, 1975a)—not at least as definitions but as a form of "use" or "disposition" (see also Wittgenstein (1953) for an anti-mentalistic approach). We take the idea of verb meanings as mental representations to be common to theories within the classic (symbolic) tradition in linguistics and cognitive science.

[10] Although this hypothesis might be more readily identified with Fodor (1975), it is also current in other theories (e.g., Jackendoff 1983, 2002) albeit there are some important distinctions. Of general concern here are the productivity and systematicity of linguistic and conceptual representations, which hinge on the characterization of the very nature of their elementary constituents.

that there are linguistically *active* aspects of verb meaning as well as linguistically *inactive* ones. The linguistically *inactive* part of verb meaning is what we called *concept*, above, and bears on the content—possibly the referential content of a verb, i.e., what sort of event/state it picks out. As Grimshaw puts it, the differences between *melt* and *freeze* are probably irrelevant from the perspective of linguistics: These verbs mean what they mean (namely, MELT and FREEZE) although linguistically they behave in similar ways—i.e., they enter into similar constructions, they have the same number of arguments, and probably assign the same thematic roles. Linguistically active aspects of verb meaning, then, are those that have an effect on the linguistic behavior of a predicate, that is, the aspects of meaning which determine the nature of argument structure and the thematic-role properties of these arguments.

The active/inactive divide raises a great number of empirical questions. For instance, if content (the linguistically inactive aspect of a verb's meaning) is opaque to the linguistic behavior of a predicate, how do they compose to form expressions in the language of thought? This issue is important vis-à-vis the hypothesis that concepts are the elements of thoughts and that, just like sentences, thoughts are compositional. One possibility is that conceptual representations inherit structural properties of linguistic predicates such that conceptual composition mirrors the linguistic structure of events. Yet, it is also possible that concepts are atomic while their labeling verbs are structurally complex, reflecting their linguistic properties such as predicate–argument relations (see, e.g., Chaps. 2, 4 for contrasting views). Sorting out which aspects of a verb's meaning are *active* and which ones are *inactive* is ultimately an empirical question requiring multiple methods—the classical methods employed in linguistic theory and the theoretical and experimental tools of other cognitive science disciplines. The characterization of what in fact constitutes the "behavior" of a given verb or verb class and how they are implemented in psychological and neurological processes might prove fruitful in further determining the active/inactive divide. While the linguistically active aspects of verb meaning have been of greater concern in linguistics and psycholinguistics, what is supposedly linguistically *inactive* is also key to understanding the nature of the conceptual system and what we encode of events and states.

We will briefly examine, next, theoretical and empirical work bearing on three types of information implicated in the representation of the meaning of a verb (or the meaning of a sentence that a verb partakes): (1) the bare specification of structural relations between the predicate and its arguments (see Argument Structure), (2) the interpretation of these arguments in terms of the roles they play in the event or state denoted by the predicate (see Thematic Roles), and (3) a discussion of the conceptual representation of a verb's meaning, traditionally studied in linguistics and psycholinguistics in terms of conceptual primitives within a conceptual or semantic template representing a verb or verb class (see Conceptual Structure).[11] Although

---

[11] We have to leave aside many other types of information contributing to the meaning of a predicate and its carrier sentence, such as tense and aspect. But see part III of the present volume for studies involving processing, representation, and impairment of tense and aspect.

these different types of representation of information regarding the meaning of a verb are rather complementary or even redundant (e.g., Jackendoff 1990; Levin and Rappaport Hovav 2005), they have constituted major areas of research in linguistics and cognitive science. We will consider arguments for and against each type of information, as well as their empirical (mostly psycholinguistic and neurolinguistic) evidence.

1.3 Argument Structure

In its standard conception, argument structure—a linguistic version of predicate logic predicate–argument relations—specifies the number and type of grammatical constituents (usually noun phrases, pronouns, and prepositional phrases) licensed by a verb. The arguments of a verb stand, in principle, for the obligatory participants in the event or state referred to by the verb and, thus, encode the hypothetically *necessary* linguistic (syntactic/semantic) properties of the event/state. A simple way of conceiving argument structure is by specifying the constituents as variables devoid of content, as in (1).

(1) drink (x, y)

Under this view, argument structure is said to be encoded with the verb. The assumption is that lexical entries—or at least verbs—are structurally complex with regards to the kinds of syntactic constituents with which they partake in grammatical sentences.

Beyond projecting the set of constituents of an event/state, argument structure can also be taken as a specification of the prominence relations of the arguments (Grimshaw 1990). The idea of prominence relations gives structural arguments a semantic life. This is so because, in order to conceive such prominence relations, arguments ought to bear information about the nature of events and states that their root verbs refer to. Thus, in (1), which represents the argument structure of the transitive variant of *drink,* the variable x stands for the external argument, the syntactic subject, while y stands for the internal argument, the syntactic object position. And moreover, given what *drink* means, x stands for the one who drinks while y stands for the thing drank. One way of further specifying the semantic life of arguments is to assign them *roles* such as *Agent* and *Theme*. In Grimshaw's (1990) theory, prominence relations are determined by thematic and aspectual properties of the predicate and thus are intertwined with the nature of thematic roles and the hypothesis of thematic hierarchy. We return to semantic or *thematic roles* in the next section. Suffice it to say that there is no agreement on how such prominence relations are established—whether by thematic properties of the arguments or whether by other purely syntactic principles blind to semantic properties of the predicate. In a theory such as Levin and Rappaport Hovav (2005), for instance, thematic role labels are not used to represent the semantic nature of the participants and thus arguments correspond to syntactic variables represented within the verb's semantic template (see Sect. 1.5).

While we have presented rather simply the view of argument structure as lexically specified information, i.e., verb-encoded structure, this is far from being a universally accepted notion. Many researchers have espoused the idea that argument structure is determined by more general syntactic principles. There are several perspectives on this camp (e.g., Åfarli 2007; Borer 2003; Goldberg 1995; Marantz 2013), but its main tenet is the idea that verbs are simplex, containing no necessary structure to be projected. Some (Åfarli 2007; Borer 2003) assume that there are syntactic–semantic *frames* which are generated independently of the verb meaning itself—i.e., not as an encoded property but as an independent frame appropriated by the verb for different uses. The idea is that the verb is *inserted* into different frames to convey different events/states, and it is then that the verb *gets* an argument structure, i.e., by being associated with a particular frame. Åfarli (2007), for instance, assumes that there are only five frames (in Norwegian) for a verb to pick from. In other approaches (see, e.g., Marantz 2013), verbs are, roughly, roots which are merged with different features to yield a variety of canonical syntactic configurations. Importantly, the syntactic structures are independent of verb meaning. Thus, for instance, the transitive and intransitive variants of *drink* as in (2) are given by independently generated syntactic structures which merge with the verbal root *drink* producing complex verb phrases.

(2) a. The man drinks
b. The man drinks beer in the morning
c. The man drinks his way out of trouble

The meaning of *drink* in these contexts remains invariant, only determined by its ontological *activity* denoting a particular concept (namely, DRINK), with semantic composition relying on the resulting syntactic structure to determine other properties of the event/state the sentence describes. Thus, meaning differences between (2a), (2b), and (2c) are given not by the root, which is constant, but by the semantic composition emerging from the different syntactic structures.

The idea that verbs can be *flexible* in their syntactic contexts (i.e., with varying internal arguments) or in the types of syntactic configurations it allows for is perhaps one of the main reasons for doubting a strict lexicalist view represented in (1). Another source of evidence against strict lexicalism comes from verbs coined anew (e.g., *to tweet*) or *verbbed* proper names such as *Justin Bieber* (as in *He Justin Biebered into the party*—whatever this means), which have no established argument structure to project. It is possible that these verbs pick up their meanings by appropriating a frame from a similar class (e.g., *to Google* for *to search*) analogically, although, as pointed out by Åfarli (2007), this just begs the question on how analogical processes work. What is relevant for the present purposes is that it seems to be the case that argument structure might not be encoded with the verb but may arise from their host frame or syntactic configuration. One advantage of separating the verb root from its structure is that it accounts for crosslinguistic variability for verb roots, which often appear in different syntactic configurations in different languages (see Marantz 2013).

There is yet another advantage of keeping roots apart from their syntactic projections and it has to do with how we might ultimately encode concepts. Although verb-encoded argument structure is said to be strongly tied to the meaning of a predicate—i.e., how the predicate encodes properties of events and states—it is not entirely clear if the idea of *necessary* constituents carries all the way to the predicate's concept. If *drink* is represented as in (1), does the very concept of drinking somehow encode these *necessary* arguments as event participants (e.g., DRINK(X, Y))? If, however, concepts are atomic or *monadic* (see, e.g., Fodor 1998; Chap. 2), they do not carry constituents by necessity and their modes of combination in the language of thought do not come from their valence but from independent syntactic (/compositional) principles akin to natural language syntax.

However, it is possible to conceive of *flexibility* as part of lexically specified argument structures as in (1) without necessarily committing to conceptual complexity, assuming flexible argument structures for the purpose of syntactic composition but atomic roots that map into concepts. Di Sciullo (2007), for instance, proposes that argument structures can shift similarly to semantic type-shifting operations (Partee 1986), with arguments occupying noncanonical positions within the default argument structures of a verb. So, for instance, assuming a canonical set of argument structures, as in (3), structures can *shift* to adjust for missing arguments or to account for different syntactic configurations (roughly, *A* stands for argument positions).

(3)

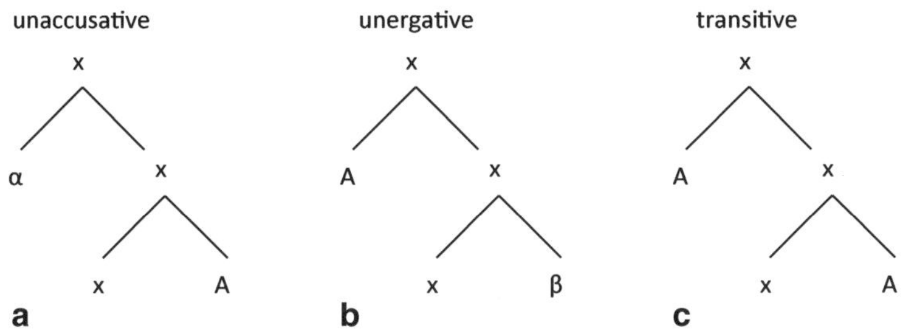

Consider, for instance, the case of *middle* constructions such as in (4).

(4) This book sells well

The verb *sell,* in its transitive sense, with a structure as in (3c) (see Keyser and Roeper 1984; Bowers 2002), specifies one internal and one external argument, similarly to (1). But, by hypothesis, *this book* is the internal argument of *sell,* while occupying the subject, external argument position. Contrast (4) with (5), which are unergative structures (i.e., they project only an external argument).

(5) a. This clerk sells well
b. This store sells well

If argument structures are lexically specified, it is not clear how the projection of arguments, as in the case of middles, would be obtained. Di Sciullo's (2007) proposal would entail shifting the structure such that the internal argument in (3c) appears in the external argument position, yielding a structure that behaves like (3b) on the surface. This solution allows for flexibility while also preserving lexically specified binary projections.

This brief introduction is certainly far from offering any verdict on which alternative carries greater validity. As it is often the case in linguistics, with its numerous competing schools, different perspectives might in fact be guided by different sets of assumptions, which in turn make different empirical predictions. It is also possible that progress might come from converging ideas towards some form of consensus. We think that for any sort of consensus to be obtained, it would also be necessary to look at what experimental approaches have informed us about the nature of argument structure.

Most evidence coming from psycholinguistics and cognitive neuroscience (aphasia and neuroimaging) are presented in support of the very idea that verb structural complexity correlates with different neuronal signals or that verbs with more arguments are more difficult to process in cases of aphasia. For instance, many psycholinguistic studies have suggested that argument structure information is automatically accessed when verbs are encountered during sentence comprehension (e.g., Boland 2005; Boland et al. 1990; Friedmann et al. 2008; MacDonald et al. 1994; Trueswell and Kim 1998; Trueswell et al. 1993). Some of these studies have also been taken as evidence for the use of thematic information brought about by the verb (see next section). Although there are numerous methodological and theoretical differences between these studies, the general findings have been interpreted as supporting the idea that argument structure is part of a verb's representation (lexicalist view). An interesting dimension of argument structure processing comes from the literature on implicit arguments. For instance, Mauner et al. (1995) showed that speakers appear to activate (and, by assumption, encode) implicit arguments during sentence processing (e.g., an *agent* in a passive sentence) and that this activation does not require additional processing time compared to explicit argument processing. Subsequent experiments on implicit agents (Mauner and Koenig 2000) extended those findings, suggesting that argument structure information is accessed immediately at the verb, yielding semantic information about arguments even when these are not overtly expressed. These studies raise several questions with regard to the nature of arguments—whether they are only structural elements or whether they are "filled" with content by default. Either way, at a minimum they show that the system is *fast* in making semantic (content) decisions driven by the structure of the predicates (see also Chaps. 3, 4, 10 for related discussions).

Studies with agrammatic patients, in several languages, have also contributed important evidence to our knowledge about argument structure. Employing a variety of techniques, most studies have showed that more complex argument structures engender greater production difficulties, suggesting that number and perhaps semantic type of arguments play a role in the representation and processing of sentences (see, e.g., Thompson 2003; Kim and Thompson 2000, 2004; Lee and Thompson 2004; Thompson and Choy 2009, for English; Luzzatti et al. 2002, for Italian; Jonkers and Bastiaanse 1998, for Dutch; Kiss 2000, for Hungarian; De Bleser and Kauschke 2003, for German; Dragoy and Bastiaanse 2010; for Russian). These studies have largely supported what Thompson (2003) termed *Argument Structure Complexity Hypothesis,* a strict lexicalist hypothesis of argument structure which postulates that (a) more complex verbs are more difficult to produce (at least in agrammatism), and

(b) complexity is a function of number and type of arguments—in which case verbs are deemed more complex if they encode more arguments or if they encode "argument structures that trigger movement operations" (p. 163). Bastiaanse and Platonov (Chap. 7) present an extension of this hypothesis, including *tense* and *aspect* as factors affecting verb retrieval in agrammatism.

The neurological implementation of argument structure has also been investigated in neuroimaging studies with two main goals: to determine specific brain areas involved in verb-argument structure representation and processing, and to investigate particular contrasts between argument-structure variables. For instance, using fMRI, Thompson et al. (2007) investigated the neural correlates of verbs with one- (*die*), two- (*kill*), and three- (*put*) argument verbs in a lexical decision task. They found that while verbs with two and three arguments did not differ from each other in terms of activation clusters, both showed greater activation than one-argument verbs mostly at the left angular and supramarginal gyri (AG, SMG, respectively; see Fig. 1.1a). A follow-up fMRI study by Thompson et al. (2010) obtained similar effects (activation near the left AG) but only in the contrast three arguments vs. one argument, in a whole-brain analysis (Fig. 1.1b). Finally, Meltzer-Asscher et al. (2013), also using fMRI with a lexical decision task obtained small but significant greater activation for verbs that can alternate (or shift) between two structures (e.g., *break,* as in *John broke the vase/ The vase broke*) compared to unergatives (e.g., *laugh*) at a similar region (see Fig. 1.1c).[12] In principle, one could claim that (a) verb-argument structure complexity engages the AG and adjacent areas, thus that is primarily where argument-structure complexity is computed, and (b) there are neurological correlates for linguistically proposed contrasts. However, it is not the case that this area is *only* involved in argument-structure processing, or that the obtained activation contrasts are univocally attributable to linguistic differences.

Clearly, these studies are important to understanding the neurological correlates of argument structure and related processes, on the assumption that different neuronal peaks of activation and different neuronal networks correlate with contrasts between linguistic/semantic variables. But as Binder et al. (2009) show, conceptual ("content") processes involve vast networks and foci of activation, with great variability due to methodological differences between studies. In their review, Binder et al. point to several studies involving knowledge of "actions" and knowledge of "artifacts" also activating areas such as the left AG implicated in the studies on verbs mentioned above. Thus, while it is possible that more fine-grained distinctions between verb types and other semantic knowledge are computed at that general area (which in fact includes portions of the anterior occipital lobe, or BA 19; see Binder et al. 2009; and Fig. 1.1b, c, in particular), other semantic processes also engage those areas. It is also important to note that the experimental studies we reviewed do not seem to dissociate argument-structure complexity from thematic roles and

---

[12] Interestingly, Meltzer-Asscher et al. (2013) also found activation for alternating verbs in a small cluster in the anterior cingulate cortex, an area that has been implicated in conflict resolution and in indeterminate sentence processing (see de Almeida et al. 2014).

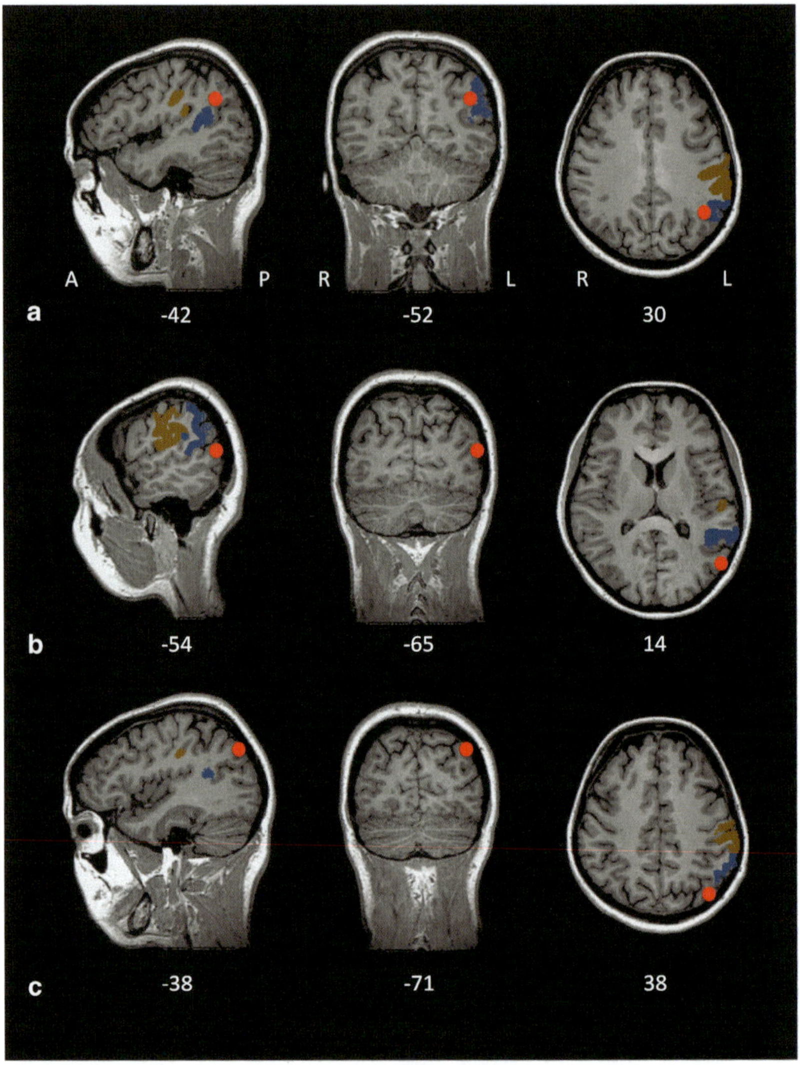

Fig. 1.1 Activation maxima for verb-argument structure contrasts based on data from (**a**) Thompson et al. (2007), (**b**) Thompson et al. 2010 (healthy control subjects), and (**c**) Meltzer-Asscher et al. (2013) shown in sagittal (*left column*), coronal (*middle column*), and transversal (*right column*) orientations. *Red dots* represent activation loci. *Colored areas* are Broadman Areas (BAs) 39 (*blue*) and 40 (*brown*). Data are plotted based on reported Talairach coordinates (in **b** and **c** these were transformed from MNI coordinates) to show approximate location of highest peak of activation for the contrasts (center of *red dots*), leaving out clusters of activation where voxels reached significant threshold. Also, choice of BAs was based on the most reported anatomical areas where contrasts occur. The activation in (**a**) represents the contrast between three- and two-argument verbs > one-argument verbs. The activation in (**b**) represents the contrast between three-argument verbs > one-argument verbs. And the activation in (**c**) represents the contrast between alternating > nonalternating (unergative) verbs. Legends: *A* anterior, *P* posterior, *R* right, *L* left

other "content" information. In fact, Thompson et al. (2010) suggest that these areas (Fig. 1.1) might not be involved in argument-structure building but in the integration with semantically selected (content) arguments. Thus, it remains to be seen whether or not there are neuronal correlates of verb-argument *structure* that are independent of content (the *linguistically inactive* aspects of meaning).

Returning to methodological issues on the investigation of argument structure, as we have seen, evidence is rarely, if ever, straightforward—be it in psycholinguistic experiments or cognitive neuroscience studies. As it happens with competing linguistic theories and their empirical support, many of the experimental studies we cited are subject to methodological and, more importantly, theoretical disputes. For instance, in a series of studies (reported preliminarily in Di Sciullo et al. 2007), we found that a sentence such as (4) takes longer to process than those in (5). In principle, this shows behaviorally that the two sentence types engender different processing routines. Clearly, the parser is sensitive to processes that go beyond sentence linear order and taps perhaps their structural differences. But it is not clear, though, if the obtained difference provides evidence for a *shift* in argument structure or for a syntactic operation like move-α. The behavioral difference between these conditions, however, could also be attributed to a reanalysis of the external, subject position argument in (4), which might be canonically interpreted similarly to those in (5). Our follow-up eye-tracking experiment with modified materials shows that the difference between (4) and (5) disappears when the middle clause (4) is preceded by adjectival predicative clauses, such as in *That novel is unpopular but this book sells well...*, suggesting a *structural priming* between clauses. This priming effect might indicate that the middle clause is parsed as an unergative (as in (3b)), supporting the argument shift hypothesis.[13] And yet there is the possibility—ever so present in psycholinguistic and cognitive neuroscience experiments—that tasks are *cognitively penetrable* (Pylyshyn 1984) and thus observed regularities might be due to "semantic" or "knowledge-level" processes: in the present case, when effects of argument structure might in fact be attributable to the overt interpretation of the sentences (including here frequency, plausibility, expectations, preference judgments, and the like) or, more specifically, when structural effects are confounded with content effects.

As we will see in the next section, there are good reasons for believing that argument structure—but in particular information about the semantic nature of arguments, their prominence as well as realization (thematic hierarchy)—seems to play an important role in verb representation and sentence processing.

[13] We offer our *middles* experiments as an example of how behavioral studies can lead to alternative theoretical accounts, much like most in the field, such as the ones we cite above. We are thus avoiding getting into a lengthy methodological and theoretical discussion on all those experimental studies. Of course, the theories that motivate such experiments are also subject to change and thus the interplay between types of evidence and theoretical proposals might lead to progress in our understanding of linguistic and conceptual phenomena.

1.4 Thematic Roles

Thematic roles have given rise to numerous controversies in linguistics. Before sampling some of these controversies, let us say roughly what "thematic roles" mean in standard usage. Simply put, thematic roles specify the semantic nature of the different arguments, which are participants in the events/states denoted by verbs. These roles are usually assigned by the verb to each of its arguments and are supposed to constitute a basic set of semantic properties—by hypothesis ontologically primitive—characterizing "who did what to whom" in the event/state. Thus, bringing back example (1) above, the specification of the roles played by the constituents represented by the variables would entail understanding the event that the verb refers to as involving two "participants," the one who drinks and the thing drunk as in (6).

(6) drink ($Agent_x$ ($Patient_y$))

The notion of thematic/semantic role has been part of modern linguistic theory at least since Gruber (1965) and Fillmore (1968), and also since Jackendoff (1972) and Katz (1972). In Chomsky's (1981) theory, thematic roles were incorporated as a form of semantic licensing of constituents (semantic selection or *s-selection*), at the interface between syntactic structures and semantic interpretation. The semantic properties assigned by lexical heads to their s-selected constituents were subject to the *theta* criterion, which governed their assignment. The *theta* criterion in fact constituted a series of principles governing the assignment of thematic roles—such as the assignment of roles to constituents in argument positions, and the assignment of only one role per argument. Notice that the thematic roles as in (6) are, in principle, independent of the actual content of the elements occupying *theta* positions. If verbs carry information about the thematic roles that they assign to *theta* positions, even semantically anomalous cases such as in (7) are assigned the thematic roles in (6).

(7) The table drank the sandwich

That is, from the perspective of syntax, the *theta* criterion is blind to the *content* of arguments, serving mainly to inform semantic interpretation about the nature of the constituents in argument positions. Several other theories have also made use of thematic roles to characterize the nature of participants in events. For instance, Parsons (1990) introduced thematic roles at the logical form (LF) representation of the event denoted by the sentence, as in (8b), with a reading such as in (8c).

(8) a. Brutus stabbed Caesar
b. ($\exists e$) [Stabbing(e) & *Agent*(e, Brutus) & *Theme*(e, Caesar) & ($\exists t$)[$t <$ now & Cul(e, t)]]
c. There is an event which is a stabbing, the *Agent* of the event is Brutus, and the *Theme* of the event is Caesar; and there is a time, the time is before now, and the event culminated at that time.

The so-called neo-Davidsonian view represented by Parsons (1990) is just one example of how event structures combine with information about thematic roles to convey the meanings of predicates and their carrier sentences. Another is Engelberg's (2004) *lexical event structures,* in which thematic roles are also included with event-structure information, although in this theory lexical conceptual information is kept separate from the verb event structure. And, as we will see in the next section, lexical semantic theories have also incorporated the notion of thematic roles by turning them into conceptual primitives of verb conceptual templates (Jackendoff 1990; Levin and Rappaport Hovav 2005; Randall 2010).

Although the use of thematic roles to convey information about event participants was and still is common to many semantic approaches to verb meaning, several proposals coming from syntax have put into question the explanatory power of lexically driven thematic assignment. For instance, the reduction of the syntactic machinery brought about by the minimalist program in syntax (Chomsky 1995) relegated thematic assignment to representations at LF. Much of what was taken to be the *theta* criterion was shown to be dealt with by syntactic computations proper without the need for a grammatical module assigning semantic roles to argument positions, thus eliminating the redundancy between lexically driven and syntactically driven structures. This of course does not mean that arguments do not bear thematic roles but thematic roles and the *theta* criterion lose their explanatory status as a module of the grammar, becoming largely a descriptive tool (see Harley (2011) for discussion on these issues).

But, even for those who assume a lexically driven form of argument structure, thematic roles have been under question. Hale and Keyser (1993) were perhaps one of the first to attempt to reduce thematic roles to syntactic positions governed by properties of their host predicates. Thus, for instance, Hale and Keyser (1993, 2002) suggested that several thematic-role effects could be attributed to lexical syntactic structures, involving operations such as "conflation" (or incorporation) of predicates. They proposed, for example, that unergative structures (akin to (4b) above) have in fact a *transitive* (4c) structure, with an argument in the object position that corresponds to the morphologically related noun. In order to illustrate this operation, consider the sentence in (9a) and the (simplified) structure in (9b), with roughly the interpretation in (9c).

(9) a. Isadora danced

b.

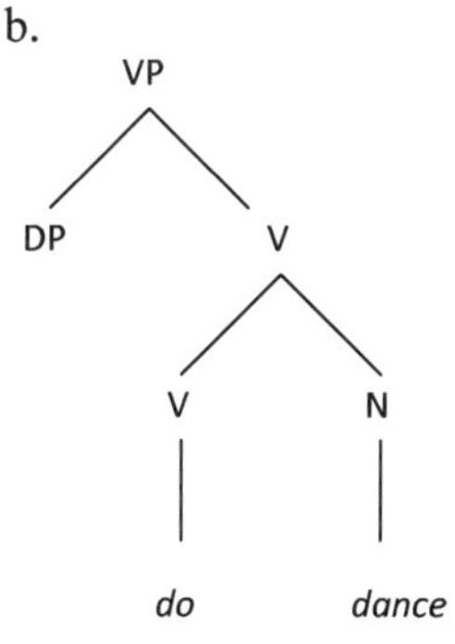

c. Isadora did a dance[14]

For Hale and Keyser, verbs that are derived from nouns (i.e., denominal verbs) incorporate a light verb (*do* in (9)). It is this hidden verb that, by hypothesis, would assign the *Agent* thematic role to the subject position (*DP, Isadora*). If all unergative verbs have such structure, then they all assign the same roles and thus thematic roles could perhaps be superseded by operations that are purely syntactic (L-syntactic or part of lexical structure). Hale and Keyser identified other similar operations that turn thematic properties into syntactic and morphological properties of predicates. Relevant for the present purposes is that, in their view, thematic roles can be dispensed with by the regularities brought about by lexical structures.

But it is not clear if we are ready to fully reject a taxonomy of roles based on the hypothesis that some verb classes conflate a covert light predicate; if we were to adopt this theoretical position, then in fact thematic roles would be reduced to syntactic operations rather than being verb-specific (or class-specific) projections of arguments bearing thematic content. One problem with the analyses that Hale and Keyser put forward is that they rest on the hypothesis that there is a hidden verb in the zero-derived denominal verb. An analysis of Hale and Keyser's arguments show that what they are arguing for is that verbs have roughly a definitional structure (viz., that *dance* means DO A DANCE or that *shelve* means PUT ON THE SHELF, among others). But as Fodor and Lepore (1999) noticed, Hale and Keyser's periphrastic (i.e., decomposed) versions of their conflated structures were not synonymous with their lexicalized structures, following many of the arguments that Fodor (1970) mounted against the decomposition of *kill* into CAUSE TO DIE. We will return to this issue in Sect. 1.5. For now, suffice it to say that we suspect that if the analyses of predicate argument structures proposed by Hale and Keyser cannot be sustained, their conclusions about the fate of thematic roles should be seen with caution.

As we have seen so far, a discussion of thematic roles cannot be entirely dissociated from predicate arguments, which are supposed to be the bearers of thematic role content. However, we can approach the nature of thematic roles from a "higher" stance, i.e., by evaluating the role they play in the content of the predicates that assign them. Let us assume that verbs do have lexically encoded argument structure, *contra* some evidence we discussed in the last section. Let us also assume that the meaning of a predicate is tied not only to the verb's content (see Sect. 1.2) but also to the very nature of the participants that the verb licenses. Taking both assumptions into account, we can say that thematic roles provide *content,* i.e., information that enables us to represent whole event/state *types,* beyond token verb meanings. That is to say that the content of an event/state lexicalized by the verb goes beyond the verb's content and spills over its arguments. In our discussion of argument structure,

[14] To be consistent with our metalanguage, the interpretation in (9c) should be roughly [ISADORA DID A DANCE], corresponding to the concepts constituents of (9a), assuming this sentence is in fact structured as in (9b). Of course, this conceptual interpretation would be the same had the natural language expression been what it is in (9c).

above, we said that arguments might be specified as variables devoid of content, corresponding to the syntactic constituents that the verb makes obligatory. If we assume that thematic roles are content-bearing entities, carrying necessary information about events/states, then arguments are not simply structural positions devoid of content, but structural or syntactic manifestations of obligatory meaning components. This position is compatible with other views of thematic roles—e.g., Reinhart's (2002) who sees thematic role *features* as an essential component of the interface between language and the conceptual system.

Assuming that some form of thematic role system still plays a role at the interface between linguistic and conceptual representations, let us now examine a few issues that have motivated experimental studies in cognitive science.

One advantage of *theta*-role grids such as in (6) is that they allow for verb classifications based on the number, kind, and positions of the roles that verbs assign. Thus, verbs such as those in (10a) may all be analyzed as having a thematic structure such as the one in (10b).

(10) a. fear, admire, despise, hate, dread, prize, deplore, appreciate
b. ($Experiencer_x$ ($Theme_y$))[15]

There is no major agreement on these labels and it is often difficult to figure out the actual role played by some arguments in events. In fact, those were some of the reasons that lead Hale and Keyser to seek alternative analyses in lexical-syntactic projections. However this plays out in terms of labels, there seems to be some agreement on a few of these labels as well as on what they stand for. Recall that when we presented the standard view of argument structure we also mentioned that in Grimshaw's (1990) theory, argument structures (such as (6) and (10b)) are representations of prominence relations among arguments, which are based in part on what has been called a *thematic hierarchy*.

There have been several proposals for thematic hierarchy, all sharing the basic assumption that meaning-to-form mapping follows some form of hierarchical relations between thematic roles. Table 1.1 presents a sample of thematic hierarchies. As observed by Levin and Rappaport Hovav (2005), thematic hierarchies have an explanatory value in accounting for the mapping between semantic roles and grammatical relations, allowing for a particular argument to be referred to in terms of its relative position in the hierarchy rather than in terms of its semantic role proper. Consider the sentences in (11a–c) with their respective thematic grids.

(11) a. The boy_x opened the $door_y$ ($Agent_x$ ($Patient_y$))
b. The key_x opened the $door_y$ ($Instrument_x$ ($Patient_y$))
c. The $door_y$ opened ($Theme_y$)

When a verb allows for these thematic alternatives, the hierarchy specifies which argument takes the external (subject) position. Fillmore (1968, p. 33), for instance,

[15] We will not dwell here on the proper labels—e.g., whether the object is a *Stimulus*, a *Causer*, or a *Theme* that the subject experiences. The same applies to example (11) below—whether the internal argument of *open* is a *Patient* or *Theme*.

Table 1.1 Sample thematic hierarchies

Study	Thematic hierarchy
Fillmore (1968)	*Ag* > *Ins* > *Th*
Jackendoff (1972)	*Ag* > *G/S/L* > *Th*
Givon (1984)	*Ag* > *Ben* > *Pat* > *L* > *Ins*
Belletti and Rizzi (1988)	*Ag* > *Exp* >*Th*
Baker (1989)	*Ag* > *Ins* > *Th/Pat* > *G/L*
Grimshaw (1990)	*Ag* > *Exp* > *G/S/L* > *Th*
Van Valin (1990)	*Ag* > *Eff* > *Exp* > *L* > *Th* > *Pat*
Jackendoff (2002)	*Ag* > *Rec* > *Th* > *L* > *Pred NP*

Ag Agent, *Exp* Experiencer, *Ins* Instrument, *Pat* Patient, *G* Goal, *S* Source, *L* Location, *Rec* Recipient, *Th* Theme, *Eff* Effector, *NP* Predicate (e.g., a genius)

suggested that the presence of an *Agent* makes it the subject of a sentence (as in (11a)), with an *Instrument* taking up this role in the absence of an *Agent* (11b). Although there is considerable variability in the rankings of thematic roles, as can be seen in Table 1.1, they all agree that whenever there is an *Agent*, it occupies the subject position.

Several studies have investigated the general hypothesis of a thematic hierarchy and, more specifically, if deviations from thematic hierarchies have a processing correlate. The goal in most cases is to understand how the processor deals with noncanonical mappings from thematic to syntactic structure and how this mapping might break down in populations with impaired semantic systems. The importance of this topic is manifested by the fact that several experimental studies have given rise to models of language comprehension/production making reference to a processing level involving the checking of thematic roles or their proper assignment to sentence constituents (e.g., Frazier and Clifton 1996; Bornkessel and Schlesewsky 2006). A common assumption is that the types of arguments required by a verb and their possible thematic roles are taken into account during the very early stages of processing. For instance, Bornkessel and Schlesewsky (2006) formulated the argument dependency model (ADM) which aims to provide an account of hierarchy mismatches in sentence comprehension. ADM is based on the incrementality of sentence comprehension assuming that hierarchical thematic dependencies are immediately set, even before the verb is encountered. As a consequence, the initial argument is interpreted as thematically higher ranking, according to hierarchical demands. In case there is a mismatch between the thematic structure and the hierarchical thematic relations, reanalysis occurs. The prediction of such model is that verbs with noncanonical argument realization, such as object-*Experiencer* verbs (e.g., *frighten*), should be harder to process and they should trigger thematic reanalysis on the assumption that *Experiencer* should be assigned canonically to an earlier argument in subject position. Indeed, it seems that various studies that have manipulated argument realization have found increased reaction times in the locus of the predicted reanalysis. (e.g., Ferreira 1994, 2003; Manouilidou and de Almeida 2013; Verhoeven 2014).

Language impairment studies examining the correspondence between thematic roles and syntactic properties have demonstrated that patients have difficulties processing sentences with deviations from canonical structure (e.g. Zurif and Swinney 1994; Burchert et al. 2008; Thompson and Lee 2009; Manouilidou et al. 2009; Dragoy and Bastiaanse 2010). Our study (Manouilidou et al. 2009) involved Alzheimer's patients, who are known to have affected semantic memory systems. Few studies have shown that Alzheimer's patients have linguistic (i.e., syntactic, argument-structure) problems other than higher-level semantic deficits (e.g., Bencini et al. 2011). We investigated whether deviations from thematic hierarchy (e.g., no *Agent*) would affect patients' production and comprehension of sentences, on the assumption that greater deviations from hierarchical order would engender worse performance. Moreover, we hypothesized that noncanonical argument realization would engender greater difficulty than canonical realization. In the main experimental conditions, we employed two classes of psychological verbs, subject-*Experiencer* (*fear*) verbs, which by hypothesis assign no *Agent* role and object-*Experiencer* (*frighten*) psych verbs which entail noncanonical argument realization (mismatch between the thematic hierarchy and the actual realization of the arguments, with *Theme* preceding *Experiencer*). For each sentence frame (e.g., *The boy____the thunder*), patients were required to select a verb, among four alternatives, that would best fill in the frame. The alternatives included the target (e.g., *feared*), a semantic competitor (*frightened*) and two distractors. Results showed that patients had difficulties completing the sentence when the target verb was a subject-*Experience* (*fear*) and even greater difficulty when the frame required an object-*Experience* (*frightened* for a sentence frame such as *The thunder____the boy*). Interestingly, patients had near-normal performance (compared to a group of age- and education-matched controls and a group of young controls) with sentences that took canonical *Agent–Theme* verbs (e.g., *kick*). When we looked at the pattern of errors for the psych verb conditions, we also found that patients selected the competitor about 70 % of the time, suggesting that while patients were able to discard the unrelated distractors, the difficulty choosing between target and competitor (e.g., the near reversible pair *fear/frighten*) could be due to the proper thematic roles assigned by these verbs. It is important to note that the difficulty was not with the linear order of constituents because the same pattern of results was obtained with passive sentences (e.g., *The thunder was ____by the boy, The boy was_____by the thunder*). See Fig. 1.2 for the Alzheimer's patient data. We suggested that the pattern of performance by patients with Alzheimer's was not entirely consistent with extant thematic hierarchy proposals (e.g., Belletti and Rizzi 1988). For instance, patients had difficulty with sentences lacking *Agent,* even when the argument realization was canonical (e.g., in *Experience–Theme* frames) but had no difficulty with some noncanonical structures (e.g., *Theme–Agent*; see also Manouilidou and de Almeida (2009) for discussion).

Although thematic roles are among the most controversial types of linguistic representations bearing on verb meaning (see, e.g., Newmeyer 2002), they also seem

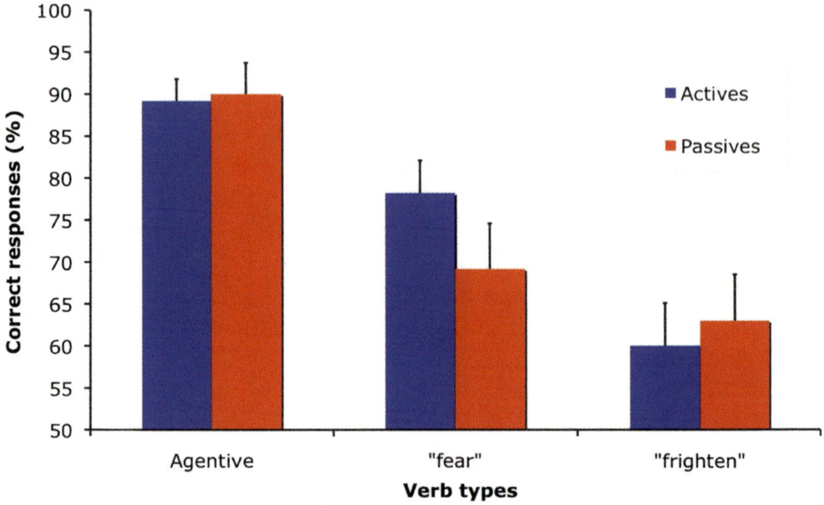

Fig. 1.2 Alzheimer's patient data from Manouilidou et al. (2009). The bars show correct selection of verb type per sentence condition (see text). *Error bars* represent standard error

to account consistently for effects in sentence processing. In fact, most of the evidence we presented for argument-structure effects cannot be easily dissociated from thematic-role effects; nor can we easily discard the role of a thematic hierarchy in accounting for the pattern of semantic impairment. What seems to be an important point of contention is the proper characterization of thematic roles and in particular their level of representation. While Newmeyer (2002, p. 71) appears to discard thematic roles from the "core" mechanisms of the language faculty due to inconsistencies in hierarchies and their realization, as well as the proliferation of thematic-role labels, we think their status is yet to be determined. What is clear is that until other grammatical or semantic constraints account for the thematic-role effects found in the linguistic and experimental data, we can claim their psychological reality. Even if it turns out that they are not part of the "core," thematic roles might have a role to play most likely at the interface between linguistic and conceptual systems, assisting in the mapping of form to meaning.

1.5 Conceptual Structure

The final and "highest" level of verb representation we would like to discuss is what is generally called *conceptual structure*. This label covers at least two possibly opposing views: one assumes that concepts are simplex (or atomic) and the other, that concepts are complex (or molecular). From this latter perspective, concepts themselves are structured representations, but the machinery responsible for combining concepts might also involve a fair amount of structuring, perhaps akin to natural language syntax or predicate logic. From the former perspective, that concepts are atomic, the basic units of representation are not themselves structured, but to a large extent what the conceptual structure does is to combine concepts, also deploying something akin to a logical or a syntactic structuring system. One could as well call it a *conceptual system,* or the *language of thought,* assuming that concepts are the elements of thoughts (Fodor 1998; Fodor and Pylyshyn 2014).[16] In Sect. 1.2, we alluded to the idea that verbs are lexicalizations of "happenings" (Levin and Rappaport Hovav 2005) and much of what we represent (and, thus, lexicalize) are properties of events and states out there in the world, including those that are beyond the perceptual circle. The conceptual representation of a verb thus conceived stands for the ultimate codes serving for other cognitive processes—not only linguistic interpretation and production but also those involving actual events and states.

A fair question at this juncture is whether what we have presented so far about conceptual structure characterizes in any sense a *linguistic* level of representation. When one surveys theories of verb meaning, it is not always clear at which level the purported representations are encoded. While Jackendoff (1983 and subsequent work) assumes that verb conceptual structure does not constitute a level of linguistic representation (see below), others such as Levin and Rappaport Hovav (2005) assume a certain degree of autonomy for verb-conceptual representations, interacting with syntax. It has been a long tradition in linguistics to attribute to the semantic system its proper level within the language faculty. The work by Katz (1972) typifies this latter perspective. For him, "... the semantic component of the grammar contains a dictionary that formally specifies the senses of every syntactically atomic constituent in the language" (p. 33). What Katz had in mind for the semantic system was akin to syntax: "... a specification of the form of the dictionary and a specification of the form of the rules that project semantic representations for complex syntactic constituents from the dictionary's representations of the senses of their minimal syntactic parts" (p. 33).

Clearly, when one thinks about issues of "semantic" representation—the level of LF and other representations bearing meaning—much of the groundwork comes from semantics conceived as a level of linguistic analysis, perhaps with its own principles, interfacing with syntax and other linguistic representations. Work on the logical properties of linguistic expressions, including lexical representations and compositional processes are properly semantic. But one could well see LF, for instance, as a system responsible for structuring representations at the conceptual level, whether these representations are linguistic—output from the language faculty—or whether they are nonlinguistic (such as in the output of visual perception). Thus, the

[16] And yet there are those who do not believe there is a conceptual system, but only "conceptual processes" (Barsalou et al. 2003) implemented by linguistic and other input/output systems, including action. We will have to restrain our discussion to those who assume there is some form of cognitive system enabling conceptual processes. But see, e.g., Chap. 9 for a discussion on how a distributed account of verb meaning might be implemented.

conceptual representation in (8) above is not about the sentence (exclusively) but an encoding of the event of Brutus killing Ceasar at a certain point in time.[17]

This semantics tradition in linguistics, in particular in generative grammar, and much of what has followed, were attempts—explicit or not—to meet Quine's challenge according to whom "...pending a satisfactory explanation of the notion of meaning, linguists in semantic fields are in the situation of not knowing what they are talking about" (Quine 1953, p. 47). For him, "the idea of the mental counterpart of a linguistic form is worse than worthless for linguistic science" (p. 48). Consistent with this view, in subsequent work, he famously defended a behaviorist view of language acquisition and use (Quine 1960). However, much of the contemporary work in semantics (and on concepts) and related fields are aimed at characterizing the very idea that meaning *is* mentally represented and that the codes represented in the mind/brain serve for other cognitive processes, including but not exclusively, language.

The current work on conceptual structure born out of linguistics has been instigated mainly by Jackendoff (e.g., 1983, 1990) as well as by researchers working in generative semantics and cognitive linguistics, broadly speaking (see, e.g., McCawley 1972; Croft 2012, Chap. 5). This work amounts to a large constellation of ideas on what the conceptual level is, and how it interfaces with other systems, such as language, vision, and action. From Jackendoff's (1983) early conception, conceptual structure is seen as a central cognitive system, operating at the outputs of diverse input systems such as vision and language. In Jackendoff's theory, the algebraic language that operates on conceptual representations is a development of Fodor's (1975) *language of thought* hypothesis; except that, contra Fodor's perspective, the algebraic language of conceptual structure also serves for structuring concepts themselves from a set of innate primitives.

There are numerous views on the representation of verb meaning, more specifically, many seemingly compatible, sharing common properties as semantic primitives and variables standing for the linguistic arguments of a verb. In order to motivate our discussion—and some of the experiments reviewed below—in (13) we present a brief example of different notations used to convey verb meanings, according to three theories of verb representation (see Engelberg (2004, 2011a) for a review of several theories of verb meaning decomposition).[18]

(13) a. Conceptual Semantics (Jackendoff 1990)
dress: [CAUSE ([$]_i$, [GO ([$]_j$, [TO [IN [CLOTHING$]]_k$])])][19]
Mary$_i$ *dressed John(/herself)*$_j$

---

[17] For a more in-depth discussion on the representation and processing of events see Chap. 6, 8. For a perspective on the encoding of events by bilinguals, see Chap. 11.

[18] Also see Engelberg (2011b) for a comprehensive review of issues involved in lexical decomposition, most of which we cannot begin to discuss here.

[19] We present a simplified version of these templates. The conceptual templates in Jackendoff's theory (at least in his 1990 work) involve also several features, assuming that even CAUSE is decomposable or that it entails different events. They also contain an *action tier* specifying whether or not the object is *affected*. We will not get into these details here.

b. Lexical Conceptual Structure (Levin and Rappaport Hovav 2005)
break: [[*x* ACT] CAUSE [BECOME [*y* < *BROKEN* >]]
John$_x$ *broke the vase*$_y$

c. Lexical Decomposition Grammar (Wunderlich 1996)
dry: $\lambda x \lambda y \lambda e^{<P,T>}$ CAUSE(x, BECOME(DRY(y))) (e)
Mary$_x$ *dried (off) her clothes*$_y$

These theories differ in significant ways (e.g., from the level of the purported representation, to the kinds of features that enter into the templates), but they are in general agreement about the very idea that surface verbs are represented more deeply (or at a "higher" level) by other, perhaps conceptually primitive, predicates. All the verbs in (13) and their corresponding decompositions are taken to be *causatives,* i.e., verbs denoting an agent's act which brings about a change of state in an entity or a patient. This form of "defining" causatives is, of course, inherent to theories that assume that the event/state that the verb labels is represented by certain regularities involving participants (namely, the fillers of arguments) and predicates making explicit what these participants do and what they cause. The very idea that causative verbs are represented by templates such as those in (13) came from generative-semantic analyses, which introduced semantic primitives into the tree-structural representations of sentences (e.g., McCawley 1972). According to this proposal, two main processes were involved in the transformation of the complex semantic expression involving predicates such as CAUSE and BECOME into a surface verb such as *close:* predicate raising and lexicalization. Transformations would successively raise predicates in the *deep* structure of a sentence and adjoin them next to the immediate higher ones, which would then be lexicalized into a verb such as *break* at surface structure. What is important to note regarding McCawley's proposal is that CAUSE and other predicates were taken to be semantic primitives that form complex structures underlying simple morphologically unmarked forms as *break*. The proposal that semantic primitives form the basic ingredients in the analysis of verb meaning was adopted by many semantic theories past and present, as can be seen in (13).

It is interesting to note that if verbs have a conceptual-structure representation such as those in (13), their representations encode, among other properties, information about the roles played by different arguments. For instance, if *break* is represented as in (13b), the role played by the argument (*x*) in the subject position is determined by a predicate *ACT*; similarly, *BECOME* represents the role played by the object (*y*) argument. Thus the mapping from syntax to conceptual-structure template can dispense with thematic roles if we encode verb meanings as templates.

The theories exemplified in (13) have many interesting characteristics. First, they account for our intuitions on the nature of the events denoted by the verb in terms of relations between participants and their roles in the events/states referred to by their carrier sentences. Second, the postulation of common predicates and similar template structures is taken into account for the categorization of "happenings," similar to that obtained for "things" (see Smith and Medin 1981); more specifically, this enables the conceptual system to represent classes for purportedly similar verb types based on the idea of shared constituents and structure the same way two concepts or categories are supposed to share their sets of features or properties. By the same token, verb meanings represented as sets of primitives and their relations allow for a typology of events/states, thus enabling verb classifications within and across languages. In addition to some of these arguments for decompositions such as in (13), there is also some experimental evidence supporting this type of theoretical account of verb meaning. We discuss this evidence below.

Despite some of the advantages of verb decomposition theories, there have been many challenges to the views represented in (13). First, there are the arguments (and distributional evidence) against the synonymy of *kill* and *cause to die*. Second, there are many reasons for rejecting (and, apparently, lack of criterion for) the analytic/synthetic distinction upon which these theories rest. And finally, there is also some experimental evidence *against* the lexical decomposition hypothesis more generally, and against the decomposition of causatives, more specifically. We will address the first two challenges before discussing experimental work.

The arguments—and evidence—against the synonymy of *kill* and *cause to die* were first developed by Fodor (1970) and rely on distributional analyses of sentences containing the verb and its periphrastic counterpart. First, we should say that the assumption is that if the lexical item *kill* is semantically represented by something like CAUSE TO DIE (with a template such as (13b)), so is the overt linguistic expression *cause to die,* unless there are reasons for assuming that, e.g., *cause* does not mean CAUSE and *die* does not mean DIE. Leaving aside this eerie possibility, Fodor's arguments were based on the effects that replacing *kill* for *cause to die* have on the resulting semantic representation of sentences as well as on their entailments. Consider two of the arguments, one exemplified in (14a–d), and the other, in (15a, b).

(14) a. John caused Mary to die and it surprised me that he did so
b. John caused Mary to die and it surprised me that she did so
c. John killed Mary and it surprised me that he did so
d. *John killed Mary and it surprised me that she did so
(15) a. John caused Bill to die on Sunday by stabbing him on Saturday
b. *John killed Bill on Sunday by stabbing him on Saturday

As can be seen in (14), while *Mary* (*she*) is the subject of the elliptical verb in (14b), *Mary* cannot be the subject of the elliptical verb *kill* in (14d) showing that at least the surface predicates in (14) have different distributional properties and that the sentences are not synonymous. In (15a), the event that ultimately led to Bill's death on Sunday could have happened on Saturday. But *kill* in (15b) allows for only one adverb because *kill,* but not *cause to die*, points to one event. While it is clear that the surface predicates have different "behaviors," what is important for the present discussion is that their semantic translations carry different properties: If we were to translate *kill* for CAUSE TO DIE, the semantic representations of those sentences would inherit the anomalies that their linguistic equivalents carry. In essence, these examples show that to assume that one-predicate sentences can be represented by two-predicate structures leads to representations that do not preserve the meaning that the sentences are supposed to express.

Although these arguments have gone virtually unchallenged (see, e.g., Jackendoff 1990, 2002 for exceptions)[20], there is an even older reason to suspect of the lexical decomposition project. We invoked Quine's "challenge" to semantics only to motivate our presentation of the current view of semantics—the discipline—as part of the endeavor aiming to understand the ultimate mental representation codes carrying conceptual information. But if on matters of semantics *representation,* Quine's view has been to a large extent deflated, some remnants of his critique remain en vogue: the lack of a clear criterion for establishing an analytic/synthetic distinction (Quine 1951). Roughly speaking, lexical semantic theories such as the ones exemplified in (13), assume a form of representation (the templates with their predicates and features) that embody ideas born out of definitional theories (e.g., Katz 1972): they require a criterion for sorting out the features (or predicates) that are *necessary* from the ones that are contingent on experience. Not having such a criterion—and in fact not being able to sort out between necessary and contingent ones—leads to a semantic dead end. Not all of those who are committed to lexical decomposition neglect these difficulties, but surprisingly many do.

Among the psycholinguistic studies investigating the nature of verb-conceptual representation, some have supported the predicate decomposition theory (Gennari and Poeppel 2003; Gentner 1981; McKoon and Love 2011; McKoon and McFarland 2000, 2002) while some others have failed to find evidence for decomposition (e.g., de Almeida 1999a; Fodor et al. 1975, 1980; Kintsch 1974; Manouilidou and de Almeida 2013; Rayner and Duffy 1986; Thorndyke 1975). These studies vary widely in terms of methods, materials, and in particular the predictions on what should count as evidence for and against decomposition.

We start off with experiments supporting the long-held idea of decomposition. To our knowledge, the first experimental support for decompositional structures came from studies by Gentner (1975, 1981) which assumed that more complex structures were deemed more memorable because they had more components upon which to create meaningful connections (e.g., verbs such as *receive* and *borrow* are supposed to share constituents such as CAUSE and CHANGE OF POSSESSION). Although her study employed a small number of materials, and the results were based on a small proportion of recall errors committed by the subjects, greater confusions were obtained with items that supposedly share more constituents. Two more recent studies supporting the decompositional hypothesis are also of note here. In one (McKoon and McFarland 2000), participants were presented with two types of change-of-state

[20] Although we cannot address all arguments posed by Jackendoff (e.g., 1990, 2002) for the decomposition of lexical causatives (or more properly *against* the view that lexical concepts *do not* decompose), it is important to note that Jackendoff assumes that the best course for semantics (or the study of conceptual structure) is to rely on the ample analytic possibilities that decomposition affords, for decomposition "... is a richly textured system whose subtleties we are only beginning to appreciate (...). It does remain to be seen whether all this richness eventually boils down to a system built from primitives, or if not, what alternative there may be. And it does remain to be seen whether lexical meaning can be exhaustively constituted by the techniques discussed here" (Jackendoff 2002, p. 377).

sentences, denoting an externally caused change-of-state event as in (16a), and an internally caused change-of-state, as in (17a) (semantic templates such as (16b) and (17b) represent their analyses). They found that the more complex type of sentence, (16a), took longer to accept than the simplex type, (17a).

(16) a. The cement crumbled
 b. [[*x* ACT] CAUSE [*y* BECOME < *crumbled* >]]
(17) a. The potatoes rotted
 b. [*x* BECOME < *rot* >]

It is important to note that these two types of verbs are usually represented by different *argument structures/transitivity properties*; and they also differ with regard to semantic properties: e.g., while many things *crumble,* only a few things *rot* (Levin and Rappaport Hovav 2005).[21] Thus, it is possible that differences in response times between these two conditions reflect other aspects of the verbs' content rather than their templates.

Another study supporting decomposition (Gennari and Poeppel 2003) found similar effects of template complexity: sentences with eventive verbs such as (18a) took longer to read (self-paced) at the verb position than sentences with stative verbs such as (19a), supposedly because these constructions are represented by templates such as those in (18b) and (19b) (based on their analyses/notation).

(18) a. The young boy bullied his parents
 b. [x CAUSE [y BECOME < *bullied* >]]
(19) a. The young boy adored his parents
 b. [*x adore y*]

With regard to evidence against decomposition, although there have been a few other studies (as early as Kintsch's 1974), perhaps the most persuasive was by Fodor et al. (1980) who employed a variety of sentence types. Relevant to the present discussion is their contrast between lexical causatives (e.g., *close*) and other verb types deemed semantically simplex (e.g., *sell*), as in (20).

(20) a. Despite protests from the manager, the owner closed the theater
 b. Despite protests from the manager, the owner sold the theater

In one of their experiments, Fodor et al. (1980) employed a related-intuitions task in which subjects had to judge how closely related the main arguments of the verb (e.g., *owner* and *theater*) were in the sentence. The task is supposed to capture the underlying semantic representation of the sentence. Under the decomposition hypothesis,

[21] As we briefly mentioned above (fn. 9), Putnam (1975) argued against meaning representation—at least against definitions—mostly because he assumed correctly that the definition of natural kind terms (*gold, tiger*) could only be given in scientific terms (viz., the tiger DNA), thus definitions could not be the representations upon which we rely when we entertain the meaning of such terms. We mention this motivated by the puzzle of the internally/externally caused distinction, which must rest on a mentally encoded knowledge of how molecules of potatoes and cement might behave upon rotting or crumbling, respectively.

owner and *theater* should be judged less related in the causative (20a) sentence than in the simple transitive case (20b). This is because there is supposed to be a "shift" in the predicate-argument relations if indeed the surface *close* is represented as something like [[*x* ACT] CAUSE [*y* BECOME < *CLOSED* >]]: namely, *x* becomes the "agent" of the causative predicate, while *y* becomes the thing undergoing a change of state. Fodor et al. found no difference between (20a) and (20b), while showing that the technique was sensitive to underlying semantic relations, using a control experiment. This effect was replicated by de Almeida (1999a) using a larger set of sentences than Fodor et al.'s and also employing the same related intuitions paradigm as well as different response–time techniques.

Experimental studies investigating the predicate decomposition hypothesis are rather few compared to the number of studies investigating argument structure and thematic roles. We do not think that this discrepancy is because most people stand against decomposition. On the contrary, it is possible that the paucity of experimental studies in this area reflect a tacit understanding—if not a general consensus—that decompositions are the standard or, as Jackendoff (2002) suggested, that lexical decompositions represent a more fertile ground for making progress in semantic/conceptual representation. It could also be that, similar to the psychological study of concepts and categorization, most researchers believe that decomposition is the *only* way to encode the meanings of verbs, or the only way to capture generalizations about verb classes as well as linguistic properties affecting the linguistic behavior of predicates (the linguistically active aspects of meaning). On the methodological side, it also be pointed out that experimental designs employed in the investigation of verb decomposition (or lack thereof) also differ substantially: Studies supporting decomposition (see above) have in general employed simple designs (e.g., simple vs. complex templates), which of course begs the question of the outcome of the studies, had differences between the two conditions been null. On the other side of the spectrum, experiments that have failed to find support for decomposition have usually employed more experimental conditions—the ones comparing the variable of interest (hypothetically simple vs. complex predicates) in addition to conditions designed to show that complexity effects would have been found in the variables of interest had they existed—i.e., that null effects are not due to methodological confounds.

While these considerations are important in the evaluation of the theories and experimental findings on both sides, there is yet a question of alternatives to verb decomposition. Perhaps one such alternative—adopted by few but perhaps one that appears to be equally powerful in terms of accounting for a wide range of phenomena that decompositional theories appear to account for—is what has been called "meaning postulates." This approach, inspired in Carnap (1956) and later supported by diverse theoretical and empirical works (e.g., de Almeida 1999a, b; Fodor et al. 1975; Fodor 1975; Partee 1995), appears to have some of the advantages of decomposition without some of its potential pitfalls. Crucial to this approach is its potential for accounting for the *entailments* between causatives and change-of-state events as well as relations between (verb) concepts belonging to different conceptual classes without committing to the conceptual constituency typical of semantic templates.

That is, the sets of entailments or inferences that a predicate (or other concept) triggers are not *analytic* entailments but constitute (nonlogical) inferential relations. This is not simply a notational variant of predicate decompositions because, contrary to templates, these entailments are not obtained by *necessity* as the constituents of templates are.

As we remarked in the previous sections, an understanding of conceptual structures should come from the interplay between diverse theoretical and experimental approaches. It is from this methodological stance typical of cognitive science—rather than from a commitment to the "richness" of decompositions—that a better understanding of verb semantic/conceptual phenomena might come.

1.6 Overview of Chapters

The chapters constituting the present volume address some of the above outlined issues but also go way beyond, either by contributing new theoretical insights (Chaps. 2, 3, 5) or by providing theoretical and experimental evidence from sentence processing (Chaps. 4, 8, 10), patient studies (Chap. 7), neuroimaging studies (Chaps. 6, 9), bilingualism (Chap. 11), as well as acquisition (Chaps. 12, 13). We present the chapters here by "method" but have organized them in the volume by proximity of topic, aware that no linear order would do justice to their intricacies.

Bill Croft's chapter proposes an analysis of event lexicalization and argument realization within the framework of force dynamics. He argues that the contributions of causal (force-dynamic) and aspectual structure can be most clearly identified by using a representation of event structure that includes both causal and aspectual structure but clearly distinguishes the two. The chapter also introduces the category of *directed change*, an aspectual category that, according to Croft, appears to play the most important role in understanding event lexicalization. Brendan Gillon focuses on the optionality of some verb complements and extends his proposal to adjectives. After providing a typology of intransitives, Gillon argues that optionally transitive verbs should not be taken for *ambiguous verbs* as previous research has considered them to be. Rather, he develops his account considering optionally transitive verbs unambiguous. Paul Pietroski, on the other hand, using a more philosophical approach stemming from formal semantics sees verbs as instructions to fetch monadic concepts which can be conjoined with others for composition. This perspective leads to a nonstandard conception of how words and the process of lexicalization are related to human thought and communication. It also helps make sense of some otherwise puzzling phenomena which suggest that lexical items do not themselves have fixed arguments. The chapter concludes by locating the specific proposal in the context of Chomsky's (1986, 1995) conception of distinctively human languages as biologically instantiated procedures, I-languages, whose expressions make contact with other cognitive systems.

1.6.1 On-Line Processing

Several chapters contribute not only experimental evidence but also theoretical analyses regarding aspect and event structure (Chap. 8), thematic roles (Chap. 4), grammar and semantic processing resources (Chap. 10). Matt Husband and Linnaea Stockall present a review of linguistic aspect from two perspectives: linguistic theory and on-line language comprehension. They touch upon issues related to event comprehension and the syntax–semantic interface. Their experiments provide a detailed look at the time course of aspectual interpretation and the processing of compositional structures more generally. Results argue for incremental commitment to aspectual interpretation, placing the commitment point for telicity at the VP, which is the first point when all the information needed to construct an aspectual interpretation has been provided to the system (i.e., both the verb and the internal argument). Gail Mauner's chapter also contributes data from processing about verb participants. The studies which she reports on (employing self-paced reading and visual world paradigms) indicate that whether participant role information is used predictively or instead is used later in the course of understanding a sentence depends upon constraints from both the linguistic and real-world contextual environments. Thus, while participant roles are rapidly activated upon verb recognition, whether participant role information is used anticipatorily depends in part on the availability of processing resources, which can be modulated by, among other things, referential contexts. Jean-Pierre Koenig and colleagues addressed the question of what causes the difference in the kind and amount of information used by the human parser and the human "grammar maker." They report on some computational models of on-line reading experiments which suggest that a distinct and much larger kind of event knowledge is used by the human parser. They propose an explanation for the difference in the use of event knowledge. Specifically, Koenig et al. conclude that grammars and parsers use different kinds of event knowledge because the tasks that listeners and grammar learners must perform are quite distinct.

1.6.2 Clinical, Electrophysiological, and Neuroimaging Studies

The study by Bastiaanse and Platonov involving data from aphasia brings evidence regarding the interaction between aspect and telicity. The authors contribute evidence from agrammatic aphasia in Russian-speaking individuals trying to delineate the observed verb deficit in agrammatic aphasia crosslinguistically. Results of a sentence–picture matching task support the predictions made by the *Aspect Assignment Model* which relates the observed difficulties with argument structure to difficulties in time reference, highlighting the role of aspectual selection.

Telicity (Chap. 6) and verb classes (Chap. 9) are the issues under investigation by two electrophysiological and neuroimaging chapters. Evie Malaia and colleagues focus on another semantic feature of verbs—*telicity*. The authors present electrophysiological and neuroimaging data on the processing of telic versus atelic verbs in spoken American English as well as in American and Croatian sign language. Combined results from both experiments point to early interaction of syntax and semantics in human languages, and suggest telicity correlates with neural resources used for language processing. David Kemmerer's chapter focuses on the idiosyncratic root-level semantic features of action verbs (*running, hitting, cutting, putting, throwing* verbs). His main goal is to show how recent developments in cognitive neuroscience have begun to illuminate the representational character of these aspects of verb meaning. By discussing fMRI, Kemmerer explores specific hypotheses within the Embodied Cognition Framework, that is, whether the visual motion features of action verbs and the motor features of action verbs depend on different cortical areas. Results from these studies suggest that distinguishing between, say, *running* verbs (e.g., *stroll, jog, run, sprint,* etc.) requires access to experience-based knowledge stored in modality-specific cortical areas. These areas partially overlap with those involved in perceiving and producing the designated types of actions.

1.6.3 Bilingualism and Acquisition

Finally, one chapter contributes new data from verb representation and processing in bilinguals, and two chapters approach the process of acquiring verbs. Vicky Lai and Bhuvana Narasimhan investigated how Spanish–English bilinguals represent and process path and manner of motion, on the assumption that different languages might encode and express these properties differently thus affecting how they are used in understanding/describing events. They provide evidence for the influence of verb-specific representations on "thinking-for-speaking." Sudha Arunachalam's chapter explores the persistent question of the relation between lexical (semantics) and syntactic structure in relation to acquisition. Arunachalam shows that any of the available theories can be more or less equally compatible with the acquisition data—this is to some extent expected as experiments are primarily designed following certain theoretical assumptions. In some cases, but not all, the same results can be made compatible with different theories, since they present alternative points of view. The challenging data then are those that are compatible with one kind of analysis but problematic for others. The study by Alexandra Marquis and Rushen Shi investigated the question of verb morphology and acquisition. The authors, who conducted their experiments in French-speaking children, argue for a decompositional view of infants' morphological development. In particular, they suggest that infants at the initial learning stage parse verb stems and affixes without relying on semantics but on the basis of high-token frequency of affixes and high-type frequency of stems (i.e., regular morphological operations).

While all these studies rely on linguistic-theoretical claims to motivate their theoretical or empirical investigations, they all employ multiple methods and draw from different disciplines constitutive of cognitive science. It is this interdisciplinary endeavor that might propel a shift—if not already happening—in the investigation of linguistic constructs as well as on the nature of the interface between linguistic and conceptual representations.

References

Åfarli, T. A. (2007). Do verbs have argument structure? In E. Reuland, T. Bhattacharya, & G. Spathas (Eds.), *Argument structure* (pp. 1–16). Amsterdam: John Benjamins.
Baker, M. (1989). Object sharing and projection in serial verb construction. *Linguistic Inquiry, 20*, 513–553.
Barsalou, L. W., Simons, K. W., Barbey, A. K., & Wilson, C. D. (2003). Grounding conceptual knowledge in modality-specific systems. *Trends in Cognitive Sciences, 7*, 84–91.
Bastiaanse, R., & Zonnevelt, R. van (2005). Sentence production with verbs of alternation transitivity in agrammatic Broca's Aphasia. *Journal of Neurolinguistics, 18*, 57–66.
Belletti, A., & Rizzi, L. (1988). Psych-verbs and θ-theory. *Natural Language and Linguistic Theory, 6*(3), 291–352.
Bencini, G. M. L., Pozzan, L., Biundo, R., McGeown, W. J., Valian, V. V., Venneri, A., & Semenza, C. (2011). Language-specific effects in Alzheimer's disease: Subject omission in Italian and English. *Journal of Neurolinguistics, 24*, 25–40.
Bever, T. G., & McElree, B. (1988). Empty categories access their antecedents during comprehension. *Linguistic Inquiry, 19*, 34–43.
Binder, J. R., Desai, R. H., Graves, W. W., & Conant, L. L. (2009). Where is the semantic system? A critical review and meta-analysis of 120 functional neuroimaging studies. *Cerebral Cortex, 19*, 2767–2796.
Boland, J. E. (2005). Visual arguments. *Cognition, 95*, 237–274.
Boland, J. E., Tanenhaus, M. K., & Garnsey, S. M. (1990). Evidence for the immediate use of verb control information in sentence processing. *Journal of Memory and Language, 29*, 413–432.
Borer, H. (2003). Exo-skeletal vs. endo-skeletal explanations: Syntactic projections and the Lexicon. In J. Moore & M. Polinsky (Eds.), *The nature of explanation in linguistic theory* (pp. 31–67). Stanford: CSLI.
Bornkessel, I., & Schlesewsky, M. (2006). Generalised semantic roles and syntactic templates: A new framework for language comprehension. In I. Bornkessel, M. Schlesewsky, B. Comrie, & A. D. Friederici (Eds.), *Semantic role universals and argument linking: Theoretical, typological and psycholinguistic perspectives* (pp. 327–353). Berlin: Mouton de Gruyter.
Bowers, J. (2002). Transitivity. *Linguistic Inquiry, 33*(2), 183–224.
Breedin, S. D., Saffran, E. M., & Schwartz, M. F. (1998). Semantic factors in verb retrieval: An effect of complexity. *Brain and Language, 63*(1), 1–31.
Burchert, F., Meiner, N., & De Bleser, R. (2008). Production of non-canonical sentences in agrammatic aphasia: Limits in representation or rule application? *Brain and Language, 104*, 170–179.
Carnap, R. (1956). Meaning postulates. In R. Carnap (Ed.), *Meaning and necessity* (2nd ed. pp. 222–232). Chicago: University of Chicago Press.
Chomsky, N. (1981). *Lectures on government and binding*. Dordrecht: Foris.
Chomsky, N. (1986). *Knowledge of language: Its nature, origin, and use*. New York: Praeger.
Chomsky, N. (1995). *The Minimalist Program*. Cambridge: MIT Press.
Croft, W. (2012). *Verbs: Aspect and causal structure*. Oxford: Oxford University Press.
Culicover, P. W., & Jackendoff, R. (2005). *Simpler syntax*. Oxford: Oxford University Press.
Davidson, D. (1967). The logical form of action sentences. In D. Davidson (Ed.) (1980), *Essays on actions and events* (pp. 105–122). Oxford: Clarendon.
de Almeida, R. G. (1999a). The representation of lexical concepts: A psycholinguistic inquiry. Unpublished PhD Dissertation, Rutgers University.
de Almeida, R. G. (1999b). What do category-specific semantic deficits tell us about the representation of lexical concepts? *Brain and Language, 68*, 241–248.
de Almeida, R. G. (2004). The effect of context on the processing of type-shifting verbs. *Brain and Language, 90*, 249–261.
de Almeida, R. G. (2006). On the status of "linguistic psychology" and the language faculty. In G. Wiebe, G. Libben, T. Priestly, R. Smyth, & S. Wang (Eds.), *Phonology, morphology, and the empirical imperative: Papers in honour of Bruce Derwing*. Taipei: Crane.

de Almeida, R. G., & Dwivedi, V. D. (2008). Coercion without lexical decomposition: Type-shifting effects revisited. *Canadian Journal of Linguistics, 53,* 301–326.
de Almeida, R. G. & Riven, L. (2012). Indeterminacy and coercion effects: Minimal representations with pragmatic enrichment. In A. M. Di Sciullo (Ed.), *Towards a biolinguistic understanding of grammar: Essays on interfaces*. Amsterdam: John Benjamins.
de Almeida, R. G., Riven, L., Manouilidou, C., Lungu, O., Dwivedi, V., Jarema, G., & Gillon, B. (2014). *Coercion or pragmatics? An fMRI study on indeterminate sentences.* Manuscript submitted for publication.
De Bleser, R., & Kauschke, C. (2003). Acquisition and loss of nouns and verbs: Parallel or divergent patterns? *Journal of Neurolinguistics, 16,* 213–229.
Di Sciullo, A. M. (2007). *Asymmetry in morphology*. Cambridge: MIT Press.
Di Sciullo, A. M., de Almeida, R. G., Manouilidou, C., & Dwivedi, V. D. (Aug 2007). This poster reads clearly: Processing English middle constructions. Poster presented at the at the Architectures and Mechanisms of Language Processing conference, Turku, Finland.
Dowty, D. (1991). Thematic proto-roles and argument selection. *Language, 67,* 547–619.
Dragoy, O., & Bastiaanse, R. (2010). Verb production and word order in Russian agrammatic speakers. *Aphasiology, 24,* 28–55.
Engelberg, S. (2004). Lexical event structures for verb semantics. *Journal of Language and Linguistics, 3,* 62–108.
Engelberg, S. (2011a). Frameworks of lexical decomposition of verbs. In C. Maienborn, K. von Heusinger, & P. Portner (Eds.), *Semantics: An international handbook of natural language meaning* (Vol. 1, pp. 358–399). Amsterdam: De Gruyter Mouton.
Engelberg, S. (2011b). Lexical decomposition: Foundational issues. In C. Maienborn, K. von Heusinger, & P. Portner (Eds.), *Semantics: An international handbook of natural language meaning* (Vol. 1, pp. 124–144). Amsterdam: De Gruyter Mouton.
Faroqi-Shah, Y., & Thompson, C. K. (2010). Production latencies of morphologically simple and complex verbs in aphasia. *Clinical Linguistics and Phonetics, 24,* 963–979.
Ferreira, F. (1994). Choice of passive voice is affected by verb type and animacy. *Journal of Memory and Language, 33,* 715–736.
Ferreira, F. (2003). The misinterpretation of noncanonical sentences. *Cognitive Psychology, 47,* 164–203.
Fillmore, C. J. (1968). The case for case. In E. Bach, & R. T. Harms (Eds.), *Universals in Linguistic Theory*. New York: Holt, Rinehart and Winston.
Fodor, J. A. (1970). Three reasons for not deriving 'kill' from 'cause to die'. *Linguistic Inquiry, 1,* 429–438.
Fodor, J. A. (1975). *The language of thought*. New York: Crowell.
Fodor, J. A. (1998). *Concepts: Where cognitive science went wrong*. Oxford: Oxford University Press.
Fodor, J. A., & Lepore, E. (1999). Impossible Words? *Linguistic Inquiry, 30,* 445–453.
Fodor, J. A., & Pylyshyn, Z. W. (2014). *Minds without meanings*. Cambridge: MIT Press.
Fodor, J. A., Garrett, M., & Bever, T. G. (1968). Some syntactic determinants of sentential complexity. II: Verb structure. *Perception and Psychophysics, 3,* 453–461.
Fodor, J. D., Fodor, J. A., & Garrett, M. F. (1975). The psychological unreality of semantic representations. *Linguistic Inquiry, 6,* 515–531.
Fodor, J. A., Garrett, M. F., Walker, E. C. T., & Parkes, C. H. (1980). Against Definitions. *Cognition, 8,* 263–267.
Frazier, L., & Clifton, C. Jr. (1996). *Construal*. Cambridge: MIT Press.
Friedmann, N., Shapiro, L. P., Taranto, G., & Swinney, D. (2008). The leaf fell (the leaf): The on-line processing of unaccusatives. *Linguistic Inquiry, 39*(3), 355–377.
Gennari, S., & Poeppel, D. (2003). Processing correlates of lexical semantic complexity. *Cognition, 89,* B27–B41.
Gentner, D. (1975). Evidence for the psychological reality of semantic components: The verbs of possession. In D. A. Norman & D. E. Rumelhart (Eds.), *Explorations in cognition* (pp. 211–246). San Francisco: Freeman.

Gentner, D. (1981). Verb semantic structures in memory for sentences: Evidence for componential representation. *Cognitive Psychology, 13,* 56–83.
Givon, T. (1984). *Syntax: A functional-typological introduction* (Vol. 1). Amsterdam: John Benjamins.
Gleitman, L., & Gleitman, H. (1992). A picture is worth a thousand words: The role of syntax in vocabulary acquisition. *Current Directions in Psychological Science, 1,* 31–35.
Gleitman, L., Cassidy, K., Nappa, R., Papafragou, A., & Trueswell, J. (2005). Hard words. *Language Learning and Development, 1,* 23–64.
Goldberg, A. (1995). *Constructions: A construction grammar approach to argument structure.* Chicago: University of Chicago Press.
Grimshaw, J. (1990). *Argument structure.* Cambridge: MIT Press.
Grimshaw, J. (1993). Semantic structure and semantic content in lexical representation. In J. Grimshaw (Ed.) (2005), *Words and structure* (pp. 75–89). Stanford: CSLI
Grossman, M., Mickanin, J., Onishi, K, & Hughes, E. (1996). Verb comprehension in probable Alzheimer's disease. *Brain and Language, 53,* 369–389.
Gruber, J. (1965). Studies in lexical relations. Ph. D. Dissertation, MIT, Cambridge.
Hale K., & Keyser, S. J. (1993). On argument structure and the lexical expression of syntactic relations. In: K. Hale & S. J. Keyser (Eds.), *View from building 20* (pp. 53–109). Cambridge: MIT Press.
Hale, K., & Keyser, J. K. (2002). *Prolegomenon to a theory of argument structure.* Cambridge: MIT Press.
Harley, H. (2011). A minimalist approach to argument structure. In C. Boeckx (Ed.), *The handbook of linguistic minimalism* (pp. 427–448). Oxford: Oxford University Press.
Hauser, M., Chomsky, N., & Fitch, W. T. (2002). The language faculty: What is it, who has it, and how did it evolve? *Science, 298,* 1569–1579.
Humphreys, G. W., & Forde, E. M. (2001). Hierarchies, similarity, and interactivity in object recognition: 'Category-Specific' neuropsychological deficits. *Behavioral Brain Sciences, 24,* 453–509.
Jackendoff, R. (1972). *Semantic interpretation in generative grammar.* Cambridge: MIT Press.
Jackendoff, R. (1983). *Semantics and cognition.* Cambridge: Cambridge University Press.
Jackendoff, R. (1990). *Semantic structures.* Cambridge: MIT Press.
Jackendoff, R. (2002). *Foundations of language: Brain, meaning, grammar, evolution.* Cambridge: MIT Press.
Jonkers, R., & Bastiaanse, R. (1998). How selective are selective word class deficits? Two case studies of action and object naming. *Aphasiology, 12,* 245–256.
Kaplan, R., & Bresnan, J. (1982). Lexical functional grammar: A formal system for grammatical representation. In J. Bresnan (Ed.), *The mental representation of grammatical relations* (pp. 173–281). Cambridge: MIT Press.
Katz, J. J. (1972). *Semantic theory.* New York: Harper & Row.
Kemmerer, D. (2006). Action verbs, argument structure constructions, and the mirror neuron system. In M. Arbib (Ed.), *Action to language via the mirror neuron system* (pp. 347–373). Cambridge: Cambridge University Press.
Kemmerer, D., & Eggleston, A. (2010). Nouns and verbs in the brain: Implications of linguistic typology for cognitive neuroscience. *Lingua, 120*(12), 2686–2690.
Kemmerer, D., Tranel, D., & Barrash, J. (2001). Patters of dissociation in the processin of verb meanings in brain-damaged subjects. *Language and Cognitive Processes, 16,* 1–34.
Keyser, S. J., & Roeper. T. (1984). On the middle and ergative constructions in English. *Linguistic Inquiry, 15,* 381–416.
Kim, M., & Thompson, C. K. (2000). Patterns of comprehension and production of nouns and verbs in agrammatism: Implications for lexical organization. *Brain and Language, 74,* 1–25.
Kim, M., & Thompson, C. K. (2004). Verb deficits in Alzheimer's disease and agrammatism: Implications for lexical organization. *Brain and Language, 88,* 1–20.
Kintsch, W. (1974). *The representation of meaning in memory.* Hillsdale: Lawrence Erlbaum.

Kiss, K. (2000). Effects of verb complexity on agrammatic aphasic's sentence production. In R. Bastiaanse & Y. Gordzinsky (Eds.), *Grammatical disorders in aphasia*. London: Whurr.
Laudan, L., Donovan, A., Laudan, R., Barker, P., Brown, H., Leplin, J., Thagard, P., & Wykstra, S. (1986). Scientific change: Philosophical models and historical research. *Synthese, 69,* 141–223.
Lee, M., & Thompson, C. K. (2004). Agrammatic aphasic production and comprehension of unaccusative verbs in sentence contexts. *Journal of Neurolinguistics, 17,* 315–330.
Levin, B., & Rappaport Hovav, M. (2005). *Argument realization*. Cambridge: Cambridge University Press.
Luzzatti, C., Raggi, R., Zonca, G., Pistarini, C., Contardi, A., & Pinna, G. D. (2002). Verb–noun double dissociation in aphasic lexical impairments: The role of word frequency and imageability. *Brain and Language, 81,* 432–444.
MacDonald, M. C., Pearlmutter, N. J., & Seidenberg, M. S. (1994). The lexical nature of syntactic ambiguity resolution. *Psychological Review, 101,* 676–703.
Manouilidou, C., & de Almeida, R. G. (2009). Canonicity in argument realization and verb semantic deficits in Alzheimer's disease. In S. Featherston & S. Winkler (Eds.), *The fruits of Empirical Linguistics, Volume I: The process* (pp. 123–150). Berlin: Mouton De Gruyter.
Manouilidou, C., & de Almeida, R. G. (2013). Processing correlates of verb typologies: Investigating internal structure and argument realization. *Linguistics, 51*(4), 767–792.
Manouilidou, C., de Almeida, R. G., Schwartz, G., & Nair, V. (2009). Thematic roles in Alzheimer's disease: Hierarchy violations in psychological predicates. *Journal of Neurolinguistics, 22,* 167–186.
Marantz, A. (2013). Verbal argument structure: Events and participants. *Lingua, 130,* 152–168.
Marr, D. C. (1982). *Vision*. San Francisco: Freeman.
Mauner, G., & Koenig, J.-P. (2000). Linguistic vs. conceptual sources of implicit agents in sentence comprehension. *Journal of Memory and Language, 43,* 110–134.
Mauner, G., Tanenhaus, M., & Carlson, G. (1995). Implicit arguments in sentence processing. *Journal of Memory and Language, 34,* 357–382.
McCawley, J. D. (1972). Lexical insertion in a transformational grammar without deep structure. In J. McCawley (Ed.), *Grammar and meaning* (pp. 155–166). New York: Academic.
McKoon, G., & Love, J. (2011). Verbs in the lexicon: Why is hitting easier than breaking? *Language and Cognition, 3*(2), 313–330.
McKoon, G., & McFarland, T. (2000). Externally and internally caused change of state verbs. *Language, 76,* 833–858.
McKoon, G., & McFarland, T. (2002). Event templates in the lexical representations of verbs. *Cognitive Psychology, 45,* 1–44.
Meltzer-Asscher, A., Schuchard, J., den Ouden, D. B., & Thompson, C. K. (2013). The neural substrates of complex argument structure representations: Processing "alternating transitivity" verbs. *Language and Cognitive Processes, 28,* 1154–1168.
Miller, G. A., & Johnson-Laird, P. N. (1976). *Language and perception*. Cambridge: Harvard University Press.
Murphy, G. L. (2002). *The big book of concepts*. Cambridge: MIT Press.
Newmeyer, F. (2002). Optimality and functionality: A critique of functionally-based optimality-theoretic syntax. *Natural Language & Linguistic Theory, 20,* 43–80.
Parsons, T. (1990). *Events in the semantics of English*. Cambridge: MIT Press.
Partee, B. (1986). Noun phrase interpretation and type-shifting principles (Reprinted in B. Partee (2004), Compositionality in formal semantics) (pp. 203–230). Oxford: Blackwell.
Partee, B. (1995). Lexical semantics and compositionality. In L. Gleitman & M. Liberman (Eds.), *An invitation to cognitive science: Language* (2nd ed., pp. 311–360). Cambridge: MIT Press.
Pietroski, P. (2005). *Events and semantic architecture*. Oxford: Oxford University Press.
Pollard, C., & Sag, I. A. (1994). *Head-driven phrase structure grammar*. Chicago: University of Chicago Press.
Pustejovsky, J. (1995). *The generative lexicon*. Cambridge: MIT Press.

Putnam, H. (1970). Is semantics possible? In H. Putnam (1975), *Mind, language, and reality: Philosophical papers* (Vol. 2, pp. 139–152). Cambridge: Cambridge University Press.

Putnam, H. (1975). The meaning of "meaning". In H. Putnam (Ed.) (1975), *Mind, language, and reality: Philosophical papers* (Vol. 2, pp. 215–271). Cambridge: Cambridge University Press.

Pylkkänen, L., & McElree, B. (2006). The syntax–semantics interface: On-line composition of sentence meaning. In M. Traxler & M. A. Gernsbacher (Eds.), *Handbook of psycholinguistics* (2nd ed.). New York: Elsevier.

Pylyshyn, Z. W. (1984). *Computation and cognition: Toward a foundation for cognitive science*. Cambridge: MIT Press.

Pylyshyn, Z. W. (2004). *Seeing and visualizing*. Cambridge: MIT Press.

Quine, W. V. O. (1951). Two dogmas of empiricism. In W. V. O. Quine (Ed.) (1961), *From a logical point of view* (pp. 20–46). Cambridge: Harvard University Press.

Quine, W. V. O. (1953). The problem of meaning in linguistics. In W. V. O. Quine (Ed.) (1961), *From a logical point of view* (pp. 47–64). Cambridge: Harvard University Press.

Quine, W. V. O. (1960). *Word and object*. Cambridge: MIT Press.

Randall, J. (2010). *Linking: The geometry of argument structure*. Boston: Springer.

Rayner, K., & Duffy, S. (1986). Lexical complexity and fixation times in reading: Effects of word frequency, verb complexity, and lexical ambiguity. *Memory and Cognition, 14*(3), 191–201.

Reinhart, T. (2002). The theta system: An overview. *Theoretical Linguistics, 28,* 229–290.

Smith, E. E., & Medin, D. L. (1981). *Categories and concepts*. Cambridge: Harvard University Press.

Thompson, C. K. (2003). Unaccusative verb production in agrammatic aphasia: The argument structure complexity hypothesis. *Journal of Neurolinguistics, 16,* 151–167.

Thompson, C. K., & Choy, J. J. (2009). Pronominal resolution and gap-filling in agrammatic aphasia: Evidence from eye movements. *Journal of Psycholinguistic Research, 38,* 255–283.

Thompson, C. K., & Lee, M. (2009). Psych verb production and comprehension in agrammatic Broca's aphasia. *Journal of Neurolinguistics, 22,* 354–369.

Thompson, C. K., Bonakdarpour, B., Fix, S., Blumenfeld, H., Parrish, T., Gitelman, D., & Mesulam M-M. (2007). Neural correlates of verb argument structure processing. *Journal of Cognitive Neuroscience, 19*(11), 1753–1767.

Thompson, C. K., Bonakdarpour, B., & Fix, S. (2010). Neural mechanisms of verb argument structure processing in agrammatic aphasic and healthy age-matched listeners. *Journal of Cognitive Neuroscience, 22,* 1993–2011.

Thorndyke, P. W. (1975). Conceptual complexity and imagery in comprehension and memory. *Journal of Vebal Learning and Verbal Behavior*, 14, 359–369.

Trueswell, J. C., & Kim, A. E. (1998). How to prune a garden path by nipping it in the bud: Fast priming of verb argument structure. *Journal of Memory and Language, 39,* 102–123.

Trueswell, J. C., Tanenhaus, M. K., & Kello, C. (1993). Verb-specific constraints in sentence processing: Separating effects of lexical preference from garden-paths. *Journal of Experimental Psychology, 19,* 528–553.

Tyler, L. K., & Moss, H. E. (2001). Towards a distributed account of conceptual knowledge. *Trends in Cognitive Science, 5,* 244–252.

Van Valin, R. D. Jr. (1990). Semantic parameters of split intransitivity. *Language*, 66, 221–260.

Verhoeven, E. (2014). Thematic prominence and animacy asymmetries. Evidence from a cross-linguistic production study. *Lingua, 143,* 129–161.

Wittgenstein, L. (1953) *Philosophical investigations*. New York: MacMillan.

Wunderlich, Dieter (1996). Models of lexical decomposition. In E. Weigand & F. Hundsnurscher (eds.), *Lexical structures and language use* (Vol. 1, pp. 169–183). Tübingen: Niemeyer.

Zurif, E., & Swinney, D. (1994). The neuropsychology of language. In M. A. Gernsbacher (Ed.), *Handbook of psycholinguistics* (pp. 1055–1074). New York: Academic.

Part II
Structure and Composition

Chapter 2
Lexicalizing and Combining

Paul M. Pietroski

Often, theorists mean different things by "meaning," and understandably so.[1] Even restricting attention to language, one might want to talk about what *speakers* mean when they communicate, or what *expressions* of a language mean. Regarding the latter, one might focus on languages that human children can naturally acquire, certain systems of animal communication, possible languages of thought, formal languages invented for purposes of computation or for modeling "ideal" thought/communication, etc. Like many words, "mean" is polysemous. So if the task is to study whatever natural phenomena we are gesturing at, it's hard to know where to begin.

On the other hand, it can seem obvious that whatever *verb meanings* are, they vary along a dimension that can be described in terms of valence, adicity, or Frege's (1892) metaphor of *saturation*. This is a tempting starting point, with implications for semantic composition that have become standard. But I'll urge a different view, according to which verbs—along with nouns, common and proper—are instructions for how to access *monadic* concepts that can be *conjoined* with others; cp. Hobbs (1985), Parsons (1990), Schein (1993, 2001), Pietroski (2005, 2006). As we'll see, adopting this perspective leads to an attractive though nonstandard conception of how words and the process of lexicalization are related to human thought.

Section 2.1 reviews some facts that motivate the view I want to challenge, and then some other facts that motivate the search for an alternative view of the sort discussed in Sect. 2.2. I'll conclude by locating my specific proposal in the context of

[1] This chapter, a written version of material presented at the Verb Concepts conference in 2008, has older descendants; see Pietroski (2010, 2011, 2012a, b). Though for various reasons, I have not revised this early presentation of my views in light of subsequent work.

P. M. Pietroski (✉)
Department of Linguistics, University of Maryland, Marie Mount Hall,
College Park, MD 20742, USA
e-mail: pietro@umd.edu

R. G. de Almeida, C. Manouilidou (eds.), *Cognitive Science Perspectives on Verb Representation and Processing,* DOI 10.1007/978-3-319-10112-5_2

Chomsky's (1986, 1995) conception of distinctively human languages as biologically instantiated procedures, I-languages, whose expressions make contact with other cognitive systems.

2.1 Fregean Verbs: Idealization and Myth

We humans can express endlessly many thoughts by linguistic means. This suggests that expressible thoughts are composed of concepts that are linked to expressions, which combine in ways that somehow mirror the ways in which the concepts combine. Frege offered a model language whose expressions reflect thoughts of a certain kind (*Gedanken*). But as Frege stressed, even if humans can have such thoughts, his *Begriffsschrift* may not be a good model of the languages that we naturally use to express the thoughts we typically entertain. Still, one can hypothesize that a verb is like a predicate of Frege's invented language in expressing a concept whose adicity determines the number of arguments the verb can/must combine with in a sentence. In this section, I note some well-known difficulties for this idea. My suspicion is that its familiarity, easily mistaken for inevitability, leads us to underestimate these difficulties and the attractions of an available alternative.

2.1.1 A Pretty Picture

In a sentence like (1) or (2), consisting of a verb and one or more names,

(1) Brutus arrived.
(2) Brutus saw Caesar.

each name is an argument of the verb. The relation a verb bears to its argument(s), in a sentence or sentential clause, is somehow asymmetric. Verbs *take* arguments. By contrast, the names in (1) and (2) do not take verbs: "saw Caesar" is a verb phrase, akin to "arrived," not a phrase that is grammatically akin to "Brutus." In some sense, the names appear as satellites of the verbs. Let's take this as given, for now, and precisify later.

One might hope to explain this grammatical asymmetry in terms of a more fundamental asymmetry exhibited by constituents of thoughts. For present purposes, let's assume that at least many thoughts can be described as the result of combining an *unsaturated* concept with one or more *saturating* concepts. Saturating concepts, like BRUTUS and CAESAR, can be used to think about things like Brutus and Caesar. Unsaturated concepts, like ARRIVED(X) and SAW(X, Y), can be saturated to form thoughts like ARRIVED(BRUTUS) andSAW(BRUTUS, CAESAR). Correlatively, an unsaturated concept can be viewed as the result of abstracting away from the specific contents of one or more saturating concepts in a thought.[2]

Given some such conception of articulable thoughts—thoughts with parts that can be systematically combined and expressed—one might suppose that verbs are argument-taking words *because* they express unsaturated concepts, while names appear in sentences as arguments *because* they express saturating concepts.[3] If verbs have unsaturated meanings in this sense, then perhaps the *number* of arguments that a verb can combine with is determined by the adicity of (i.e., the number of variable positions in) the concept expressed with the verb.

One can go on to hypothesize that this determination is often transparent: "arrived" takes a single argument because it indicates the monadic concept ARRIVED(X); "saw" takes two arguments, at least in active voice, because it indicates the dyadic concept SAW(X, Y). On this view, "saw Caesar" indicates the complex monadic concept SAW(X, CAESAR), which is like ARRIVED(X) in being saturatable by BRUTUS. One can also say that "gave" indicates the triadic concept GAVE(X, Y, Z) and so takes three arguments, as in (3);

(3) Brutus gave Caesar a sandwich.

where "a sandwich" reflects existential quantification over (as opposed to saturation of) the conceptual "slot" for the thing given to the recipient by the giver, as shown in (3a).

(3a) ∃Z:SANDWICH(Z)[GAVE(BRUTUS, CAESAR, Z)]

In this thought, the complex monadic concept GAVE(BRUTUS, CAESAR, Z) saturates the second-order concept ∃Z:SANDWICH(Z)[Φ(Z)], which is the result of saturating a dyadic concept, ∃Z:Ψ(Z)[Φ(Z)], with the monadic concept SANDWICH(Z). The idea is that an unsaturated concept can saturate suitable concepts of a higher order; by contrast, BRUTUS and CAESAR are said to be inherent *saturaters*. Correlatively,

---

[2] I take concepts to be composable mental symbols of a special sort; see Margolis and Laurence (1999), especially their introduction. In Fregean terms, starting with ARRIVED(CAESAR) and abstracting away from the specific content of CAESAR yields the monadic concept ARRIVED(X). Starting with SAW(CAESAR, BRUTUS) and abstracting away from the contents of both saturating concepts yields the dyadic concept SAW(X, Y). I assume that concepts have contents, which need not be linguistic meanings. I follow the usual conventions of using small capitals to indicate concepts, with variables ("*x*," "*y*," ...) indicating the number and logical order of saturaters: SAW(CAESAR, BRUTUS) implies that Caesar saw Brutus; SAW(X, BRUTUS) is a monadic concept that applies to anything that saw Brutus, while SAW(CAESAR, Y) is a monadic concept that applies to any entity that Caesar saw. But as discussed below, I do not assume that the contents of unsaturated concepts are functions, or that ARRIVED(CAESAR) denotes the value of some function with Caesar in its domain.

[3] I assume that talk of lexical items expressing concepts is to be understood, eventually, in terms of how concepts are indicated in speech and/or accessed in comprehension. But I do not assume that each lexical item λ is paired with a single concept C: if only because of polysemy, and the possibility of different perspectives on the things thinkers think about, a speaker might indicate one concept with a word that fetches a related but distinct concept in a hearer. For me, saying that λ expresses C is a simple way of saying that λ is linked, in a special indicating/fetching way, to one or more concepts that share a certain form and perhaps a common root; see Sect. 2.2.

"surface syntax" need not reflect the order of saturation. In (3), "a sandwich" is a grammatical constituent of a verb phrase headed by "gave."[4]

Given a "saturationist" conception of semantic composition, a verb's valence may *exceed* its overt arguments, at least in some sentences. Perhaps "ate" is fundamentally transitive/dyadic, as suggested by (4), and (5) somehow involves a covert argument.

(4) Caesar ate a sandwich.
(5) Caesar ate.

I'll return to some complications for this suggestion. But first, let me stress that saturationists can and should posit event variables, following Davidson (1967) and much subsequent work. For example, the untensed verb "arrive" can be treated as an indicator of the formally dyadic concept ARRIVE(E, X), which applies to an ordered pair of things just in case the first is an arrival of the second. Correspondingly, theorists can represent the thought expressed with (1) as in (1a).[5]

(1) Brutus arrived.
(1a) ∃E[PAST(E) & ARRIVE(E, BRUTUS)]

Eventish analyses of this sort account for the pattern of entailments and nonentailments exhibited by (the thoughts expressed with) sentences like (6) and (7).

(6) Brutus poked Caesar with a red stick sharply.
(7) Brutus poked Caesar with a blue stick softly.

Note that while (6) implies each of (8–10), and (7) implies each of (10–12),

(8) Brutus poked Caesar with a red stick.
(9) Brutus poked Caesar sharply.
(10) Brutus poked Caesar.
(11) Brutus poked Caesar softly.
(12) Brutus poked Caesar with a blue stick.

the conjunction of (6) and (7) implies neither (13) or (14).

(13) Brutus poked Caesar with a red stick softly.
(14) Brutus poked Caesar with a blue stick sharply.

---

[4] This divergence can be described in terms of "covert" movement or type-adjustment; see, e.g., May (1985) and Jacobson (1999).

[5] But if events of arriving are not independent of arrivers, no value of the variable in ARRIVE(E, BRUTUS) is independent of Brutus, and so ARRIVE(E, X) is not a concept of a genuine *relation*. Compare AFTER(E, F), ABOVE(X, Y), and ARRIVE-AT(T, X), whose first variable ranges over *times*, which are independent of arrivers. Likewise, while SEE(E, X, Y) is formally triadic, the corresponding relation does not hold among three independent entities. In this sense, hypothesizing that verbs indicate concepts like ARRIVE(E, X) and SEE(E, X, Y)—as opposed to ARRIVED(X) and SAW(X, Y)—adds one to the posited adicities, allowing for adverbial modification of event variables, without changing much else.

This pattern is expected if (6) and (7) have the logical forms displayed in (6a) and (7a).[6]

(6a) ∃E{PAST(E) & POKE(E, BRUTUS, CAESAR) & ∃X[RED(X) & STICK(X) & INSTRUMENT(E, X)] & SHARP(E)}
(7a) ∃E{PAST(E) & POKE(E, BRUTUS, CAESAR) & ∃X[BLUE(X) & STICK(X) & INSTRUMENT(E, X)] & SOFT(E)}

Moreover, a tenseless version of (10) can appear as the direct object of certain verbs, as in (15). This suggests that the perceptual verb "saw" does *not* express the dyadic SAW(X, Y).

(15) Antony saw Brutus poke Caesar.

For "Brutus poke Caesar" does not name or describe any particular seeable thing. Brutus may have poked Caesar many times, in many ways, with sticks of varied colors; cp. Ramsey (1927). Instead, one can say that "saw" expresses SAW(E, X, Y), where values of the last variable include events as well as people; see Higginbotham (1983). On this view, the thought expressed with (15) has the form shown in (15a).

(15a) ∃E{PAST(E) & ∃F[SEE(E, ANTONY, F) & POKE(F, BRUTUS, CAESAR)]}

And if the adverbial phrase in (16) is understood as a conjunct of a complex event description,

(16) Antony saw Brutus poke Caesar with a telescope.

then the *ambiguity* of (16) can be represented as in (16a) and (16b).

(16a) ∃E{PAST(E) & ∃F[SEE(E, ANTONY, F) & POKE(F, BRUTUS, CAESAR)] & ∃X[TELESCOPE(X) & INSTRUMENT(E, X)]}
(16b) ∃E{PAST(E) & ∃F[SEE(E, ANTONY, F) & POKE(F, BRUTUS, CAESAR)] & ∃X[TELESCOPE(X) & INSTRUMENT(F, X)]}

On one reading, Antony does his seeing (of a poke) with a telescope; on the other, Brutus does his poking (of Caesar) with a telescope.

I don't know how to account for such facts, in any systematic way, without appealing to event variables. So I assume that saturationists will allow for such variables, and maintain that a verb typically expresses a concept whose adicity *exceeds* the number of arguments that the verb takes in a sentence. (And covert existential closure is not limited to event variables.) But let me note one more reason for positing event variables in the concepts expressed with verbs.

---

[6] If the adverbial phrases correspond to conjuncts of a complex monadic concept, closed by existential quantification, the valid inferences are instances of conjunction reduction:∃E[Φ(E) & Ψ(E) & Δ(E)] implies ∃E[Φ(E) & Ψ(E)], which implies ∃E[Φ(E)]. But an instance of ∃E[Φ(E) & Ψ(E) & Δ(E)] & ∃E[Φ(E) & Γ(E) & Θ(E)] need not imply ∃E[Φ(E) & Ψ(E) & Θ(E)] or ∃E[Φ(E) & Δ(E) & Γ(E)]. See Taylor (1985), expounding an argument due to Gareth Evans. The example also shows that values of event variables are not ordered *n*-tuples consisting of participants and a moment in time; a sharp hit (of *y* by *x*) with a red stick can occur at the same time as a soft hit with blue stick.

If verbs like "poked" express dyadic concepts like POKED(X, Y), which applies to poker–pokee pairs, it is hard to describe (much less explain) the thematic asymmetry that these verbs exhibit; see, e.g., Dowty (1991), Carlson (1984), Baker (1997). Consider the possible concept KOPED(X, Y): when saturated by CAESAR and then BRUTUS, the result—KOPED(BRUTUS, CAESAR)—is true just in case Caesar poked Brutus; KOPED(X, Y) applies to pokee–poker pairs. Human children do not naturally acquire verbs that express such "thematically inverted" concepts. If they did, there would be sentences with verbs whose direct objects indicate agents and whose subjects indicate patients of the relevant events. This suggests that "poke" expresses a concept with an event variable, *and* that if this concept also includes variables for a poker and pokee, then this concept has a thematic decomposition along the lines shown in (17).[7]

(17) $\forall$E$\forall$X$\forall$Y[POKE(E, X, Y) $\equiv$ POKE(E) & AGENT(E, X) & PATIENT(E, Y)]

One can maintain that monadic concepts like POKE(E)—concepts of events that may be expressed with nouns—are abstracted from the polyadic concepts expressed with verbs. So one can embrace generalizations like (17) while saying that intransitive, transitive, and ditransitive verbs express concepts that exhibit distinct adicities. Nonetheless, appeal to event variables can feed doubts about the saturationist picture of semantic composition for verb phrases.

2.1.2 *Messy Facts*

Some of these doubts are specific to the introduction of event variables. Others are often set aside as puzzles for any account. Though as we'll see, the relevant facts are not so puzzling if verbs express monadic concepts like ARRIVE(E) and POKE(E).

If "arrive" and "poke" express ARRIVE(E, X) and POKE(E, X, Y), respectively, then one needs some explanation for why (18) and (19) cannot have the indicated meanings.

(18) That Brutus arrived.
(18a) #That was an event of Brutus arriving.
(19) The witnessed event Brutus poked Caesar.
(19a) #The witnessed event was one of Brutus poking Caesar.

[7] Or perhaps $\forall$E$\forall$X[POKE(E, X, Y) $\equiv$ POKE(E, Y) & AGENT(E, X)]; where POKE(E, Y) applies to event–pokee pairs (cp. Kratzer 1996, but also note 9 below). See Parsons (1990) on "subatomic" semantics. Schein (1993, 2001) extends arguments for "thematic separation" to plural constructions; see also Pietroski (2005) on action descriptions, including causative and serial verb constructions. Note that while thematic concepts are formally dyadic, like AFTER(E, F) and ABOVE(X, Y), the corresponding relation does not hold between independent entities; cp. note 5.

Why can't the event variable correspond to an overt grammatical argument? If a verb cannot be combined with an overt argument for *each* variable that the verb introduces, then perhaps verbs do not take arguments *because* they express unsaturated concepts.

I'll return to the actual meaning of (18), which casts doubt on the idea that names appear as arguments *because* they express saturating concepts. For now, recall (5) and consider its relation to (20–22). Note that (5) does not follow from (20); these sentences are not synonymous.

(5) Caesar ate.
(20) Caesar ate something.
(21) Caesar dined.
(22) Caesar dined on pencils.

Suppose that Caesar ate a pencil, but Caesar is a normal human for whom pencils are not nutritious. Then an utterance of (20) can be true while an utterance of (5) is false. In this respect, (5) is more like (21). Prima facie, the implications go from (22) to (21) to (5) to (20). So even if (5) has a covert argument, and "eat" always expresses the polyadic concept EAT(E, X, Y), one needs to say why (5) implies that the unspecified thing eaten is food for the eater. And the concept expressed with "dine" presumably does not have a *lower* adicity.[8]

On the contrary, one might think this concept adds something about the manner of the eating and/or the food eaten. Yet "Caesar dined something" is not a sentence of English—as if the concept expressed with "dine" does *not* have a variable for the food eaten, and describing this (essential) event participant requires a *grammatically optional* prepositional phrase. But then perhaps the concept expressed with "eat," which does take a direct object, also lacks a variable for the food eaten. Perhaps "eat" and "dine" express EAT(E) and DINE(E), respectively.

As discussed in Sect. 2.2, this is compatible with speakers *having* the polyadic concepts EAT(E, X, Y) and DINE(E, X, Y). Indeed, these concepts may be related to the verbs in a way that helps capture the intuition that events of eating/dining require eaters and things eaten. But in any case, "eat" and "dine" differ: the former can take a direct object that specifies whatever was eaten; the latter requires use of a prepositional phrase to specify what was dined on. This difference must be encoded somehow, whatever concepts the verbs express. And as we'll see, it is easily encoded if the concepts expressed are monadic. So in my view, the interesting questions here concern the *kinds* of concepts that verbs indicate/fetch for purposes of semantic composition. Do the thoughts expressed with (20–21) have the forms shown in (20a–21a),

(20a) ∃E{PAST(E) & ∃X[EAT(E, CAESAR, X)]}

(21a) ∃E{PAST(E) & DINE(E, CAESAR)}

[8] By contrast, (20) has a more permissive construal; cp. "There is something that Caesar ate." So perhaps "eat" can express INGEST(E) or REFUEL(E), and that for whatever reason, a covert direct object forces the second choice. Perhaps events of ingestion are represented as having agents and patients, without any necessary connection to nourishment, while events of refueling need not be represented as having patients.

with thematic information represented elsewhere, or the forms shown in (20b–21b)?

(20b) ∃E{PAST(E) & AGENT(E, CAESAR) & EAT(E) & ∃X[PATIENT(E, X)]}
(21b) ∃E{PAST(E) & AGENT(E, CAESAR) & DINE(E)}

Similar questions arise in the context of much discussed examples like (23–25).

(23) Brutus gave a museum a painting.
(24) Brutus donated a painting.
(25) Brutus donated a painting to a museum.

If "give" takes three arguments because it expresses GIVE(E, X, Y, Z), one wants to know why "donate" does not express DONATE(E, X, Y, Z) and also take three arguments. So perhaps "give" expresses a concept of *lower* adicity. The synonymy of (23) and (26)

(26) Brutus gave a painting to a museum.

invites the hypothesis that "give" expresses GIVE(E, X, Y), and that (23) is used to express thoughts of the form shown in (23a), as opposed to (23b); cp. Larson (1988).

(23a) ∃E{PAST(E) & ∃Y[PAINTING(Y) & GIVE(E, BRUTUS, Y)] & ∃Z[MUSEUM(Z) & RECIPIENT(E, Z)]}
(23b) ∃E{PAST(E) & ∃Y[PAINTING(Y) & ∃Z[MUSEUM(Z) & GIVE(E, BRUTUS, Y, Z)]]}

And upon reflection, the mere availability of ditransitive *constructions* like (23) does not favor the second analysis.

Examples like (27) do not lead us to say that "kick" expresses KICK(E, X, Y, Z).

(27) Brutus kicked Caesar a bottle.

For plausibly, (27) and (28) are both used to express thoughts of the form shown in (28a).

(28) Brutus kicked a bottle to Caesar.
(28a) ∃E{PAST(E) & ∃Y:BOTTLE(Y)[KICK(E, BRUTUS, Y) & RECIPIENT(E, CAESAR)]}

But if "give" and "donate" are like "kick" in expressing concepts with no variable for recipients, we must consider the possibility that these verbs express concepts with no variables for Agents, as in (23c); cp. Kratzer (1996).

(23c) ∃E{PAST(E) & AGENT(E, BRUTUS) & ∃Y[PAINTING(Y) & GIVE(E, Y)] & ∃Z[MUSEUM(Z) & RECIPIENT(E, Z)]}

The existence of passive constructions like (29)

(29) Caesar was kicked.

is puzzling if "kick" expresses KICK(E, X, Y). One can posit a process of introducing a related concept—KICK(E, Y)—that has no variable for kickers, yet still has a saturatable variable for kickees: ∀E∀Y{KICK(E, Y) ≡ ∃X[KICK(E, X, Y)]}. This goes some way toward the view urged here. But why should "passivization" be available at all? Why not understand "kicked Caesar" with a covert subject, or always require an overt quantificational subject as in (30)?

(30) Someone kicked Caesar.

Such considerations can help motivate the idea that "kick" expresses KICK(E, Y). But then we must also consider "objectless" examples like (31) and nominal constructions like (32).

(31) The baby kicked.
(32) I get no kick from champagne.

Especially in light of the pressure to say that "dine" can express a concept with no variable for the food eaten, perhaps we should say that "kick" expresses KICK(E), with no variable for kickees.[9]

Likewise, given passive uses of "give" and the possibility of giving at the office, perhaps we should say that "give" expresses GIVE(E), with no variables for event participants. Moreover, if "give" expresses GIVE(E, X, Y, Z), one might expect "sell" to express a concept with an *additional* argument, SELL(E, X, Y, Z, W). For selling differs from giving, in that the seller gets something from the buyer: x sells y to z for w. Likewise, one might expect "buy" to express BUY(E, X, Y, Z, W). So if combining verb V with argument A signifies saturation/binding of the concept expressed with V by the concept expressed with A, one might expect "sell" and "buy" to combine with *four* arguments (ignoring any event variable). But prima facie, neither verb can take four arguments. Note that (33) only has a bizarre meaning,

(33) *Brutus sold/bought Caesar the car a dollar.

according to which Caesar is a car *for whom* Brutus sold/bought a dollar; cp. (40) below. So if SELL(E, X, Y, Z, W) and BUY(E, X, Y, Z, W) are expressible concepts, we face the question of why they aren't expressed with "sell" and "buy."

One can say that syntax somehow forbids tritransitive constructions. But this is to grant that linguistic constraints may require a process of lexicalization that results in verbs with adicities that are *lower* than those of the concepts expressed. Examples like (34) and (35)

(34) Brutus sold the car.
(35) Caesar bought the car.

suggest that "buy" and "sell" express concepts with no more than two variables for participants—buyers/sellers and things bought/sold—in the relevant events. Especially given the facts concerning "give"/"donate"/"kick," noted above, the synonymy of (36) with (37)

---

[9] Again, see Parsons (1990) and Schein (1993, 2001). One can say that (31) has a covert direct object, and that it means something like "The baby did a kick"; cp. Hale and Keyser (1993). But if anything, this supports the idea that "kick" expresses KICK(E) in both (31) and (32). And if one has already posited the concept KICK(E, Y), one might use it to introduce a monadic concept of events: $\forall$E{KICK(E) $\equiv$ $\exists$Y[KICK(E, Y)]}. Kratzer (1996) offers a few reasons for not going this far, and instead leaving themes/patients semantically "unsevered" from verbs that apply to pairs of events and their "internal" participants; see note 7. But Williams (2007) argues that Kratzer's arguments are not decisive for English, and that they seem less plausible for Igbo and Mandarin.

(36) Brutus sold Caesar the car.
(37) Brutus sold the car to Caesar.

suggests that "sell" expresses a concept with no variable for recipients. And note that while (35) follows from (38), much as (34) follows from (36), (38) is not synonymous with (39).

(38) Caesar bought Antony the car.
(39) Caesar bought the car from Antony.

Rather, (38) has a benefactive meaning like (40),

(40) Caesar bought the car for Antony

which differs from (41), which follows from (42), which employs two prepositional phrases.

(41) Caesar bought the car for a dollar.
(42) Brutus sold the car to Caesar for a dollar.

But if "Antony" does not indicate a saturater of the concept expressed by the verb in (38), then prima facie, "Caesar" does not indicate a saturater of the concept expressed by the verb in (36).

If "sell" does not require more arguments than "give" or "donate," and "buy" does not require more arguments than "take," perhaps that is because no verb expresses a concept with more than two variables for the relevant event participants. If so, we want to know the source of this constraint, which would follow from the stronger constraint that all verbs express monadic concepts of things that can have participants. But in any case, once saturationists adopt the weaker constraint, this reduces the interest of the hypothesis that verbs inherit adicities from the concepts they express. Moreover, if saturationists posit processes that *introduce* concepts like GIVE(E, X, Y) in terms of concepts with *higher* adicities, they can hardly complain if other theorists do the same and extend this strategy in light of examples like (43) and (44).

(43) Brutus gave/donated at the office.
(44) Caesar wants to buy low and sell high.

One can call these cases of "coercion" and set them aside for special treatment. But we shouldn't suppose that we have any clear conception of how a concept can *have* an adicity that (if coerced) *changes*. We can, however, posit processes of using polyadic concepts to introduce concepts of lower adicity—even if this leads us in surprising directions.

2.2 A Conjunctivst Picture

Let's assume that lexicalizers have many polyadic concepts like GIVE(X, Y, Z) or GIVE(E, X, Y, Z).

We can describe lexicalization as a process that uses available mental representations, over time and given experience, to make atomic linguistic expressions that can be combined in certain ways that correspond to certain ways in which concepts can be combined. If constraints on the available modes of combination create pressure for lexical items that fetch monadic concepts, and lexicalization can be a process of using polyadic concepts to introduce fetchable monadic concepts, then one expects to find lexical items that fetch monadic analogs of (prelexical) polyadic concepts. And if the methods of introduction often make use of event variables, which can appear in both monadic concepts like GIVE(E) and thematic concepts like RECIPIENT(E, X), one might expect to find lexical items that fetch monadic concepts *and* invoke thematic concepts—via functional elements like prepositions, or certain grammatical relations that verbs can bear to their arguments. So if the available modes of combination create pressure to treat phrasal composition as an instruction to *conjoin monadic concepts,* as opposed to an instruction to *saturate one concept with another,* the facts illustrated with (1–44) are unsurprising.

2.2.1 Possible Minds

One can imagine minds that simply pair combinable concepts with perceptible signals, yet manage to communicate tolerably well by producing the signals in a linear order. Producing a string of atomic signals $S_1 \ldots S_k$ could be interpreted as the expression of a thought whose atomic components are the corresponding concepts $C_1 \ldots C_k$, at least one of which must be unsaturated. Given a few conventions to reduce ambiguity—e.g., put the signal for a dyadic concept between the signals for its saturaters, and associate the first signal with a particular argument position—short sentences, pronouns, occasional parataxis, and lists can go a long way. (Hemingway wrote novels. He liked newspapers. People understood him. He won a prize.)

Of course, humans are not so limited.[10] We acquire lexical items that can be combined to form phrases, of unbounded length, that exhibit a nontrivial syntactic typology. But if our lexical items signal concepts that are independently combinable because of their valences, and the semantic role of syntax is basically to determine order of saturation, one wonders why we have the syntax we do. So perhaps lexicalization and syntax conspire in a less obvious way, with a restricted form of conjunction as the primary mode of semantic composition.

I have pursued the technical details—especially concerning the composition principles governing verb/determiner/prepositional phrases with various kinds of nominal constituents—in other places; see Pietroski (2005, 2006, 2008). So here, let me simply present the main ideas in the context of an example that initially seems unfriendly.[11] Suppose the sound of "gave" is initially paired with a triadic concept GAVE(X, Y, Z) with no event variable.

[10] I am indebted to Norbert Hornstein for a series of conversations on these topics.

[11] See also note 1. But I have no firm commitments about any particular example. It is *very* hard to know the adicity of any prelexical concept. Even the classically monadic "mortal" may express a concept that relates individuals to events of death. Indeed, this should make us wary of hypotheses according to which some feature of verbs *matches* the adicity of the concept expressed. How does

One can envision a process of first introducing an event variable along the following lines: ∀x∀y∀z{∃e[GAVE(e, x, y, z) ≡ GAVE(x, y, z)]}. This assumes the apparatus required for such introduction; see Horty (2007) for related discussion of Frege's notion of definition. But one can at least imagine a mind that can use n-place concepts to define $n+1$-place concepts in this way; cp. Davidson (1967). The added variable can then be used like a variable for times:∀e∀x∀y∀z[GIVE(e, x, y, z) & PAST(e) ≡ GAVE(e, x, y, z)]. And given n thematic concepts, a monadic concept can be introduced: ∀e∀x∀y∀z[GIVE(e) & AGENT(e, x) & PATIENT(e, y) & RECIPIENT(e, x) ≡ GIVE(e, x, y, z)]; cp. Castaeda (1967).

As one would expect, this is a "contextual" introduction of GIVE(e), which applies to certain events that occur when three individuals exhibit the relation that GIVE(x, y, z) is a concept of. And the biconditionals in question need not be logical truths. The hypothesis is that in lexicalizing GIVE(x, y, z), we effectively assume an equivalence:∀x∀y∀z{GAVE(x, y, z) ≡ ∃e[PAST(e) & GIVE(e) & AGENT(e, x) & PATIENT(e, y) & RECIPIENT(e, x)]}; where the right side implies each of its conjunct-reducing variants. But this generalization need not hold—like ∀x∀y∀z [GAVE(x, y, z) & GAVE(x, y, z)], which is an instance of noncontradiction—as a matter of logic. There is much more to be said here about the relations among logic, meaning, and psychology. But for present purposes, let me bracket these larger issues.

2.2.2 *Recapturing Distinctions*

Let's assume that for any given speaker, finitely many concepts can be fetched with lexical items. Call these lexically fetchable concepts, which can be combined via operations corresponding to phrasal syntax, *L-concepts*. For any given lexicalizer, let her *P-concepts* be those available independent of lexicalization, with "P" connoting "prior" and "prelexical." This leaves room for the hypothesis that all or most L-concepts are P-concepts, and it does not require that L-concepts be atomic. It also leaves room for the hypothesis that many of our L-concepts are not P-concepts, but rather, concepts introduced in the course of lexicalization: a P-concept like GIVE(x, y, z) might be used to introduce an L-concept like GIVE(e) that would otherwise be unavailable for fetching with a lexical item.[12]

If all L-concepts fetched with open-class lexical items are monadic, this has implications for names as well as verbs. But before turning to this point, let me stress that words can differ formally while expressing concepts of the same adicity. In

one tell if such a hypothesis is correct, absent a reliable independent means of discerning the relevant conceptual adicity?

[12] This is an instance of a more general idea: P-concepts may exhibit certain formal distinctions that L-concepts do not; L-concepts may, by design, abstract away from certain respects in which P-concepts differ. For example, each P-concept may be essentially singular or essentially plural, while at least many L-concepts are neither; see Pietroski (2006), drawing on Boolos (1998), and Schein (1993).

particular, traditional ideas about subcategorization/selection can be recast in terms of hypotheses about which thematic concepts a verb invokes along with the monadic concept it expresses.

We can grant that "put" requires both an object and a prepositional phrase as in (45),

(45) Brutus put a book on a table.

without saying that "put" expresses PUT(E, X, Y, L), with a variable for locations—or PUT(E, Y, L), without a variable for agents; cp. Hale and Keyser (1993). We can say instead that "put" expresses PUT(E), or perhaps PLACE(E), but that "put" *also* imposes a lexical requirement on the verb phrases it heads: they must invoke the thematic concepts corresponding to a thing placed and its location when placed; cp. Levin and Rappaport (1995), Levin and Rappaport Hovav (2005). If the grammatical relation between "put" and its direct object invokes the concept PATIENT(E, X), and the preposition invokes LOCATION(E, X), then "put a book on a table" meets this requirement by expressing concepts like PUT(E) & ∃X[BOOK(X) & PATIENT(E, X)] & ∃X[TABLE(X) & LOCATION(E, X)].

If "put" lexicalized a polyadic concept of making something be in a place, locations might be "conceptually tied" to puttings in a way they are not tied to eatings, even if we know a priori that every eating occurs in a place. For the concept lexicalized with "put" might have a variable for locations, while the concept lexicalized with "eat" does not. And this can be so, even if "put" and "eat" are on a semantic par in the sense that both verbs fetch monadic concepts of events.

Let the Semantic Composition Adicity Number (SCAN) of a verb be the adicity of the concept it expresses: the SCAN of a verb V reveals how many saturaters/binders are required to convert the concept expressed with V into a complete thought. Let a verb's Property of Smallest Sentential Entourage (POSSE) be the number of "satellite" expressions—arguments or adjuncts, be they noun, determiner, prepositional, or complementizer phrases—that must accompany the verb in a clause with active voice: the POSSE of a verb V reveals how many satellites are needed to make V into an active voice sentence. A verb's SCAN need not determine its POSSE, and its POSSE need not determine its SCAN. One can say, for example, that "put" has a SCAN of 1 and a POSSE of 3. One can also define a verb's Lexicalized Adicity Number (LAN) as the adicity of the concept initially lexicalized with the verb. And one can speculate that a verb's POSSE is determined by, or at least interestingly related to, its LAN. This speculation seems plausible; though is hard to evaluate, absent independent and reliable ways of discerning LANs.

In one sense, this simply recodes the facts. But that is no objection, absent good reasons for coding the facts in terms of *diverse* SCANs, as opposed to POSSEs and/or LANs. Of course, if one posits diverse SCANs *in addition to* diverse POSSEs, one might be accused of needlessly introducing an unwanted degree of freedom into our theories. But the hypothesis here is that SCANs are uniform: all verbs express monadic concepts, even if the concepts lexicalized vary in adicity. And this at least avoids the need to explain particular SCAN/POSSE mismatches. For example, if the verb "jimmy" (as in "jimmy the lock with a knife") has a SCAN greater than 2, one needs some explanation for why it (unlike "put") has POSSE of 2; see Williams (2005, 2007). Otherwise, one *is* positing various SCANs and various POSSEs.

If SCANs greater than 1 are possible, one also needs some explanation for why apparently simple concepts like BETWEEN(X, Y, Z), TALLER(X, Y), and FROM(X, Y) are not lexicalized with monomorphemic verbs—yielding constructions like "Brutus betweened Antony Caesar," "Caesar talls Antony," and "Brutus froms Rome." The intended thoughts are expressible, with circumlocution, by using functional expressions: Brutus *is* between Antony *and* Caesar; Caesar *is taller than* Antony; Brutus hails *from* Rome. This suggests some kind of block on directly fetching the relevant nonmonadic concepts. And it invites the hypothesis that functional vocabulary lets us find circumlocutory ways to express essentially relational thoughts, despite our massively monadic lexicons, when the thematic concepts invoked by grammatical relations (like being the subject or object of a verb) are inadequate.

2.2.3 Weather Reports and Names

Verbs that can apparently take *no* arguments, as in (46) and (47),

(46) It is snowing in Rome.
(47) Brutus saw it rain today.

are often set aside for special treatment. From a saturationist perspective, such examples are puzzling. Given the need for event variables, the verbs in "It rained/snowed/poured/drizzled" cannot be treated as devices for expressing thoughts with no unsaturated elements, even if there are such thoughts; cp. Montague (1974). But if "rain" expresses RAIN(E), and the argumentless verb corresponds to an argumentless concept modulo the event variable, we need some explanation for why (48) is acceptable and why it implies (49).

(48) Rocks rained down on the village.
(49) Rocks fell on the village.

An obvious initial suggestion is that RAIN(E) is an essentially plural variant of FALL(E), which is introduced via FALL(E, X), a concept that relates falls to fallen; cp. Boolos (1998). If some events satisfy RAIN(E), they are falls; if their patients were rocks that ended up on the village, they were falls of rocks that ended up on the village. And if we typically use "rain" to think/talk about waterdrops, we might add a nominal use as in (50).[13]

[13] Note that "Cats and dogs rained down on Rome" does not have the idiomatic meaning of "It rained cats and dogs in Rome," which is roughly that it rained heavily in Rome. One might argue that "snow" expresses SNOW(E, L), with a variable for locations. But even if this is correct, it is little comfort to saturationists. For unlike the variable for the fallen in FALL(E, X), the location variable is not saturated by the concept expressed with any argument of the verb. We can say "Snow fell" and "Rome fell," but not "Rome snowed." And if one insists that "It snowed" has a

(50) Brutus watched the rain fall.

My aim is not, however, to provide a theory of weather reports. It is rather to highlight two points. First, if verbs express monadic concepts of things that can have participants, then "argument optionality" is not surprising. If "rain" expresses RAIN(E), then absent lexical restrictions of the sort imposed by "put," (46–50) do not present puzzles. Likewise, if "kick" expresses KICK(E), its appearance in the range of constructions repeated below is unsurprising.

(27) Brutus kicked Caesar a bottle.
(29) Caesar was kicked.
(31) The baby kicked.
(32) I get no kick from champagne.

In short, a verb can take arguments without expressing a polyadic concept; and a verb can have mandatory satellites, of whatever kind, without expressing a polyadic concept. Second, support for alternatives to the saturationist picture can come from considering words that do *not* take arguments. This leads to the last set of reminders I want to offer.

For purposes of this chapter, I have focused on verbs. But the saturationist conception of semantic composition is motivated in part by the idea that names like "Brutus" and "Caesar" express singular concepts like BRUTUS and CAESAR. As Russell and Montague showed, this is hypothesis is not required: one can analyze names as quantificational expressions of the same higher-order type as "every logician." But if names don't express saturating concepts, yet children have many such concepts (*pace* Russell), that would be surprising—absent some reason for thinking that concepts like BRUTUS and CAESAR can be P-concepts but not L-concepts. On the other hand, if all (open class) L-concepts are monadic, it follows that names do not express singular concepts: lexicalizers would have to use a concept like CAESAR to introduce a monadic concept—perhaps CALLED(X, PF: CAESAR), where PF: CAESAR is a concept of the phonological form associated with the singular concept—that can be combined with others.

This predicts that examples like (51) are not as simple as they appear.

(51) Caesar left.

If the lexical item "Caesar" fetches a monadic concept likeCALLED(X, PF:CAESAR), the subject of (51) is presumably a complex expression consisting of the lexical item and a covert functional item of some kind.[14] For present purposes, the details are not important. The idea is that, one way or another, (51) is used to express a thought like the following: ∃E{PAST(E) & ∃X[D(X) & CALLED(X, PF: CAESAR) & AGENT(E, X)] & LEAVE(E)}; where D(X) is a monadic concept, perhaps demonstrative in character,

covert *saturating location argument,* as opposed to a covert *conjoining location adjunct,* one needs appropriate analyses of (46) and (47).

[14] See Burge (1973) and many others, e.g., Katz (1994), Longobardi (1994), Elbourne (2005), Matushansky (2006); cp. Segal (2001).

expressed by the posited covert element. As noted by Burge (1973) and many others, there is abundant evidence that lexical proper nouns are like common nouns with respect to distribution and the kind of concept expressed. Consider (52–55).

(52) Every Caesar I saw was a politician.
(53) Every politician I saw was a Caesar.
(54) There were three Caesars at the party.
(55) That Caesar stayed late, and so did this one, but the other Caesar left early.

As shown in (52) and (53), "Caesar" can appear where other nouns can. Like common nouns, "Caesar" can take a plural form, as in (54). Examples like (55) show that "Caesar" can combine with "that" to form complex demonstratives; and "one," modifiable with "other," is ordinarily a pro-form for nouns that are *not* singular terms. It would be very puzzling if a lexical item with this distribution expressed a singular concept like CAESAR. By contrast, if "Caesar" expresses a monadic concept, then (52–55) are expected. Even if such constructions are special (or "coerced") in English, they remain grammatically possible. And in other languages, including Greek and many dialects of Romance, such constructions are quite normal.

Note that proper nouns are not only pluralizable, they can be used generically as in (56).

(56) Politicians lie, and Caesars steal.

They can also be used to make claims about some people who share a surname.

(57) The Smiths are coming to dinner.

And as surnames remind us, names can be overtly complex, as in (58).

(58) At noon, I saw Caesar Smith.

Prima facie, "Caesar Smith" is *semantically* related to "Caesar" and "Smith," roughly as "red stick" is to"red" and "stick": a Caesar Smith is both a Caesar and a Smith. A random Smith need not be a Caesar Smith. But in a context where the only Caesar is also the only Smith, one can use (59) or (60) to say what one says with (58), suggesting that "Caesar" fetches a monadic concept.

(59) I saw Caesar at noon.
(60) I saw Smith at noon.

Titles, as in (61), raise similar issues.

(61) Professor Caesar Smith and Doctor Caesar Smith are both republicans.

One might insist that a speaker who uses "Caesar" to talk about two people, who we might call "Sid" and "Romero," has two homophonous lexical names. On this view, the sound of "Caesar" is associated with potentially many singular meanings *in addition to* its monadic meaning. (Positing the latter seems unavoidable.) I find it hard to believe that a speaker who knows and distinguishes n Caesars has $n+1$ meanings for "Caesar." But it is hard to establish that an ambiguity hypothesis is false: that is why such hypotheses are, methodologically, not options of first resort; see Kripke (1979). Still, this particular hypothesis faces serious difficulties.

Given that many languages allow for grammatically complex names, with predicative nouns as constituents, it is a substantive assumption that English *forbids* such an analysis of (51).

(51) Caesar left.

But if such an analysis is possible, positing even one singular meaning for "Caesar" makes (51) ambiguous; and any posited singular meanings seem theoretically otiose. Lexical nouns with such meanings would also be theoretically unattractive, since they would not head any noun phrases. There are, to be sure, subclasses of nouns. Indeed, as (52) shows,

(52) *Politician left.

common nouns do not combine with a covert name maker in English. But one can distinguish common nouns from proper nouns, especially if the latter do not correspond to language-independent concepts, without positing unmodifiable nouns. Accounting for any grammatical categories is hard enough, without needlessly positing odd subcategories; see Baker (2003).[15]

My aim is not, however, to provide a theory of names. It is to highlight a potential source of support for the idea that open-class lexical items (and hence verbs) express monadic concepts, in contrast to the saturationist picture of verbs often expressing polyadic concepts saturated by the singular concepts expressed with names.

[15] A similar point applies to acquisition. We must ask if the faculty that supports the acquisition of languages that allow for complex names—names composed of lexical proper nouns and overt determiners—*also* supports the acquisition of "singular" names. For example, in Greek, names may and typically *must* be complex: a bare proper in a context like (51) is anomalous, like (52) in English; see Giannakidou and Stavrou (1999). Any child can acquire a "G(reek)-style" language. And *if* languages like English allow for lexical singular names, any child can acquire such an "E-style" language; in which case, experience with E-style languages must differ from experience with G-style languages, in a robust way that leads *every* normal child to acquire a lexicon of the right sort: in cases of acquiring English, a lexicon with *enough* entries, despite homophony and the possibility of complex-name analyses that would shorten the lexicon; in cases of acquiring Greek, a lexicon with *fewer* entries, despite the possibility of ambiguity and lexical-name analyses that would lengthen the lexicon. Usually, children treat lexical sounds as ambiguous only given reason to do so. So what would lead children to conclude that English name sounds are ambiguous? One can conjecture that not hearing the determiner, in examples like (51), lets children know that English has singular names. But on this view, children use "negative" evidence to *disconfirm* that English names are complex; and the use of such evidence in acquisition remains unattested (see Crain and Pietroski 2001). Worse, an unwanted lexical type must be posited to allow children to use negative evidence to acquire a grammar that admits theoretically superfluous ambiguities.

2.3 Meaning and the Language Faculty

At the outset, I noted that the saturationist picture of verb meanings can seem to offer an attractive starting point for theorizing about meaning more generally. If verbs express unsaturated concepts of varying adicities, then, presumably, combining verbs with arguments signifies saturation of one concept by another. And this conception of verbs coheres with the idealization that lexical items express prior concepts, which were available for lexicalization. One might hope to develop detailed accounts of meaning, and its relation to human psychology, within these framework assumptions. But if the saturationist picture is inadequate, even as a description of how verbs do and don't combine with arguments, we need a different point of departure. So let me suggest that Chomsky's (1986, 1995) focus on I-languages provides a congenial setting for the idea that lexical meanings are instructions to fetch monadic concepts. We can view I-languages as cognitive tools that let humans use prior concepts to make distinctive recursively combinable concepts. But the distinctiveness may lie more with the monadic building blocks than with the modes of recursive combination.

2.3.1 I-Languages

Starting with the ancient conception of languages as pairing signals with interpretations, we can distinguish *sets* of signal–interpretation pairs from *procedures* that pair signals of some sort with interpretations of some sort. Since we can describe the former as extensions of functions, and the latter as intensions, Chomsky speaks of E-languages and I-languages.[16] While "I-" also connotes the internalistic and idiolectic character of the procedures that interest Chomsky, the basic distinction is simpler and less tendentious. We can distinguish the set of input–output pairs determined by "$x - 1$" from the indicated procedure, which differs from the procedure indicated with "$+\sqrt{(x^2 - 2x + 1)}$," which determines the same set of input–output pairs. Likewise, even if a speaker's linguistic competence can be partly characterized by a set of signal–interpretation pairs, we can distinguish any such set from the procedure the speaker *implements* in pairing signals with interpretations as she does; cp. Marr (1982) on the distinction between functions computed and implementable algorithms for computing them.

Using this terminology, let's say that human I-languages are naturally acquirable procedures that pair distinctively human linguistic signals—like the sounds of spoken English or signs of American Sign Language (ASL)—with the corresponding interpretations, whatever they are.[17] Human I-languages are biologically implemented

[16] Compare Church (1941), who was echoing Frege (1892). Given homophony and synonymy, a signal may be paired with an n-tuple of interpretations, and an interpretation may be paired with more than one signal.

[17] This leaves room for the externalist idea that interpretations are individuated by features of the environment (see, e.g., Burge 1989), even if these interpretations are themselves concepts; cp.

procedures that normal children can acquire, given an ordinary course of experience. But depending on what we mean by "signal" and "interpretation," the procedures in question may be indirect. For example, my "I-English" need not be a procedure that directly links acoustic vibrations to interpretations of any kind. Human I-languages may be procedures that pair instructions to generate signals/percepts of a certain sort with instructions to generate interpretations/concepts of a certain sort. From this perspective, the human faculty of language consists of whatever aspects of human cognition are responsible for generating such instructions, via the acquisition of human I-languages. And following Chomsky (1995), we can think of phonological forms (PFs) and logical forms (LFs)—or more neutrally, PHONs and SEMs—as instructions at two "interfaces" between the human faculty of language and other cognitive systems: the articulatory/perceptual systems germane to the production/perception of signals, and the conceptual/intentional systems germane to the construction/expression of concepts.

Let's assume that in acquiring a human I-language, a child lexicalizes available concepts, which are symbols of one or more mental languages that children may share with nonhuman animals. Once acquired, human I-languages can be used (via the expressions they generate) in thought and communication. Indeed, we may use "human I-expressions" mostly in thought. The significance of such expressions is presumably *inherited* from the significance of lexicalizable concepts; see, e.g., Fodor (1975, 2003). But this inheritance may be indirect, since lexical expressions of a human I-language may do more than merely label lexicalizable concepts.

If the phonological form of "poked" is initially paired with a polyadic concept, this may initiate a process that results in a lexical item that connects the phonological form of "poke" with a monadic concept POKE(E); where the monadic concept is henceforth the one fetched with "poke," and "poked" is treated as complex instruction to fetch the concept PAST(E) and conjoin it with a concept fetched via "poke." Likewise, if the phonological form of "Brutus" is paired with a singular concept like BRUTUS, this may initiate a process that results in a lexical item that connects the sound with a monadic concept like CALLED(X, PF:BRUTUS); where the monadic concept is henceforth the one fetched with the lexical item "Brutus," which might be combined with an overt or covert determiner. Given a polysemous word, there is presumably more than one fetchable concept. But a lexical item that is, early on, linked to at least two concepts—one lexicalized and one fetchable—might become linked to several.

Of course, lexicalization and composition must dovetail. If lexical items can be combined to form a phrase, the concepts fetched with those lexical items must be combinable via the operations (conjunction, saturation, or whatever) invoked by the relevant syntax. Correlatively, operations of semantic composition must be applicable to the concepts fetched by lexical items that can be combined to form expressions of a human I-language. This raises chicken/egg issues. Are certain operations of

---

Pietroski (2006, 2008) for discussion drawing on Chomsky (2000). And of course, the point is not to deny that humans can have languages of thought that are independent of public signals. But these languages may be neither acquired nor distinctively human.

conceptual combination invoked, as correlates of phrasal syntax, because lexically fetchable concepts have the formal character they do? Or do the fetchable concepts have the formal character they do because of constraints on which operations of conceptual combination are available as correlates of phrasal syntax?

I suspect that in the end, the answer to both questions is affirmative, but that the second is especially important. In my view, there are independent empirical and theoretical reasons for thinking that in human I-languages, the core operation for combining expressions recursively yields expressions that are instructions to conjoin monadic concepts that may have thematic constituents in which one variable has been closed; see Hornstein and Pietroski (2009). But even restricting attention to the facts noted above, we have seen reasons for taking some such conception of semantic composition seriously. And if one were to begin anew, without adopting a Fregean/saturationist conception by default, one might well explore the idea that combining human I-expressions often signifies an operation that is *reversible* and in some sense *additive*. If one focuses on considerations of easy computability, and not mere recursive specifiability, one might see patterns of conjunction reduction as manifestations of the core operation invoked by the human faculty of language to combine lexically fetchable concepts.

From this perspective, the distinctively human aspect of this faculty may lie with lexicalization and the concepts it delivers, as opposed to (i) the composition operations applied to these concepts, or (ii) the concepts lexicalized; cp. Hauser, Chomsky, and Fitch (2002). Indeed, lexically fetchable concepts may be special in various respects. We know that human children are distinctive primates who lexicalize with a vengeance. So perhaps we should explore the hypothesis that lexicalization was the new trick that somehow let humans exploit extant operations of conceptual combination to new effect. If our ancestors were already saturating polyadic concepts with singular concepts, at least within various local domains, lexicalization may not have added much. But perhaps monadicizing, massively, lets us do a lot—with simple operations of conjunction and existential closure—that we couldn't otherwise do; see Pietroski (2005, 2006, 2011) for further discussion.

2.3.2 Reprise

Imagine a mind that has unsaturated concepts, of various adicities, and singular concepts that can saturate them. This mind can express thoughts by means of intransitive, transitive, and ditransitive constructions in which verbs combine with one, two, or three referential expressions. One might expect this mind to express its concepts in accord with the saturationist picture: intransitive verbs express concepts like ARRIVED(X) or eventish analogs like ARRIVE(E, X); transitive verbs express concepts like KICKED(X, Y)/KICK(E, X, Y); ditransitive verbs express concepts like GAVE(X, Y, Z)/GIVE(E, X, Y, Z); and names express singular concepts like BRUTUS.

One can imagine such a mind having a transitive verb "poke" that expresses POKE(E, X, Y), with no variable for instruments, or a ditransitive verb that expresses POKE(E, X, Y, Z). But one wouldn't expect both. Likewise, given transitive "kick," one wouldn't expect an intransitive or ditransitive version. Other things equal, one wouldn't expect passives or nominalizations. Nor would one expect names to figure in sentences like (53)

(53) Every Antony saw a Brutus poke a Caesar while it rained.

But given ditransitive constructions and some concepts like SELL(E, X, Y, Z, W), one might expect "tritransitive" constructions like (54), but with the meaning that Brutus sold a car for a dollar.

(54) *Brutus sold a car a dollar.

Given concepts like JIMMY(E, X, Y, Z) and BETWEEN(X, Y, Z), one would expect verb phrases like "jimmied the lock a knife" and "betweened Brutus Caesar." Given BIGGER(X, Y), and FROM(X, Y), one would expect constructions like "Antony bigged Caesar" and "Brutus froms Rome."

Put another way, if a mind stocked with such concepts *could* acquire a language that conforms to the saturationist picture, one would expect it to acquire such a language—especially given evidence that local adults had acquired such a language. So if human children have the concepts, but they *don't* acquire the expected verbs, this suggests that kids *can't* acquire languages that conform to the saturationist picture. Defending this suggestion requires far more evidence, technical detail, and consideration of potential saturationist replies. But a first step is to recognize that a host of well-known facts can be seen as symptoms of the massive monadicity of lexical meanings, and verb meanings in particular.[18]

References

Baker, M. (1997). Thematic roles and grammatical categories. In L. Haegeman (Ed.), *Elements of grammar* (pp. 73–137). Dordrecht: Kluwer.

Baker, M. (2003). *Verbs, nouns, and adjectives*. Cambridge: Cambridge University Press.

Boolos, G. (1998). *Logic, logic, and logic*. Cambridge: Harvard University Press.

Burge, T. (1973). Reference and proper names. *Journal of Philosophy, 70,* 425–439.

Burge, T. (1989). Wherein is language social. In A. George (Ed.), *Reflections on Chomsky*. New York: Blackwell.

Carlson, G. (1984). Thematic roles and their role in semantic interpretation. *Linguistics, 22,* 259–279.

Castaneda, H. (1967). Comments. In N. Rescher (ed.), *The logic of decision and action*. University of Pittsburgh Press, Pittsburgh.

Chomsky, N. (1986). *Knowledge of language*. New York: Praeger.

Chomsky, N. (1995). *The minimalist program*. Cambridge: MIT Press.

Chomsky, N. (2000). *New horizons in the study of language and mind*. Cambridge: Cambridge University Press.

Church, A. (1941). *The calculi of lambda conversion*. Princeton: Princeton University Press.

[18] My thanks to the conference organizers and participants for very helpful discussions.

Crain, S., & Pietroski, P. (2001). Nature, nurture, and universal grammar. *Linguistics and Philosophy, 24,* 139–186.
Davidson, D. (1967). The logical form of action sentences. In N. Rescher (ed.), *The logic of decision and action*. University of Pittsburgh Press, Pittsburgh.
Dowty, D. (1991). Thematic proto-roles and argument selection. *Language, 67,* 547–619.
Elbourne, P. (2005). *Situations and individuals*. Cambridge: MIT Press.
Fodor, J. (1975). *The language of thought*. New York: Crowell.
Fodor, J. (2003). *Hume variations*. Oxford: OUP.
Frege, G. (1892). Function and concept. In P. Geach & M. Black (trans.), *Translations from the philosophical writings of Gottlob Frege*. Oxford: Blackwell (1980).
Giannakidou, A., & Stavrou, M. (1999). Nominalization and ellipsis in the Greek DP. *The Linguistic Review, 16,* 295–332.
Hale, K., & Keyser, J. (1993). On argument structure and the lexical expression of syntactic relations. In K. Hale & J. Keyser (Eds.), *The view from building 20*. Cambridge: MIT Press.
Hauser, M., Chomsky, N., & Fitch, T. (2002). The faculty of language: What is it, who has it, and how did it evolve? *Science, 298,* 1569–1579.
Higginbotham, J. (1983). The logical form of perceptual reports. *Journal of Philosophy, 80,* 100–127.
Hobbs, J. (1985). *Ontological promiscuity*. Proceedings, 23rd Annual Meeting of the Association for Computational Linguistics (pp. 61–69). Chicago, Illinois, July 1985.
Hornstein, N., & Pietroski, P. (2009). Basic operations. *Catalan Journal of Linguistics, 8,* 113–139.
Horty, J. (2007). *Frege on definitions: A case study of semantic content*. Oxford: OUP.
Jacobson, P. (1999). Variable free semantics. *Linguistics and Philosophy, 22,* 117–184.
Katz, J. (1994). Names without bearers. *Philosophical Review, 103,* 1–39.
Kratzer, A. (1996). Severing the external argument from its verb. In J. Rooryck & L. Zaring (Eds.), *Phrase structure and the lexicon*. Dordrecht: Kluwer.
Kripke, S. (1979). Speaker's reference and semantic reference. *Midwest Studies, 2,* 255–276.
Larson, R. (1988). On the double object construction. *Linguistic Inquiry, 19,* 335–391.
Levin, B., & Rappaport, M. (1995). *Unaccusativity: At the syntax-semantics interface*. Cambridge: MIT Press.
Levin, B., & Rappaport Hovav, M. (2005). *Argument realization*. Cambridge: CUP.
Longobardi, G. (1994). Reference and proper names. *Linguistic Inquiry, 25,* 609–665.
Margolis, E., & Laurence, S. (1999). *Concepts: Core readings*. Cambridge: MIT Press.
Marr, D. (1982). *Vision*. New York: WH Freeman.
Matushansky, O. (2006). Why rose is the rose: On the use of definite articles in names. *Empirical Issues in Syntax and Semantics, 6,* 285–307.
May, R. (1985). *Logical form: Its structure and derivation*. Cambridge: MIT Press.
Montague, R. (1974). *Formal philosophy*. New Haven: Yale University Press.
Parsons, T. (1990). *Events in the semantics of English*. Cambridge: MIT Press.
Pietroski, P. (2005). *Events and semantic architecture*. Oxford: Oxford University Press.
Pietroski, P. (2006). Interpreting concatenation and concatenates. *Philosophical Issues, 16,* 221–245.
Pietroski, P. (2008). Minimalist Meaning, Internalist Interpretation. *Biolinguistics, 4,* 317-341.
Pietroski, P. (2010). Concepts, meanings, and truth. *Mind and Language, 25,* 247–278.
Pietroski, P. (2011). Minimal semantic instructions. In C. Boeckx (Ed.), *The Oxford handbook of linguistic minimalism*. Oxford: Oxford University Press.
Pietroski, P. (2012a). Semantic monadicity with conceptual polyadicity. In M. Werning, W. Hinzen, & E. Machery (Eds.), *Oxford handbook of compositionality*. Oxford: Oxford University Press.
Pietroski, P. (2012b). Language and conceptual reanalysis. In A. DiSciullo (Ed.), *Towards a biolinguistic understanding of grammar: Essays on interfaces*. Amsterdam, John Benjamins (2012).
Ramsey, F. (1927). Facts and propositions. *Proceedings of the Aristotelian Society* (Supplementary), *7,* 153–170.

Schein, B. (1993). *Plurals*. Cambridge: MIT Press.
Schein, B. (2001). Adverbial, descriptive reciprocals. In R. Hastings, et al. (Eds.), *Proceedings of semantics and linguistic theory XI*. Ithaca: CLC.
Segal, G. (2001). Two theories of names. *Mind and Language, 5,* 547–563.
Taylor, B. (1985). *Modes of occurrence*. Oxford: Blackwell.
Williams, A. (2005). *Complex causatives and verbal valence*. Doctoral dissertation, University of Pennsylvania.
Williams, A. (2007). Patients in Igbo and Mandarin. In J. Dölling & T. Heye-Zybatow (Eds.), *Event structures in linguistic form and interpretation*. Berlin: Mouton de Gruyter.

Chapter 3
Optional Complements of English Verbs and Adjectives

Brendan S. Gillon

3.1 Introduction

With the development of Chomsky's second version of transformational grammar (Chomsky 1965), generative linguists studying English syntax have puzzled over how best to handle verbs whose complements are optional. Such complements include not only noun phrasal and clausal complements but also prepositional phrasal complements. Much less studied are English adjectives with optional complements. To date, no one has provided a model theoretic account of such verbs which does not take them to be ambiguous. In this chapter, I shall provide such an account and show how it extends to English adjectives with optional complements. To do this, I shall proceed in three steps. First, I shall review the essential facts pertaining to English verbs with optional complements. I shall then outline how such verbs can be treated as unambigous and given a model theoretical treatment. Finally, I shall review the corresponding facts pertaining to English adjectives and show how adjectives with optional complements yield to essentially the same treatment.

3.2 Verbs

As is well-known, some English verbs take no complements, while others take a variety of complements. Sometimes the complements are obligatory and sometimes they are optional. Verbs which resist all complements, traditionally called *intransitive verbs*, include *to bloom*, *to die*, *to disappear*, *to elapse*, *to expire*, *to fall*, *to faint*, *to laugh*, *to sleep*, *stroll*, and *to vanish*. Though traditional grammar calls verbs which require a noun phrase complement *transitive verbs*, it has no general term for verbs requiring or admitting other kinds of complements, such as clauses, prepositional

B. S. Gillon (✉)
Department of Linguistics, McGill University, 1085 Dr. Penfield Ave.,
Montreal, Quebec H3A 1A7, Canada
e-mail: brendan.gillon@mcgill.ca

R. G. de Almeida, C. Manouilidou (eds.), *Cognitive Science Perspectives on Verb Representation and Processing,* DOI 10.1007/978-3-319-10112-5_3

phrases, adjectival phrases and adverbial phrases. Our attention here will be confined to English verbs taking either noun phrase complements or clausal complements. As it happens, what is true for them is true for verbs taking other complements or combinations of complements.

As we are about to see, verbs with optional noun phrase complements are liable to four kinds of construal: an indefinite, or existential, construal, a reflexive construal, a reciprocal construal and a definite, or pronoun like, construal. These labels are given in virtue of the kind of paraphrase the verb without a complement can be given by the same verb used with a complement.

Consider, for example, the verb *to read*. It may, but need not, have a noun phrase complement. To a good first order approximation, its use without a complement can be paraphrased by the very same verb with a general existentially quantified noun phrase, as illustrated in (1).

(1.1) Bill read (something).
(1.2) Bill perused *(something).

The parentheses in (1.1) indicate that the noun phrase complement is optional. The expression enclosed in parentheses provides the complement which, when it appears, yields a sentence nearly synonymous with the sentence in which it does not appear. In (1.2), the asterisk in front of the left parenthesis indicates that the omission of the complement yields an unacceptable sentence. Here, and in several other examples below, the second sentence is obtained from the first sentence by replacing the verb in the first sentence with a near synonym. This pairing of verbs, which, in the case of the sentences in (1), is due to Huddleston (2002, p. 404), shows that whether or not a verb's complement is optional is not a matter of the verb's meaning.

Other verbs giving rise to similar paraphases are the verbs *to eat*, *to drink*, *to bake*, *to carve*, *to cook*, *to write*, *to embroider*, *to knit*, *to sew*, *to weave*, *to whittle*, *to plow*, *to weed*, *to sweep*, *to iron*, *to hunt*, *to crochet*, *to file*, *to type*, etc. I call them *implicit indefinite object verbs*.

Another class of verbs admitting an optional noun phrase complement is what I call *implicit reflexive object verbs*. They include such verbs as *to bathe*, *to disrobe*, *to dress*, *to undress*, *to shave* and *to shower*. These verbs give rise to a paraphrasal equivalent which uses a form of the reflexive pronoun. The following pair of sentences is due to an observation of Huddleston (2002, p. 302.)

(2.1) *Bill* dressed (*himself*).
(2.2) *Bill* clothed *(*himself*).

(Underlining indicates the relata of the antecedence relation.)

A third class of English verbs admitting an optional noun phrase complement comprises verbs such as *to divorce*, *to fight*, *to kiss*, *to marry*, *to match*, *to meet*, and *to touch*. I find it convenient to call them *implicit reciprocal object verbs*. They fall under a broader class of expressions which Langendoen (1978) called *covert reciprocals*. Their paraphrasal equivalents require the reciprocal pronoun.

(3.1) *Bill and Carol* met (*each other*).
(3.2) *Bill and Carol* encountered *(*each other*).

Verbs which permit the omission of their noun phrase complements are liable to still a fourth construal. Here I shall call these verbs *implicit definite object verbs.*[1] Among the ones identified by Fillmore are *to approach*, *to follow*, *to leave*, *to lose*, and *to win*.

(4.1) Mary walked to *the store*. Her dog followed (*her*).
(4.2) Mary walked to *the store*. Her dog pursued *(*her*).

Here, the construal is determined by the cotnext of the verb *to follow*. However, like a pronoun, its construal can be determined by the context of its utterance. This is perhaps best illustrated by the verb *to leave*.

(5.1) Bill left (here) an hour ago.
(5.2) Bill departed *(here) an hour ago.

We now turn from verbs with optional noun phrase complements to those with optional clausal complements. Such verbs were dubbed by early transformational linguists as *null complement anaphora verbs*. As suggested by this name, these are verbs which, when their complements are omitted, seem to require an antecedent. They are thus like implicit definite object verbs. Just as, when the noun phrase complement is absent, the verb is construed as if it had a pronominal complement with an antecedent noun phrase in the context, so when the clausal complement is absent, the verb is construed as if it had a pronominal complement with an antecedent clause in the context.

Clausal complements of verbs fall into three principal categories: those whose complements clauses are finite, those whose complements clauses are infinite, and those whose complement clauses are gerundial clauses. The first two categories can be either declarative or interrogative. Below are examples where a suitable context is furnished for such verbs, together with a contrasting verb. The facts here are drawn from Fillmore (1986), Grimshaw (1979), and Stirling and Huddleston (2002).

(6.0) Bill was walking with a limp.
(6.1) Carol noticed (that he was walking with a limp).
(6.2) Carol noted *(that he was walking with a limp).

(7.0) Bill told Carol that Bradley was sleeping.
(7.1) But when Brian asked her (whether or not Bradley was sleeping), she said that she could not remember (whether or not Bradley was sleeping).

(8.0) Bill suggested to Carol that she drive the car.
(8.1) But she would not even try (to drive the car).
(8.2) But she would not even attempt *(to drive the car).

[1] In Gillon (2012) and in other work, I have called them *implicit ambiphoric object verbs*. Fillmore (1986), who seems to have the first to identify them, has referred to them as having *implicit definite arguments*.

(9.0) Carol told Bill where to find the car.
(9.1) But Bill could not remember (where to find the car).
(9.2) But Bill could not recollect *(where to find the car).

(10.0) Carol is building a deck.
(10.1) This morning Carol finished (building it).
(10.2) This morning Carol completed *(building it).

3.3 Model Theory

The question arises: how does one capture their syntactic and semantic properties? One common idea is to signal the optionality of the noun phrase complement in the strict subcategorization frame of the verb. This is customarily done by enclosing the symbol for the subcategorized phrasal category in parentheses. Thus, while an obligatorily transitive verb, such as *to greet*, is assigned the strict subcategorization frame $\langle NP \rangle$, which signals that it requires a direct object, an optionally transitive verb, such as *to read*, is assigned the strict subcategorization frame of $\langle (NP) \rangle$. Though this notation adequately indicates the optionality of the verb's noun phrase complement, it fails to distinguish the variation in construal which results when the noun phrase complement is omitted. Consider the sentences in (11). Their construals are not uniform.

(11.1) Bill read.
(11.2) Bill dressed.
(11.3) Bill and Carol met.
(11.4) Bill left.

Thus, for example, *Bill dressed* does not mean that Bill dressed someone or something; it means only that Bill dressed himself.

To distinguish the construals, one might index the parentheses, a different index for different construals. This, however, turns out to be cumbersome, and ultimately, not sufficiently general. Instead, one can introduce four diacritics: rec for the cases where the construal is that of a reciprocal pronoun, ref for the cases where the construal is that of a reflexive pronoun, ind for the cases where the construal is that of a suitable indefinite noun phrase and def for the cases where the construal is that of a definite noun phrase, typically a suitable pronoun. We shall therefore write the strict subcategorization frame for *to read* as $\langle \{NP, ind\} \rangle$, the one for *to dress* as $\langle \{NP, ref\} \rangle$, the one for *to kiss* as $\langle \{NP, rec\} \rangle$ and the one for *to leave* as $\langle \{NP, def\} \rangle$.[2]

[2] Should the word strictly subcategorize for more than one sister constituent, the different constituents can be identified by adding further complement specifications enclosed within braces, the braced specification being separated by commas. For a complete presentation of the notation, see Gillon (2012) § 2.

While the presence of the diacritics assures the optionality of the noun phrase complements for the verbs so marked, the difference among the diacritics assures a difference in interpretation. To see how this works, recall that transitive verbs are assigned a set of ordered pairs of members of the domain of the model in which they are interpreted. More generally, a verb with n complements is assigned a set of $n + 1$-tuples from the domain in the model. Under the proposal set out here, this assignment is made, even if the complements are optional. Thus, verbs with optional noun phrase complements are also assigned a set of ordered pairs, just like those with obligatory noun phrase complements. Nonetheless, what value is assigned to the verb phrase node depends on whether or not there is a complement; and if there is no complement, on which diacritic appears associated with the omitted complement.

Thus, if a verb with a complement, whether optional or not, is followed by a complement, the usual rule of assigning a value to the dominating verb phrase node applies. If, however, the verb admits no complement, then the diacritic will determine which value will be assigned to the verb phrase node. If the verb is an implicit indefinite object verb, such as *to read*, then the value assigned to the verb phrase node is the set of members of the domain which are found in the first coordinate of the set of ordered pairs associated with the verb. In other words, the verb phrase node has associated with it the set of members of the domain such that each member has read something. If the verb is an implicit reflexive object verb, such as *to dress*, then the value assigned to the verb phrase node is the set of members of the domain which are found in the first coordinate of the set of ordered pairs associated with the verb and which are identical with the members in the second coordinate. That is to say, the verb phrase node has associated with it the set of members of the domain such that each member dressed himself or herself. If the verb is an implicit reciprocal object verb, such as *to kiss*, then the value of its verb phrase node is the set of members of the domain which kiss each other.[3] More formally stated, the verb phrase node has associated with it the set of members of the domain such that each member kissed some other member which in turn kissed him/her. Finally, if the verb is an implicit definite object verb, such as *to leave*, then the value assigned to the verb phrase node is the set of members of the domain which are found in the first coordinate of the set of ordered pairs associated with the verb and each of which has some contextually determined member in the second coordinate. In other words, the verb phrase node has associated with it the set of members of the domain such that each member left some contextually determined place or thing.[4] This also holds for verbs with optional clauses.

---

[3] This informal statement does not take into account fully the complexity which arises from the fact that the subject noun phrase is plural.

[4] The complete model theoretic details are found in Gillon (2012) § 2.3.

3.4 Adjectives

This treatment extends easily to adjectives with optional complements. I start with a brief outline of the relevant facts. English adjectives may occur either attributively, that is, as a modifier within a noun phrase, or predicatively, that is, as the complement of a copular verb. While the overwhelming majority of adjectives occur both attributively and predicatively, some occur only predicatively and others occur only attributively. Adjectives which occur attributively may either precede or succeed the head noun of the noun phrase in which they occur. But the distribution is not free. English adjectives, whether occurring attributively or predicatively, like English verbs, may have complements. Now while most English adjectives, like many English verbs, take complements optionally, some resist any complement and some require their complements.

Our attention will be confined to adjectival complements of predicative adjectives. Complements to adjectives are of two major kinds: prepositional phrases and clauses. We shall start with prepositional phrase complements.

Not all predicative adjectives admit complements. Here are some which resist any complement: *ambulatory*, *bald*, *dead*, *enormous*, *farcical*, *gigantic*, *hasty*, *lovely*, *main*, *nefarious*, *ostentatious*, *purple*, *quiet*, *regular*, *salty*, *tentative*, *urban*, *vivid*, *wild*, and *young* (Pullum and Huddleston 2002, p. 543).

When prepositional phrases serve as complements to predicative adjectives, the prepositions heading such phrases are more or less confined to these: *about*, *at*, *by*, *for*, *from*, *in*, *of*, *on*, *upon*, *to*, *toward*, and *with*.

A few English adjectives have been identified as requiring prepositional phrase complements. They are: *averse to*, *contingent on*, *dependent on*, *due to*, *fond of*, *incumbent on*, *intent on*, *liable to*, *loath to*, *mindful of*, *reliant on*, and *subject to*. Thus, for example, Quirk et al. (1985) (16.69) point out the following:

(12) Max is averse *(to games).

Most English adjectives seem to permit complements but also to permit their omission. However, things are not always what they appear to be. The principal empirical problem is how to distinguish adjectives whose complements are optional from homophonic adjectives with very similar meanings, one of which takes no complement and the other which does.

Let us begin with a case where the meanings are easily distinguished, the adjective *sick*. In the first sentence below, *sick* has no complement and it is synonymous with the adjective *ill*; in the second, it has a complement yet it not synonymous with the adjective *ill*.

(13.1) Bill is sick.
Cp.: Bill is ill.
(13.2) Bill is sick of school.
Cp.: Bill has a strong distaste of school.

Notice that Bill can be sick, or ill, without being sick of anyone or anything and Bill can be sick of someone or something without being sick, or ill. Thus, neither sentence in (13) entails the other.

Another adjective with distinguishable meanings is the adjective *proud.*

(14.1) Bill is proud.
Cp.: Bill is arrogant.
(14.2) Bill is proud of his success.
Cp.: Bill is highly satisfied with his success.

Once again we note the lack of entailment: Bill can be proud, or arrogant, without being proud of, or highly satisfied with, anyone or anything, and Bill can be proud of, or highly satisfied with, his success without being proud, or arrogant.

It need not be the case that an adjective, on one sense, require a complement and, on another, exclude a complement. The adjective *familiar* has two different senses, indeed, one being the converse of the other. These senses are distinguished by whether the prepositional phrase complement is headed by the preposition *to* or *with.*

(15.1) These facts are familiar to the expert.
Cp.: these facts are known to the expert.
(15.2) The expert is familiar with these facts.
Cp.: The expert knows these facts.

But observe that, when the preposition *to* heads the complement prepositional phrase, the complement may be omitted, but not when the preposition *with* heads it.

(16.1) These facts are familiar.
Cp.: These facts are known.
(16.2) *The expert is familiar.
Cp.: The expert knows.

Of course, the second sentence is acceptable, provided the adjective *familiar* be construed as *is known*.[5]

To my knowledge, no counterpart to implicit reflexive object verbs is to be found among the adjective of English. In other words, English has no adjectives with optional complements whose construal is reflexive when the complement is omitted. But the counterpart to implicit reciprocal object verbs are to be found among the English adjectives. They include such adjectives as *compatible* (*with*), *distinct* (*from*), *divergent* (*from*), *equivalent* (*to*), *identical* (*to* or *with*), *incompatible* (*with*), *parallel* (*to* or *with*), *perpendicular* (*to*), *similar* (*to*), *simultaneous* (*with*), and *separate* (*from*).[6] The adjective *similar* is especially interesting, as it forms one of a minimal triple. Below are three synonymous words, the adjectives *similar* and *alike* and the preposition *like*.

---

[5] I thank Andrew Reisner for bringing this example to my attention.
[6] Of course, the subcategorization frames must be enriched so as to specify the choice of preposition.

(17.1) Bill and Carol are similar (to each other).
(17.2) Bill and Carol are like *(each other).
(17.3) Bill and Carol are alike (*each other).

Yet they differ regarding their allowing, requiring, or excluding a complement. These possibilities are easily handled by the notation introduced here: *like*, which requires a complement, has the strict subcategorization frame of $\langle\{NP\}\rangle$; *similar*, which permits its complement to be omitted and, when it is omitted, has a reciprocal construal, has the strict subcategorization frame of $\langle\{PP, rec\}\rangle$; *alike*, which excludes any complement but has a reciprocal construal, has strict subcategorization frame of $\langle\{rec\}\rangle$. This threefold distinction cannot be handled by the parentheses notation.

Also found among the adjectives of English are the counterparts of implicit definite object verbs, that is, English adjectives whose proper construal requires that a suitable value be found either in the context or in the setting. Here we have such adjectives as *close*, *faraway*, *foreign*, *local*, and *near*. Consider the adjective *faraway*.

(18) Bill lives faraway (from here).

Below, kindly brought to my attention by Ernie Lepore.

(19) Although Bill lives faraway (from *them*), he visits *his parents* regularly.

Such adjectives have the strict subcategorization frame of $\langle\{PP, def\}\rangle$.

We now turn to clausal complements of English adjectives. To a first approximation, English adjectives used predicatively fall into three major categories: the clausal complement is a finite clause, an infinite clause or a gerundial clause. The finite and infinite clauses can be either noninterrogative or interrogative.

(20.0) Carl told me that Bill had left,
(20.1) But Carl was wrong (that Bill had left).
(20.2) But Carl was not sure (whether or not Bill had left).

(21.0) Carl invited Bill to attend the ceremony
(21.1) Carl was unwilling (to attend the ceremony).
(21.2) Carl was unsure (whether or not to attend the ceremony).

(22) Carl was washing the dishes. Jill asked me to set the table. But Carl refused (to set the table), because he was too busy (washing the dishes).

As the strict subcategorization frames provide the relevant syntactic type and since the strict subcategorization frames used with the adjectives are a subset of those used with verbs, the model theoretic interpretation of the adjectives is the same as that of the corresponding verbs.

3.5 Conclusion

In the preceding pages, I have sketched out a model theoretic treatment of verbs and adjectives which take optional complements in which the verbs and adjectives are taken as unambiguous. As it happens, this treatment extends to prepositions which permit their object noun phrases to be omitted as well as to relational nouns whose complements need not be expressed. (See Gillon 2012 § 3 for such extensions.)

Acknowledgments This research was supported by the Social Sciences and Humanities Research Council of Canada (SSHRC 410-2010-1254).

References

Chomsky, N. (1965). *Aspects of the theory of syntax*. Cambridge: The MIT Press.

Fillmore, C. (1986). Pragmatically controlled zero anaphora. In *Proceedings of the Annual Meeting of the Berkeley Linguistics Society*, Vol. 12, 95–107.

Gillon, B. S. (2012). Implicit complements: A dilemma for model theoretic semantics. *Linguistics and Philosophy, 35*(4), 313–359.

Grimshaw, J. (1979). Complement selection and the lexicon. *Linguistic Inquiry, 10*(2), 279–326.

Huddleston, R. (2002). The clause: Complements (Chapter 4) . In R. Huddleston, G. K. Pullum, et al. (Eds.), *The Cambridge grammar of the English language*. Cambridge: Cambridge University Press.

Langendoen, D. T. (1978). The logic of reciprocity. *Linguistic Inquiry, 9*(2), 177–197.

Pullum, G. K., & Huddleston, R. (2002). Adjectives and adverbs (Chapter 6). In R. Huddleston, G. K. Pullum, et al. (Eds.), *The Cambridge grammar of the English language*. Cambridge: Cambridge University Press.

Quirk, R., Greenbaum, S., Leech, G., & Svartik, J. (1985). *A comprehensive grammar of the English language*. London: Longman.

Stirling, L., & Huddleston, R. (2002). Deixis and anaphora (Chapter 17). In R. Huddleston, G. K. Pullum, et al. (Eds.), *The Cambridge grammar of the English language*. Cambridge: Cambridge University Press.

Chapter 4
The Representation and Processing of Participant Role Information

Gail Mauner

When we recognize a verb, much of what we retrieve is information about the situation types that verb describes. Each situation type has associated with it a set of entities specified in terms of their mode of participation in the described situation. I refer to these entities and their associated modes of participation as participant roles or participant role information. An ongoing debate regarding participant roles revolves around two questions: (1) How is participant role information represented? and (2) How is participant role information used in sentence comprehension? The first part of this chapter explores attempts to answer the first question. The second part is devoted to delineating how and when different types of participant role information are used in online language comprehension.

4.1 The Representation of Participant Role Information

Intuitively, some participant roles seem to be more central to the meaning of verbs than others. This intuition is often captured in the linguistics literature as a distinction between arguments, which will be operationalized here as participant role information that is activated when a verb is recognized, and adjuncts, which is operationalized as participant role information that is only weakly activated upon verb recognition. The distinction between arguments and adjuncts has played an important role in linguistic theorizing. This is especially true for frameworks that posit lexically projected structure. In these frameworks, argument structure drives clause construction. The argument–adjunct distinction also underlies explanations of many psycholinguistic phenomena, including the interpretation of implicit arguments (Mauner et al. 1995, 2002; Mauner and Koenig 1999, 2000), the processing of sentences with long-distant dependencies (e.g., Boland et al. 1989, 1995; Conklin

G. Mauner (✉)
Department of Psychology, Center for Cognitive Science, University at Buffalo,
The State University of New York, Buffalo, NY 14260-4110, USA
e-mail: mauner@buffalo.edu

R. G. de Almeida, C. Manouilidou (eds.), *Cognitive Science Perspectives on Verb Representation and Processing,* DOI 10.1007/978-3-319-10112-5_4

et al. 2004; Garnsey et al. 1989; Koenig et al. 2003; Stowe et al. 1991; Sussman and Sedivy 2003; Traxler and Pickering 1996), sense selection (e.g., Hanna et al. 1996; Hare et al. 2003; Liversedge et al. 1998), syntactic attachment preferences (e.g., Blodgett and Boland 2004; Boland and Boehm-Jernigan 1998; Britt 1994; Clifton et al. 1991; Liversedge et al. 1998; Schütze and Gibson 1999; Speer and Clifton 1998), and syntactic ambiguity resolution (e.g., Christianson et al. 2001; McRae et al. 1997; Schmauder and Egan 1998; Trueswell et al. 1994).

Despite the centrality of the argument–adjunct distinction in linguistic theorizing (e.g., Fillmore 1968; Jackendoff 1990) and in explaining many psycholinguistic phenomena, there is widespread disagreement regarding what arguments are, and consequently, how to determine the argument status of role bearing elements. One common approach has been to use syntactic criteria to determine argument status. Many researchers who have adopted this approach place great weight on the fact that some participant roles that are associated with syntactic dependents do not pass all syntactic diagnostics for argument status. For instance, Carlson and Tanenhaus (1988) argued that only those participant roles that correspond to constituents that are sisters of a verb and are governed by it should be accorded argument status. Because instrument and beneficiary participants are typically associated with optional prepositional phrases (PPs) that fail standard verb phrase (VP)-constituency tests, Carlson and Tanenhaus argued that instrument and beneficiary participant roles are never arguments (for similar arguments, see Spivey-Knowlton and Sedivy 1995). In contrast, the fact that syntactic dependents sometimes pass most, but not all, syntactic diagnostics for argument status have led other researchers to consider instrument participants to be either arguments (e.g., Schütze and Gibson 1999) or quasi-arguments (e.g., Van Valin and LaPolla 1997). Yet others who have advocated syntactic criteria for determining argument status have placed greatest emphasis on diathesis alternations, noting that participant roles associated with syntactic dependents that participate in these alternations often correspond to arguments (e.g., Levin 1993; Kim and Thompson 2004). On this syntactic approach, beneficiary but not instrument participant roles, will sometimes correspond to arguments.

Problems with using only syntactic criteria to determine argument status have been identified by a number of scholars (Koenig et al. 2003; Miller 1997; Schütze 1996; Schütze and Gibson 1999; Van Valin and LaPolla 1997). Some have argued that syntactic criteria for identifying arguments do not uniformly reflect structural dichotomies. Others have argued that many of the syntactic diagnostics used to determine argument status rely on rare structures about which there is considerable disagreement regarding grammatical acceptability. Finally, there is disagreement about how to interpret cases in which a participant role passes some but not all syntactic tests.

A number of nonsyntactic alternatives for determining argument status have been proposed. One proposal from the psycholinguistics literature posits that argument status depends on how frequently explicitly mentioned linguistic elements co-occur with specific verbs or verb forms (MacDonald et al. 1994). On this view, argument and adjunct status is a continuum with arguments corresponding to the linguistic elements that are most likely to frequently co-occur with a verb and thus be strongly activated when that verb is recognized. One practical issue that immediately arises on this view is that it is not clear how frequently a linguistic element must co-occur with a verb before the two are reliably associated and activation is assured. To understand the import of this point, consider how we interpret so-called short (also known as agentless or truncated) passives. Readers routinely interpret short passive sentences like *the ship was sunk,* but not intransitive sentences such as *the ship sank,* as including an unexpressed agent (Mauner et al. 1995, 2002; Mauner and Koenig 2000). We might ask: How do we learn to associate an *unexpressed* agent with a passive participle or short passive construction on a frequency of co-occurrence view? Learning that a passive participle is strongly associated with an agent participant role would be the consequence of frequently encountering a passive participle co-occurring with an *explicitly* mentioned agent *by*-phrase. But passives are not all that frequent. While estimates vary, it has been reported that 2–3 % of spoken sentences in corpora of spoken language and 9–12 % of written sentences in corpora of nonscientific writing are passives (Gahl et al. 2004; Svartvik 1966). More importantly, only 20–24 % of the time do written passives include *by*-phrases (Quirk et al. 1985; Svartvik 1966). Thus, co-occurrence frequencies for passive participles and *by*-phrases represent, at best, 2 % of the sentences a language learner might encounter in print and even fewer of the sentences they might encounter in spoken language. Crucially, most of the instances of passive participles we encounter will not have co-occurred with an agent *by*-phrase. Because the overwhelming majority of passives we encounter do not include an agent *by*-phrase, it is unclear how the correspondence between an agent *by*-phrase and a passive participle could be relied upon to learn that passive participles encode an implicit agent argument. As a consequence, it is difficult for the frequency of co-occurrence view to account for the routine inclusion of implicit agents in our understanding of short passive sentences.

Another nonsyntactic approach to determining the argument status of participant role fillers avoids the problems associated with requiring the presence of explicit constituents to learn associations between specific participant roles and verbs. On this view, participant roles are considered to be verb-specific concepts (Ferretti et al. 2001; McRae et al. 1997). Specifically, participant roles are considered to be concepts that are modeled as feature-based prototypes computed over the set of encountered situations described by a verb. What a given verb makes strongly available when it is recognized are the features of the participant role fillers in the situations a verb labels. Because participant role concepts are conceived of as prototypes, a consequence of this view is that there are no semantically necessary features for any participant role type. Instead, verb-specific participant role concepts are argued to have properties similar to nominal concepts, that is, they are supposed to exhibit graded structure and typicality effects. The verb-specific concepts approach is similar in spirit to the co-occurrence frequency view in two ways. First, the association of participant roles and verbs arises from co-occurrences. However, in this case, the relevant co-occurrences are between specific situations that a verb labels and the most typical or frequent participants in those situations. Second, presumably neither approach makes a qualitative distinction between a verb's arguments and adjuncts. For both, the argument status of a participant role is a matter of the degree or strength of association between a verb that labels a set of situations and either the explicit constituents mentioned in a sentence that bear particular participant roles or the features of role participants in the representations of situations themselves.

One practical challenge for the verb-specific concepts view is determining the features of participant role concepts for each verb. This is usually accomplished by asking participants to, for instance, *list features that come to mind for someone who serves someone something* (McRae et al. 1997). While elicitation is a reasonable approach, it is also subject to biases. Instructions can unintentionally limit the features that participants generate. Consider the example instruction for generating features for the verb *serve*. It requires participants to consider features for only animate fillers of the agent and patient roles of *serve*. Yet there are situations in which inanimate entities (e.g., computers, robots, and vending machines) or aggregate entities whose features might be quite diffuse (e.g., markets, businesses, the public) can be served or do the serving. Situations involving these kinds of participants will not be considered. Consequently, the features associated with inanimate and aggregate agent and patient role fillers of serving situations will not be elicited. Feature elicitation may also tap only the most accessible types of situation knowledge. There is considerable evidence to suggest that much of our conceptual knowledge about situations is organized in terms of schemata (Rumelhart 1980), scripts (Schank and Abelson 1977), scenarios (Sanford and Garrod 1981), and/or frames (Minsky 1975) and that access to this situation-based conceptual knowledge makes available information about typical situation participants (Sanford and Garrod 1981; Schank and Abelson 1977). When participants generate features for participant role fillers for verbs such as *serve,* they likely access script-based situation knowledge (e.g., a restaurant script). To see why this is so, consider the agent features that McRae et al.'s participants must have generated for the verb *serve*. While we do not have access to these features, we do know that *waitress* was judged to be a very typical agent of *serve,* and importantly, that many of the features generated for the agent of *serve* were judged to be central to the concept of *waitress*. The fact that waitress had a high role/filler similarity rating for *serve* suggests that participants must have activated a restaurant script when generating agent features for *serve*.

There are three points I want to make about the participant role features that are likely to be generated if situation scripts are activated by elicitation. First, if feature elicitation taps situation knowledge organized into scripts and such like, then only those features that are associated with those scripts will be generated. Second, features generated from script-based situation knowledge do not always correspond to features associated with the most frequent role fillers of events described by verbs. If McRae et al.'s participants are at all like me, they have been served many more meals by their mothers than by waitresses. Thus, they should have generated many features associated with mothers. Yet, I would be willing to wager that *mother* would have much lower role-filler similarity for *serve* than *waitress*. Third, the features that are generated from script-based situation knowledge rarely if ever correspond to features posited by linguistic proposals regarding the arguments of verbs (e.g., Dowty 1991; Koenig et al. 2003). One might argue that this is a rather damning result for these linguistic proposals. However, while < is scary >, < is mean >, and

< is big > may be the sorts of features that individuals reliably and easily access, it is clear that argument representations must also include grammatically relevant role properties. To illustrate, consider the sentences in (1). Sentence (1a) shows that situations that are accurately labeled by the verb *assassinate* require their agents to act volitionally, hence the acceptability of the adverb *intentionally* and the unacceptability of the adverb *accidentally*. Sentence (1b) illustrates that rationale clauses (e.g., *in order to evaluate it*) felicitously modify only sentences describing situations that involve agents whose actions are volitional. The rationale clause in (1b) is acceptable because inspecting something requires volition. The rationale clause in (1b) is unacceptable because knowing something does not require volition. The examples in (1) demonstrate that some participant role features are grammatically necessary and must be encoded in the semantic representations of verbs to license other linguistic elements, even though properties like < act volitionally > are not mentioned in elicitation tasks.

(1) a. The government-sponsored spy assassinated the politician *intentionally/ *accidentally*.
b. The art expert *inspected/*knew* the painting well in order to evaluate it.

Both linguistic judgments and online evidence support the contention that verbs encode grammatically necessary participant role features. Mauner et al. (1995) used rationale clauses like *to collect a settlement from the insurance company* when they followed short passive (*the ship was sunk*), full passive (*the ship was sunk by the captain*), active (*the captain sank the ship*), and intransitive (*the ship sank*) clauses to test the hypothesis that the passive participle verbs in short passive sentences encode an implicit agent with the feature of volition. Mauner et al. found that readers easily interpreted rationale clauses following active and full passive clauses since they provided explicit agents for interpreting the understood subjects of adjoining rationale clauses. In contrast, readers rejected rationale clauses following intransitive clauses that introduced neither an implicit nor explicit agent for the interpretation of their rationale clauses. Most importantly, there were no differences in make-sense judgments or reading times when rationale clauses followed short passives compared to when they followed clauses that introduced explicit agents. The absence of differences in judgments demonstrates that the verbs in short passives must have encoded a implicit volitional agent, while the absence of reading time differences shows that implicit agents were not inferred just to interpret rationale clauses. Had they been inferred, reading times at the verbs in rationale clauses would have been longer in the rationale clauses following short passives than following sentences with explicit agents.

The most compelling evidence for the encoding and automatic activation of abstract participant role features (e.g., volition) comes from two studies by Mauner and Koenig (2000) and Mauner et al. (2002). They compared the processing of rationale clauses such as the example in (2c) when they followed short passives like (2a) or middles like (2b). Crucially, the situations described by these short passive and middle sentences both logically require an agent (e.g., some sort of seller). Mauner and Koenig conjectured that while the passive participle and middle forms of verbs like *sell* both tap the same script-based situation knowledge in conceptual structure, they differ in whether they activate grammatically necessary agent role features like volition. They hypothesized that grammatically required participant role features stored in conceptual structure are activated only if they are marked as being *syntactically active* by a verb, where syntactically active means capable of licensing syntactic dependents. The lexical representations of passive participles but not middle verbs can be thought of as including a pointer to grammatically required, agent role feature information in conceptual structure. In support of this conjecture, Mauner and Koenig (2000) found that readers rejected rationale clauses following middle clauses whose verbs label situations that logically require an agent as soon as possible, that is, at the verb in the rationale clauses. In contrast, readers found those same rationale clauses acceptable when they followed short passives whose verbs not only logically require an agent, but additionally, activate agent features that are necessary for licensing syntactic dependents. Mauner and Koenig's implicit agent studies demonstrate that grammatically required aspects of participant role knowledge that are not captured by the verb-specific concepts approach are, nonetheless, encoded in verbs and immediately accessed and used in online sentence comprehension.

(2) a. The antique vase was sold quickly
  b. The antique vase had sold quickly
  c. to raise money for the charity.

Although the verb-specific concepts view characterizes important aspects of participant role knowledge, it fails to capture more abstract, grammatical role feature properties that are often necessary for licensing grammatical constituents. In the remainder of this section, I discuss a complementary approach to participant role/verb argument representation called the lexical encoding hypothesis (Koenig et al. 2003). This approach is not meant to replace verb-specific participant role knowledge, but rather to supplement it by capturing important aspects of participant roles and role properties are not accessed when script-based knowledge about situation participants is retrieved. The lexical encoding hypothesis approach to determining the argument status of participant roles depends entirely on semantic criteria rather than syntactic criteria, lexical co-occurrence frequencies, or the most frequently mentioned role properties. My collaborators and I have argued that the argument status of any participant role for any verb can be determined on the basis of two criteria: (1) semantic obligatoriness, or whether the participant is required of all situations described by a verb and (2) specificity, or how characteristic or distinctive the participant role or role property is of a situation type being described by a verb.

The first criterion, semantic obligatoriness, which is similar to proposals made by Dowty (1982), plays a crucial role in classifying situation types. For example, all eating situations require things that are eaten. They may also involve instruments, but they do not require them. After all, eating is possible even if one does not use a fork, a spoon, or chopsticks, as anyone who has witnessed a competitive pie-eating contest can attest. In contrast, eating is impossible if nothing is eaten. The obligatoriness criterion captures this difference. Only participants that are required of all situations (given the laws of nature as we know them) that a verb labels are semantically obligatory. Because all situations that *eat* labels require a theme participant, theme participants are semantically obligatory for the verb *eat*. But since not all situations of eating require an instrument, the instrument participants are not semantically obligatory for the verb *eat*.

Many participant roles are semantically obligatory. Consider the sentences: *The barbarian hacked someone with an axe in the glen* and *The barbarian injured someone with an axe in the glen*. Let us focus on the instrument and location participants introduced by the syntactically optional postverbal PPs. Since all situations involving hacking or injuring must occur at some location, locations are obligatory participants for *hack* and *injure*. This is not the case for the instrument participant role. It is easy to conceive of injuring events that do not involve an instrument (e.g., falling and spraining one's ankle). But, there are no hacking situations (in the noncoughing or computer programming senses) that do not require an instrument. So, an instrument participant is obligatory for the verb *hack* but not for the verb *injure*.

Semantic obligatoriness, by itself, is insufficient for establishing the argument status of a participant role for a given verb. For a participant role to be a semantic argument of a verb, it must also be highly characteristic of the situation types that require it. It must have features that are strongly associated with few situation types. Participant roles that are not semantic arguments of a verb will, in contrast, have features that are associated with many situation types. Perhaps the clearest way to illustrate the specificity criterion is with the three location participant roles exemplified by the italicized PPs in (3). In Example (3a), the PP *in the oval office* describes the location in which the seeing event occurred. This kind of location participant role is called an event location. It indicates where the event occurs. Event locations are characterized by the fact that all participants in the event must be at that location at all times throughout the event. For example, both Barack and Michelle must be in the oval office for the duration of the seeing event. Koenig et al.'s (2003) participant role survey revealed that event locations are true of almost all event types (i.e., approximately 98 % of English verbs require an event location). Intuitively, this means that most situation types must occur at some location. As a consequence of being characteristic of almost all situation types, event locations are never arguments of verbs, even when they are semantically obligatory. Another way of thinking about the specificity criterion is that participant roles that are specific help individuate the meaning of one verb from another. Event locations, being true of situations described by almost all verbs, do not help individuate the meaning of one verb from another.

(3) a. Barack saw Michelle *in the oval office*.
b. Barack hid his blackberry *in a desk drawer*.
c. The guard expelled the stalker *from the grounds*.

Now consider the location participant role expressed by the PP *in a desk drawer* in (3b). In contrast to (3a), this PP does not describe where the event of hiding as a whole occurs, but rather, where Barack's blackberry is located at the end of the hiding event. This role is called a participant location. It describes where some, but crucially not all, of the event participants are located at the end of an event. Unlike event location participant roles, participant location roles are specific to approximately 7 %

of some 4000 English verbs known by Koenig et al.'s linguistically trained raters. Participant locations have additional properties that distinguish situations types that require participant locations (e.g., hiding, depositing, concealing) from those that require event locations (e.g., seeing, eating, walking). In our examples, hiding situations are distinguished from seeing situations, in part, because they require an additional location for the object that is hidden. Under the lexical encoding hypothesis, participant locations are arguments because they are both obligatory and specific to few situation types.

Finally, consider the PP *from the grounds* in (3c). This PP expresses a source location role, i.e., a location in which one participant was present at the beginning of the event, but moved away from and was not present at the end of the event. Source locations indicate where some of the participants are no longer present at the end of an event. For instance, the stalker mentioned in (3c) is not on the grounds at the end of the expelling event. In contrast to participant locations, source locations are defined negatively by the fact that one, but not necessarily all, participants must not be at the source location at the end of the event. Conklin et al. (2004) found that only 11 % of the 3615 verbs known to their trained raters required a source location. Thus, the source location participant role meets both the obligatoriness and specificity criteria for argument status for the verb *expel.*

Constituents bearing source and participant location roles are almost always arguments of the verbs with which they co-occur because these role types are usually both semantically obligatory and true for only a restricted number of situation types. The argument status of instrument roles is both more complex and controversial with scholars arguing both for (Ferretti et al. 2001; Schütze and Gibson 1999) and against (e.g., Boland 2005; Carlson and Tanenhaus 1988; Spivey-Knowlton and Sedivy 1995) the argument status of instruments participants. In contrast to other proposals, under the criteria proposed by the lexical encoding hypothesis, instruments are arguments of some verbs but not others. When an instrument participant is not required of all situations a verb describes (e.g., *eat, make, examine, kill*), it fails the obligatoriness criterion, and thus, is not an argument. However, for other verbs (or verb senses), all the situations they describe require an instrument (e.g., *behead, mow, write* (in the sense of making marks on a page), *etch*). Instrument participants for these verbs pass the obligatoriness criterion. Koenig et al. (2003) have argued that because only 11 % of English verbs require an instrument in all of the situations they describe, there are relatively few situation types of which instruments must be true. Thus, instrument participants for these verbs would be arguments (see Koenig et al. (2008) for a more detailed analysis of instrument verb semantic classes). Results of studies examining the integration of instrument, source, participant, and event location WH fillers provides support for the semantic criteria for arguments proposed by Koenig et al.

Evidence from online comprehension studies supports the argument–adjunct distinction drawn by the lexical encoding hypothesis. Koenig et al. hypothesized that WH fillers corresponding to an argument of a verb would be more quickly integrated into sentences than WH fillers that did not correspond to an argument of a verb.

Using this logic, they examined reading times for sentences like those in Example (4) (presentation regions are delineated with a pipe (|)).

(4) a. Which sword | did the knight | behead | the rebel king with | during the rebellion?

b. Which sword | did the knight | kill | the rebel king with | during the rebellion?

The WH filler-gap sentences[1] in (4) are identical except for their verbs. The verb *behead* in (4a) lexically encodes an instrument argument while the verb *kill* in (4b) does not. Crucially, instrument WH fillers across instrument argument and adjunct sentence pairs like the pair in (4) were normed to be implausible as patient role fillers but equally plausible as instrument role fillers. Self-paced reading times confirmed the instrument argument hypothesis. Direct object + *with* regions were read significantly faster in sentences with verbs like *behead*, whose verbs encoded instrument arguments, than in sentences with verbs like *kill*, whose verbs did not encode an instrument.[2] Similar semantic argument–adjunct results have been reported for sentences whose verbs are argued to lexically encode source and participant location arguments when compared to sentences with event location adjuncts that are not lexically encoded by verbs (Conklin et al. 2004). In sum, the results of filler-gap studies that have tested the lexical encoding hypothesis support not only the semantic distinction that Koenig et al. have drawn for semantic arguments and adjuncts, they have also demonstrated that participant role properties that are not typically considered to be a part of verb-specific participant roles are used during online comprehension.

My discussion thus far has been concerned with how participant role information is represented. While the verb-specific concept view captures many important aspects of our participant role knowledge, I have presented evidence that our participant role representations cannot be reduced entirely to prototypical verb-specific conceptual information (see Boland (2005) for a similar proposal). Our participant role representations must also include participant role information such as whether roles or role features are (1) semantically obligatory, (2) specific in the sense that they associated with few rather than many situation types, and (3) syntactically active, that is, grammatically necessary for licensing syntactic operations.

[1] Conventionally, the term *filler* is used to indicate a constituent that is not in the canonical position where it would be assigned a participant role, while the term *gap* is used to indicate that now unoccupied canonical syntactic position. While these terms are most consonant with a government and binding framework, no commitment to this framework is implied. Other frameworks (e.g., Categorical Grammar, Head-Driven Phrase Structure Grammar (HPSG)) make similar predictions about when a WH-constituent will be integrated into a sentence without any commitment to syntactic gaps.

[2] One might be tempted to think that because the critical region included the preposition that the argument–adjunct reading time differences were due to *with* PPs occurring more frequently with instrument argument verbs than instrument adjunct verbs. Two factors rule this possibility out. First, reading times in the critical region did not correlate with the frequency with which verbs co-occurred with instrument PPs. Second, Hongoak Yun, a former graduate student, has replicated the WH-instrument sentence-reading time differences at the direct object when the direct object and preposition were in separate presentation regions.

4.2 The Role of Participant Role Information in Language Processing

I have argued that that there are at least two types of participant role information—verb-specific participant role properties that are derived from situation-based conceptual knowledge that is easy to consciously access, and more abstract, often linguistically required semantic role properties that help individuate the meaning of one verb from another but are less likely to be mentioned in elicitation tasks. The aim of this next section is to answer two questions about how these aspects of participant role information influence language processing. The first question is concerned with the time course of activation. Specifically, I ask: Is either aspect of participant role information automatically activated when a verb is encountered? The second question is concerned with the ways in which participant role information is used during language comprehension, specifically: How do these two aspects of participant role information influence sentence comprehension? In addressing these questions, I place considerable weight on data from studies that have examined the processing of sentences with instrument role participants. Focusing on instrument participants has a number of advantages. One is that because the argument status of instruments has been controversial, evidence regarding which aspects of participant role information are immediately activated, guide parsing decisions, facilitate integration, or lead to anticipation of as yet-to-be-mentioned elements should be particularly informative in evaluating theories of verb argument structure that disagree about the argument status of instruments. Second, because it is possible to hold role filler typicality constant while manipulating the semantic argument status of instruments, it should be possible to determine whether these two aspects of participant role representation play different roles in language processing. Third, because instruments are syntactically optional and rarely mentioned (Koenig et al. 2003, Boland 2005), and are atypical when they are mentioned (Brown and Dell 1987), the influence of participant role information is less likely to be confounded with any syntactic or lexical co-occurrence frequencies with verbs.

4.2.1 What Participant Role Information Is Automatically Activated by Verbs?

The verb-specific concepts view predicts that verbs automatically prime their prototypical participant roles (McRae et al. 1997, p. 145). Initial support for the automatic activation prototypical role fillers by verbs came from Ferretti et al. (2001). Using a paired-word priming procedure with a lexical decision task, they found that verbs primed nouns that were typical fillers of their agent, patient, and instrument (but not location) roles. The claim that instrument role priming was automatic was based on the commonly held assumption that short stimulus onset asynchronies (SOA) and low relatedness proportions (RP) deter strategic/controlled attentional processing (e.g., Neely 1977; Tweedy et al. 1977). Chung-I Erica Su, in her dissertation conducted in my laboratory, tested this claim (Su 2013). Her first experiment was an exact replication of Ferretti et al.'s Experiment 2 that examined instrument and location role priming. She observed significant instrument priming, with a large effect size. Su then tried to replicate the instrument priming effect using a continuous priming/single word presentation format in one experiment (e.g., McNamara and Altarriba 1988; Shelton and Martin 1992) and masked priming with a very short (i.e., 49 ms) SOA (e.g., Perea and Gotor 1997; Perea and Rosa 2002a, b; Bodner and Masson 2003) in another. These methodologies have been shown to be sensitive to semantic relationships but they make it harder for participants to notice prime-target relationships that could serve as the bases for developing response strategies. In neither of these two experiments was significant instrument priming observed. In three additional experiments, Su sought to more directly determine whether her replication of Ferretti et al.'s instrument priming effect was due to strategic processing. She noted that instrument priming could have been due to verbs facilitating responses to their prototypical instrument role fillers, or instead, to inhibition of those same instrument targets when they were preceded by an unrelated instrument verb. More specifically, inhibition could have resulted from generating target expectations from instrument verb primes that were then violated when instrument targets were not among the target expectation set. Su reasoned that replacing *unrelated* instrument verb primes with neutral primes should have no effect on the magnitude of instrument priming for instrument targets preceded by related instrument verb primes if instrument role priming was automatic. In contrast, if replacing unrelated instrument verb primes with neutral primes significantly reduced the magnitude of instrument priming, then Ferretti et al.'s instrument priming effect must have been due, at least in part, to strategic processing. Across three experiments each using a different neutral baseline (a blank screen, the word *readied,* and psychological and intransitive verbs that did not describe situations involving instruments), the instrument priming effect failed to achieve significance and resulted in negligible priming effect sizes. These results strongly suggest that the activation of a verb's prototypical role fillers is not automatic.

There are no semantic priming studies investigating whether lexically encoded role-filler properties are automatically activated. However, there is suggestive evidence from sentence processing studies indicating that this sort of participant role knowledge is rapidly accessed. Mauner and Koenig (1999, 2000) used fronted rationale clauses like (2c) to engender an expectation for a volitional agent in an adjoining short passive or intransitive clause. In this experiment, intransitive clauses began to elicit significantly more does-not-make-sense judgments than short passive clauses right at the verb in these clauses. This strongly suggests that volitional agent information was accessed immediately. Mauner et al. (2002) extended this finding by monitoring eye movements while participants read short passive (2a) and middle (2b) clauses that followed rationale clauses (2c). Recall that the verbs in short passive and middle clauses both logically require an agent. By hypothesis, they differ only in that the agent associated with passive participles activates grammatically required agent properties such as volition. In this experiment, differences between syntactically active and inactive agents emerged in the region immediately after the verb.[3] Thus, while Su's data indicate that at least some situation-based prototypical role-filler information is not automatically activated by verbs, Mauner and Koenig's (2002) results suggest that the grammatically required properties of agent roles are accessed when a verb is recognized.

4.2.2 How Do Different Aspects of Participant Role Information Influence Sentence Comprehension?

Participant role information has been argued to influence online sentence comprehension in three ways. It is argued to (1) guide the selection of syntactic alternatives associated with particular participant roles, (2) facilitate the integration of participant role fillers into sentence representations, and (3) contribute to processes underlying the anticipation of referents of participant role fillers. I will discuss each of these potential influences in turn.

Guidance The results of a number of studies investigating the attachment of structurally ambiguous PPs strongly suggest that participant role information is used to guide syntactic attachment preferences. While the results of early studies suggested that parsing mechanisms initially ignore participant role information and favor parsing heuristics (Rayner et al. 1983), later research showed that PP attachment preferences are influenced by a number of factors, including semantic role expectations generated from the conceptual content of verb + noun + preposition sequences (Taraban and McClelland 1988), referential constraints (Britt 1994; Spivey-Knowlton and Sedivy 1995), and the argument status of PPs (Britt 1994; Schütze and Gibson 1999). Spivey-Knowlton and Sedivy additionally confirmed the influence of another semantic factor for the attachment of *with* PPs. They showed that attachment preferences for *with* PPs are strongly associated with a verb's semantic verb class (i.e., action verbs vs. psychological or perception verbs). Given that instruments are rarely mentioned in sentences, it is likely that the VP-attachment preference for *with PPs* in sentences with action verbs is driven by semantic expectations for an instrument. However, it is not possible to determine from Spivey-Knowlton and Sedivy's data whether semantic expectations for instruments in sentences with action verbs are the result of situation-based participant role knowledge or semantically required participant role information. This issue is more directly addressed in the next section of this discussion.

Integration A number of researchers have used WH-filler-gap sentences to explore the influence of participant role information on the integration of WH fillers into sentence representations. For example, in an important series of experiments,

[3] Mauner and Koenig suggested that the slight delay in detecting the absence of volitional agent properties following middle verbs may have been due to differences in expectations about upcoming constituents following the auxiliary verbs in the middle and short passive sentences.

Boland et al. (1995) employed WH-filler-gap sentences to determine when readers detected the implausibility of WH phrases as the fillers of transitive verbs' theme participant roles, dative verbs' goal/recipient participant roles, and object control verbs' "generalized theme" roles, where the timing of implausibility detection depended on when a WH filler had been assigned a participant role. To illustrate, consider their Experiment 5, which used sentences that had alternating dative verbs and implausible (e.g., Example 5a) or plausible (e.g., Example 5b) recipient role fillers. Boland et al. found that readers began to reject as not making sense sentences with implausible fillers early in the direct object region, i.e., at *maternity*. They argued that because implausibility effects emerged at the first content word in the direct object region, readers must have already provisionally assigned a recipient role to the WH filler at the verb. Boland et al.'s results indicate that participant role information is rapidly activated when verbs are encountered and then quickly used for semantic integration or gap detection.[4] More importantly for the current discussion, they demonstrate that situation-based role property information is activated quickly once a verb is recognized, since it is this type of participant role information that is relevant for evaluating the plausibility of WH fillers. However, this does not mean that only situation-based role property information is used in WH-filler integration. Recall that Koenig et al. (2003) found that instrument WH fillers were more rapidly integrated into filler-gap sentences when the verbs in these sentences required an instrument than when they did not. Because the instrument WH fillers in these sentences were equated for plausibility, which I take to be a reasonable approximation of role-filler typicality, it is unlikely that Koenig et al.'s instrument argument–adjunct differences were due to the activation of situation-based participant role information. Rather, these differences appear to be the consequence of whether or not verbs made available more abstract semantically required participant role information. Thus, it appears that both situation-based and abstract, semantically required participant role information is used in WH-filler integration, although the circumstances under which these sources of participant role information are used may depend on filler plausibility.

(5) a. Bob wondered which bachelor Ann granted a maternity leave to this month.
    b. Bob wondered which secretary Ann granted a maternity leave to this month.

Although the results of Boland et al. (1995) and Koenig et al. (2003) demonstrate that participant role information derived from situation-based conceptual knowledge and more abstract, semantically required participant role properties both influence the integration of WH fillers, there are a number of unresolved issues. One issue is that it is not clear whether the time course for activating these two aspects of participant

---

[4] I am inclined to agree with Kamide's (2008) assessment that the results of filler-gap self-paced reading studies are better interpreted as evidence of rapid semantic integration of WH fillers rather than anticipation of the syntactic gaps that they are associated with, given evidence that gap anticipation is predicated on a filler having first been provisionally assigned a particular participant role (Boland et al. 1995).

role information differ. While it is clear that both sources of participant role information are accessed quickly, Su's results suggest that situation-based participant role information may not be automatically activated. Her data raise the possibility that this situation-based participant role information, while computed quickly, may still take longer to access than participant role properties that do not require the same sorts of computations to be retrieved. Another issue is that even if there are no differences in how quickly these two aspects of participant role information are accessed, they may not be used in the same way under all circumstances. This issue is addressed to some degree in the next section.

Anticipation Over the past decade, there has been increasing interest in the idea that language processing might be, to a considerable degree, predictive in nature. The most compelling evidence that participant role information might be used in forming predictions during language comprehension has come from studies using the visual world eye-monitoring paradigm originated by Cooper (1974) and reintroduced by Tanenhaus et al. (1995). In this paradigm, participants listen to sentences while simultaneously looking at a visual display containing a number of objects or pictures. For this discussion, what is of primary interest is whether participants launch looks to the referents of targets prior to their being mentioned, and especially whether those looks are launched soon after hearing a sentence's verb. The primary focus of this section is on visual world studies whose results may help determine if and when participant role information derived from situation-based conceptual knowledge and more abstract semantically required role information contribute to anticipatory processing during language comprehension.

Altmann and Kamide (1999) were the first to demonstrate that participant role information is used to anticipate the mention of depicted objects corresponding to plausible role fillers of a verb's patient role and that anticipation began as early as the verb. Kamide et al. (2003) demonstrated that listeners also combined participant role constraints from the verb with real-world knowledge about the likelihood that specific pairs of agent and patient role fillers would engage in the activity mentioned by a verb. The results of these studies, while immensely important in demonstrating that participant role information combined with real-world knowledge is used anticipatorily during language comprehension, the separate influences of these two sources of information cannot be disentangled. Moreover, the fact that these studies focused on the anticipation of patient role fillers leaves open the possibility that patients/direct objects may be the locus of anticipatory processing because of their frequency of occurrence and proximity to the verb (see Sussman (2006) for arguments along this line). A study by Boland (2005) addresses these issues.

Boland (2005) used a visual world paradigm to examine whether there are differences in how participant role information derived from real-world knowledge (i.e., prototypical role fillers) and more abstract participant role properties (i.e., argument status) is used to guide listeners' visual attention to depicted entities prior to their being mentioned. In one experiment, participants listened to sentences with postverbal PPs that introduced a typical or atypical location, instrument, or recipient participant in the VPs headed by their respective intransitive (6a), transitive action (6b), or dative (6c) verbs. While listening to these sentences, participants also viewed an array of four pictures, of which one was always the filler of the location, instrument, or recipient role in the sentence. Boland hypothesized that if visual attention during online language comprehension is only guided by abstract participant role information, then recipient argument role fillers should elicit more anticipatory looks than instrument or location adjunct role fillers,[5] and moreover, that there should be no preference for typical over atypical recipients. In contrast, if situation-based participant role representations guide visual attention during language comprehension, then typicality effects should be observed for recipient, instrument, and location role fillers, regardless of their argument status.

(6) a. The girl slept for a while on the *bed/bus* this afternoon.
(Pictures: GIRL, BED/BUS, PILLOW, TOY CAR)
b. The donkey would not move, so the farmer beat it vigorously with a *stick/hat* every day.
(Pictures: DONKEY, FARMER, STICK/HAT, GRASS)
c. The newspaper was difficult to read but the mother suggested it anyway to her *teenager/toddler* last week.
(Pictures: NEWSPAPER, MOTHER, TEENAGER/TODDLER, DICTIONARY)

In this experiment, Boland found that participants did indeed launch more anticipatory looks to recipients than to instruments or locations, and that they showed no preference for typical recipients. This result suggests that abstract participant role properties (i.e., being semantically required by a verb) that are associated with syntactic positions not immediately proximate to verbs guides visual attention and underlies anticipation during language comprehension. Boland also found that instrument and location adjunct role fillers elicited anticipatory looks, although not as many as recipients, and only if they were typical role fillers. This result suggests that participant role information derived from situation-based conceptual knowledge also guides visual attention during language processing, but not as quickly or to the same degree as argument role information. A different pattern of results was observed in a second experiment when visual arrays contained typical and atypical recipient, instrument, or location role fillers. Not only were anticipatory looks launched earlier when displays contained contrasting role fillers than when they contained only a typical or atypical role filler, recipient arguments no longer elicited more anticipatory looks than instruments or locations. Instead, participants launched more looks to typical role fillers, regardless of their argument status. This result suggests that participants may adopt different visual attention strategies when visual referential contrast sets are present. At this point, it is not known whether this is characteristic of language processing in general, or instead, is something that occurs in visual world experiments where repeated exposure to referential contrast sets in displays may to

[5] Boland assumed that instruments were adjuncts and in fact few, if any, of her instrument verbs met Koenig et al.'s criteria for encoding an instrument argument.

lead strategy formation. It is plausible that language comprehenders might adopt attentional strategies that are specific to particular visual contexts. Recent work by Ferreira et al. (2013) has demonstrated that when it is easy to detect, participants use contrast set information in a visual array to predict what will be mentioned or the way in which it will be mentioned. It is quite possible that the participants in Boland's second experiment adopted a strategy in which they first focused on the contrast set in the visual array, and then selected the most typical of the two elements consistent with the spoken input. Because the pictures in the array would be narrowed down to two items before the verb was heard, this would tend to wash out effects of abstract participant role information. In sum, Boland's findings suggest that both situation-based and semantically required participant role knowledge is used in anticipating the mention of participant role fillers and that the locus of anticipation influenced by participant role information extends beyond patients associated with the direct objects of verbs. Her results also suggest that the degree to which these two aspects of participant role knowledge are used to facilitate the anticipation of participant role fillers may depend on attentional strategies that are adopted in response to the features of contextual environments.

The fact that Boland (2005) failed to observe any evidence that instruments behaved like arguments stands in marked contrast to Koenig et al.'s finding of argument–adjunct differences in their instrument filler-gap study. There are two possible explanations for this discrepancy. One, acknowledged by Boland herself, is that the probability of a dative verb co-occurring with a recipient PP is much higher than the probability of an action verb co-occurring with an instrument PP. Higher expectations for recipients than instruments could have been driven by these higher co-occurrence frequencies. The second possible explanation is that most of the action verbs in Boland's study did not semantically require instrument arguments.

Two visual world experiments, conducted by my former student Breton Bienvenue, address the discrepancy between Boland's finding that instrument participants are only represented in situation-based conceptual knowledge and Koenig et al.'s finding that some instrument participants, like recipients participants, are also lexically encoded as the arguments of verbs. Bienvenue employed active declarative sentences that differed only in whether or not their verbs lexically encoded an instrument argument. His sentence materials were largely adapted from Koenig et al. The instruments mentioned in these sentences were normed to be equally plausible in pairs of sentence frames whose verbs either required (e.g., *hacked*) or did not require (e.g., *injured*) an instrument. In his first experiment, participants heard sentences like *The barbarian hacked someone with a sword during the attack* and *The barbarian injured someone with a sword during the attack* while having their eye movements monitored as they viewed an array of pictures that included of the agent (e.g., barbarian), instrument (e.g., sword), and two scenario-relevant distractors (e.g., cottage and shield). The results, presented in Table 4.1, indicate that participants made more anticipatory looks to instruments than distractors in both the verb and direct object regions, but only when verbs lexically encoded an instrument argument. When listening to sentences whose verbs did not lexically encode an instrument argument,

Table 4.1 Proportions of trials for six sentence regions in which there were saccades to a depicted instrument (e.g., sword), distractor 1 (e.g., shield), and distractor 2 (e.g., hut) when sentences included a verb that either lexically encoded (*hack*) or did not encode (*injure*) an instrument

Picture	Verb type	Hacked/injured	Someone	With	A sword	During the attack
Instrument	*Hack*	0.30	0.31	0.12	0.23	0.44
	Injure	0.21	0.22	0.08	0.33	0.47
Distractor 1	*Hack*	0.12	0.25	0.15	0.26	0.38
	Injure	0.20	0.18	0.12	0.28	0.40
Distractor 2	*Hack*	0.16	0.20	0.06	0.30	0.47
	Injure	0.20	0.22	0.11	0.22	0.49

participants' looks to instruments were significantly delayed and did not exceed looks to distractors until after instruments were mentioned.

The results of Bienvenue's first experiment are important for three reasons. First, they indicate that instrument semantic argument information clearly plays a role in anticipatory processing during language comprehension. Second, because instrument argument and adjunct role fillers were equally plausible across sentence pairs, the fact that anticipatory looks were launched only when instruments arguments suggests that abstract semantic role information may be accessed more quickly than situation-based participant role knowledge. Had situation-based participant role knowledge been used to initially guide visual attention, there should have been no differences in when participants launched anticipatory looks to potential referents of instrument arguments and adjuncts. This interpretation is consistent with both Boland's results and Su's instrument priming results. Finally, these results resolve a confound in the interpretation of Boland's results by showing that semantic information about a verb's arguments is used to guide visual attention even when the probability of being mentioned is quite low (Boland 2005; Koenig et al. 2003). While Bienvenue's results do not rule out the possibility that the probability of being mentioned did not also guide visual attention to recipients in Boland (2005), they do suggest that looks to recipients were probably not driven entirely by syntactic expectations.

Two important features of Bienvenue's first experiment were that the patient role filler (i.e., *someone*) was not contentful and the visual array never contained a picture that *someone* could plausibly refer to. This was done to test, across two experiments, Sussman's (2006) hypothesis that when participants hear the verb of a sentence in a passive-listening visual world paradigm, attentional resources are devoted primarily to anticipating the filler of the unassigned participant role that is most likely to be mentioned next. For verbs that only encode a patient, visual attention will be directed toward potential patient role fillers. For verbs that encode a patient and an instrument, visual attention will be focused primarily on finding a referent for the

Table 4.2 Proportions of trials for six sentence regions in which there were saccades to a depicted instrument (e.g., sword), patient (e.g., villager), and distractor (e.g., hut) when sentences included a contentful patient and a verb that either lexically encoded (*hack*) or did not encode (*injure*) an instrument argument

Picture	Verb type	Hacked/ injured	A villager	With	A sword	During the attack
Instrument	*Hack*	0.21	0.36	0.12	0.24	0.46
	Injure	0.20	0.38	0.14	0.27	0.49
Patient	*Hack*	0.21	0.41	0.09	0.29	0.42
	Injure	0.22	0.35	0.09	0.27	0.46
Distractor	*Hack*	0.15	0.29	0.09	0.30	0.44
	Injure	0.17	0.19	0.10	0.27	0.40

patient, although depicted referents for instrument roles might still draw some attention.[6] Bienvenue's second study was conducted to determine whether providing a contentful patient role filler and depicting a potential referent for it would reduce anticipatory looks to instruments at the verb. To do this, sentences and picture arrays from the first study were altered so that sentences now included a contentful filler for the patient role (e.g., a villager) and the visual array included a corresponding picture (i.e., a female villager) in place of one of the distractors. If Sussman's conjecture is correct, patients and instruments should compete for visual attention. This should result in fewer anticipatory looks to instruments at the verb than in Bienvenue's first experiment. As can be seen from Table 4.2, Sussman's hypothesis seems to be borne out. Although looks to patients and instruments in the verb region exceeded looks distractors, the numbers of looks to patients and instruments did not differ. A somewhat surprising finding is that there were as many looks to instruments in the verb region when verbs did not require an instrument as when they did. Anticipatory looks to instruments increased in the patient region, but again, there were no differences as a function of argument status. One possible explanation for this is that the presence of referents for both patients and instruments may have been sufficient to activate situation-based knowledge associated with the verbs in these sentences. If this were the case, looks to patients and instruments might be expected to be roughly equivalent and more likely than looks to scenario relevant distractors (e.g., hut) which are likely to be less strongly associated with the situations described by the verbs in the sentences participants heard.

At this juncture, it is worth considering whether the anticipatory processing seen in these visual world studies represents the typical state of affairs in language comprehension or whether instead anticipation occurs only when constraints are high and the possibility making an error by anticipating the wrong element are relatively low. There is some evidence from my laboratory that is relevant to this issue. Hongoak

[6] This issue does not arise in Boland (2005) because patients were fronted.

Yun (another former student) and I conducted two self-paced, region-by-region reading studies using the sentence materials adapted from Bienvenue's first experiment. Participants read sentences like *The mother* | *wiped/dried* | *the tiny baby's hands* | *with* | *a paper* | *towel* where the verb either required (*wiped*) or did not require (*dried*) an instrument argument. If readers routinely use semantically required participant roles for prediction, then reading times should be faster at *with* or *a paper* when an instrument is obligatory. When participants were required to make a secondary make-sense judgment while reading, the only reading time differences we observed were on the instrument noun itself (e.g., *towel*). Reading times at the noun were faster when instruments were required by the verb than when they were not. That is, we saw evidence of integration facilitation. However, when there was no secondary task, we found no evidence of anticipation or facilitation. These data suggest that anticipation may not be the default mode of processing in language comprehension. Instead, the likelihood that a participant role filler will be anticipated during language comprehension appears to depend on the strength of constraints imposed by the linguistic and nonlinguistics information that is available during comprehension. Thus, Altmann and Kamide's (1999) participants anticipated the mention of the patient of *eat* because only one eatable object was depicted. When the verb did not constrain the range of potential referents for its patient, no anticipation was observed. Participants failed to anticipate the patients of *move* because the visual array contained more than one movable object. In Boland's (2005) study, participants made use of verb argument information (and perhaps information about likelihood of being mentioned) to launch more anticipatory looks to typical and atypical recipient arguments than to typical instrument or location adjuncts when there was only one picture in the display corresponding to the target role filler. In contrast, when Boland's displays depicted both a typical and atypical role filler, semantic argument information was no longer constraining enough to anticipate a recipient target. In these cases, participants relied on role-filler typicality participant role knowledge as the basis for anticipation. Finally, Bienvenue observed anticipatory looks only to instruments that corresponded to the arguments of verbs. However, this was only the case when the patient was not contentful. When the patient was contentful, there was enough combinatory information available from the subject NP, verb, and direct object NP to activate situation-based participant role information such that participants anticipated instruments regardless of their argument status. In this case, clear evidence for anticipatory looks emerged later, at the direct object, rather than at the verb.

Summary

In this chapter, I have suggested that participant role information is represented both as situation-based verb-specific role-filler concepts, as suggested by the verb specific concepts view advocated by Ken McRae and his colleagues, and as more abstract semantically required participant roles or role properties, as suggested by the lexical encoding hypothesis developed by me and my collaborators. I have also presented evidence that at least one of these types of participant role information does not appear to be automatically activated when a verb is encountered. Su (2013) found no evidence of facilitatory instrument priming when the possibility of strategic responding was eliminated. Whether semantically required participant role information is automatically activated is as yet unclear. Finally, I discussed how these two aspects of participant role representation might influence language comprehension. While it is clear that participant role information is used to guide syntactic attachment decisions, the extant data are silent on which type of participant role information is used. Evidence from filler-gap studies provides clear evidence that participant role information is used to facilitate the semantic integration of WH fillers into developing sentence representations. Evidence from Boland et al. (1995) suggests that situation-based participant role information is used since it is required for evaluating the plausibility of WH fillers. Evidence from Koenig et al. (2003) demonstrates that semantically required participant role information also facilitates the semantic integration of WH fillers above and beyond the influence of role-filler typicality or probability of being mentioned. Finally, both sources of participant role information are used to anticipate the mention of participant role fillers. However, when these two sources of information are used depends on how quickly they become available and the type and availability of other sources of constraint.

References

Altmann, G. T. M., & Kamide Y. (1999). Incremental interpretation at verbs: Restricting the domain of subsequent reference. *Cognition, 73,* 247–264.

Blodgett, A., & Boland, J. E. (2004). Differences in the timing of implausibility detection for recipient and instrument prepositional phrases. *Journal of Psycholinguistic Research, 33,* 1–24.

Bodner, G., & Masson, M. (2003). Beyond spreading activation: An influence of relatedness proportion on masked semantic priming. *Psychonomic Bulletin & Review, 10*(3), 645–652.

Boland, J. (2005). Visual arguments. *Cognition, 95,* 237–274.

Boland, J., & Boehm-Jernigan, H. (1998). Lexical constraints and prepositional phrase attachment. *Journal of Memory and Language, 39,* 684–719.

Boland, J. E., Tanenhaus, M. K., Carlson, G., & Garnsey, S. M. (1989). Lexical projection and the interaction of syntax and semantics in parsing. *Journal of Psycholinguistic Research, 18,* 563–576.

Boland, J., Tanenhaus, M., Garnsey, S., & Carlson, G. (1995). Verb argument structure in parsing and interpretation: Evidence from wh-questions. *Journal of Memory and Language, 34,* 774–806.

Britt, M. A. (1994) The interaction of referential ambiguity and argument structure. *Journal of Memory and Language, 33,* 251.

Brown, P., & Dell, G. (1987). Adapting production to comprehension: The explicit mention of instruments. *Cognitive Psychology, 19,* 441–472.

Carlson, G., & Tanenhaus, M. (1988). Thematic roles and language comprehension. In: W. Wilkins (Ed.), *Thematic relations* (pp. 263–288). New York: Academic.

Christianson, K., Hollingworth, A., Halliwell, J., & Ferreira, F. (2001). Thematic roles assigned along the garden path linger. *Cognitive Psychology, 42,* 368–407.

Clifton, C., Jr., Speer, S., & Abney, S. (1991). Parsing arguments: Phrase structure and argument structure as determinants of initial parsing decisions. *Journal of Memory and Language, 30,* 251–271.

Conklin, K., Koenig, J.-P., & Mauner, G. (2004). The role of specificity in the lexical encoding of participants. *Brain and Language, 90,* 221–230.
Cooper, R. (1974). The control of eye fixation by the meaning of spoken language. A new methodology for the real-time investigation of speech perception, memory, and language processing. *Cognitive Psycholoyg, 6,* 84–107.
Dowty, D. (1982). Grammatical relations and Montague grammar. In P. Jacobson & G. Pullum (Eds.), *The nature of syntactic representations* (pp. 79–130). Dordrecht: Reidel.
Dowty, D. (1991). Thematic proto-roles and argument selection. *Language, 67*(3), 547–619.
Ferreira, F., Foucart, A., & Engelhardt, P. (2013). Language processing in the visual world: Effects of preview, visual complexity, and prediction. *Journal of Memory and Language, 69,* 165–182.
Ferretti, T., McRae, K., & Hatherell, A. (2001). Integrating verbs, situation schemas, and thematic role concepts. *Journal of Memory and Language, 44,* 516–547.
Fillmore, C. (1968). The case for case. In E. Bach & R. Harms (Eds.), *Universals in linguistic theory* (pp. 1–88). New York: Holt, Rinehart, and Winston.
Gahl, S., Jurafsky, D., & Roland, D. (2004). Verb subcategorization frequencies: American English corpus data, methodological studies, and cross-corpus comparisons. *Behavior Research Methods, Instruments, & Computers, 36,* 432–443.
Garnsey, S., Tanenhaus, M., & Chapman, R. (1989). Evoked potentials and the study of sentence comprehension. *Journal of Psycholinguistic Research: Special Issue on Sentence Processing, 18,* 51–60.
Hanna, J., Spivey-Knowlton, M., & Tanenhaus, M. (1996). Integrating discourse and local constraints in resolving lexical thematic ambiguities. *Proceedings of the 18th annual conference of the Cognitive Science Society*. Mahwah: Lawrence Erlbaum.
Hare, M., McRae, K., & Elman, J. (2003). Sense and structure: Meaning as a determinant of verb subcategorization preferences. *Journal of Memory and Language, 48,* 281–303.
Jackendoff, R. (1990). *Semantic structures*. Cambridge: MIT Press.
Kamide, Y. (2008). Anticipatory processes in sentence processing. *Language and Linguistics Compass, 24,* 647–670.
Kamide, Y., Altmann, G., & Haywood, S. (2003). The time-course of prediction in incremental sentence processing: Evidence from anticipatory eye movements. *Journal of Memory and Language, 49,* 133–156.
Kim, M., & Thompson, C. (2004). Verb deficits in Alzheimer's disease and agrammatism: Implications for lexical organization. *Brain and Language, 88,* 1–20.
Koenig, J. P., Mauner, G., & Bienvenue, B. (2003). Arguments for adjuncts. *Cognition, 89,* 67–103.
Koenig, J.-P., Mauner, G., Bienvenue, B., & Conklin, K. (2008). What with? The anatomy of a (proto)-role. *Journal of Semantics, 25,* 175–220.
Levin, B. (1993). *English verb classes and alternations: A preliminary investigation*. Chicago: The University of Chicago Press.
Liversedge, S., Pickering, M., Branigan, H., & Van Gompel, R. (1998). Processing arguments and adjuncts in isolation and in context: The case of by-phrase ambiguities in passives. *Journal of Experimental Psychology: Learning, Memory, and Cognition, 24,* 461–475.
MacDonald, M. C., Pearlmutter, N. J., & Seidenberg, M. S. (1994). The lexical nature of syntactic ambiguity resolution. *Psychological Review, 101,* 676–703.
Mauner, G., & Koenig, J.-P. (1999). Lexical encoding of event participant information. *Brain and Language, 68,* 178–184.
Mauner, G., & Koenig, J.-P. (2000). Linguistic vs. conceptual sources of implicit agents in sentence comprehension. *Journal of Memory and Language, 43,* 110–134.
Mauner, G., Tanenhaus, M., & Carlson, G. (1995). Implicit arguments in sentence processing. *Journal of Memory and Language, 34,* 357–382.
Mauner, G., Melinger, A., Koenig, J-P., & Bienvenue, B. (2002). When is schematic participant information encoded? Evidence from eye-monitoring. *Journal of Memory and Language, 47,* 386–406.
McNamara, T. P., & Altarriba, J. (1988). Depth of spreading activation revisited: Semantic mediated priming occurs in lexical decisions. *Journal of Memory and Language, 27,* 545–559.

McRae, K., Ferretti, T., & Amyote, L. (1997). Thematic roles as verb-specific concepts. *Language and Cognitive Processes: Special Issue on Lexical Representations in Sentence Processing, 12,* 137–176.
Miller, P. (1997). Complèments et circonstants: une distinction syntaxique ou sèmantique? In J.-C. Souesme (Ed.), (pp. 91–103). *Actes du 37ème Congrès de la SAES*. Nice: Presses Universitaires de Nice.
Minsky, M. (1975). A framework for representing knowledge. In: P. H. Winston (Ed.), *The psychology of computer vision*. New York: McGraw-Hill.
Neely, J. H. (1977). Semantic priming and retrieval from lexical memory: Roles of inhibition less spreading activation and limited-capacity attention. *Journal of Experimental Psychology: General, 106,* 226–254.
Perea, M., & Gotor, A. (1997). Associative and semantic priming effects occur at very short stimulus-onset asynchronies in lexical decision and naming. *Cognition, 62,* 223–240.
Perea, M., & Rosa, E. (2002a). Does the proportion of associatively related pairs modulate the associative priming effect at very brief stimulus-onset asynchronies? *Acta Psychologica, 110,* 103–124.
Perea, M., & Rosa, E. (2002b). The effects of associative and semantic priming in the lexical decision task. *Psychological Research, 66,* 180–194.
Quirk, R., Greenbaum, S., Leech, G., & Svartvik, J. (1985). *A comprehensive grammar of the English language*. London: Longman.
Rayner, K., Carlson, M., & Frazier, L. (1983). The interaction of syntax and semantics during sentence processing. *Journal of Verbal Learning and Verbal Behavior, 22,* 358–374.
Rumelhart, D. E. (1980). Schemata: The building blocks of cognition. In R. J. Spiro, B. C. Bruce, and W. F. Brewer (Eds.), *Theoretical issues in reading comprehension* (pp. 33–58). Hillsdale: Erlbaum.
Sanford, A. J., & Garrod, S. C. (1981). *Understanding written language*. Chichester: Wiley.
Schank, R., & Abelson, R. (1977). *Scripts, plans, goals and understanding: An inquiry into human knowledge structures*. New Jersey: Lawrence Erlbaum.
Schmauder, A. R., & Egan, M. (1998). The influence of semantic fit on on-line sentence processing. *Memory & Cognition, 26,* 1304–1312.
Schütze, C. (1996). *The empirical base of linguistics: Grammaticality judgments and linguistic methodology*. Chicago: The University of Chicago Press.
Schütze, C., & Gibson, E. (1999). Argumenthood and English prepositional phrase attachment. *Journal of Memory and Language, 40,* 409–431.
Shelton, J. R., & Martin, R. C. (1992). How semantic is automatic semantic priming? *Journal of Experimental Psychology: Learning, Memory, and Cognition, 18,* 1191–1210.
Speer, S. R., & Clifton, C., Jr. (1998). Plausibility and argument structure in sentence comprehension. *Memory and Cognition, 26,* 965–978.
Spivey-Knowlton, M., & Sedivy, J. (1995). Parsing attachment ambiguities with multiple constraints. *Cognition, 55,* 227–267.
Stowe, L. A., Tanenhaus, M. K., & Carlson, G. N. (1991). Filling gaps on-line: Use of lexical and semantic information in sentence processing. *Language and Speech, 34,* 319–340.
Su, C. (2013). *When is semantic priming automatic? Instrument and location participant role priming as a case study*. Unpublished doctoral dissertation, University at Buffalo, The State University of New York, Buffalo.
Sussman, R. (2006). *Verb-instrument information during on-line processing*. Unpublished doctoral dissertation, University of Rochester, Rochester.
Sussman, R., & Sedivy, J. (2003). The time course of processing syntactic dependencies: Evidence from eye movements during spoken narratives. *Language and Cognitive Processes, 18,* 143–163.
Svartvik, J. (1966). On voice in the English verb, *Janua Linguarum, Series Practica 63*. The Hague: Mouton.
Tanenhaus, M. K., Spivey-Knowlton, M. J., Eberhard, K. M., & Sedivy, J. E. (1995). Integration of visual and linguistic information in spoken language comprehension. *Science, 268,* 1632–1634.

Taraban, R., & McClelland, J. L. (1988). Constituent attachment and thematic role assignment in sentence processing: Influences of content-based expectations. *Journal of Memory and Language, 27,* 597–632.

Traxler, M., & Pickering, M. (1996). Plausibility and the processing of unbounded dependencies: An eye-tracking study. *Journal of Memory and Language, 35,* 454–475.

Trueswell, J. C., Tanenhaus, M. K., & Garnsey, S. M. (1994). Semantic effects in parsing: Thematic role information in syntactic ambiguity resolution. *Journal of Memory and Language, 33,* 285–318.

Tweedy, J., Lapinski, R., & Schvanenveldt, R. (1977). Semantic context effects on word recognition. Influence of varying the proportion of items in an appropriate context. *Memory and Cognition, 5,* 84–89.

Van Valin, R., & Lapolla, R. (1997). *Syntax: Form, meaning, and function*. Cambridge: Cambridge University Press.

Part III
Events: Aspect, and Telicity

Chapter 5
Force Dynamics and Directed Change in Event Lexicalization and Argument Realization

William Croft

5.1 Introduction

Two theories of event lexicalization and argument realization are currently widely proposed: aspectual and causal (Levin and Rappaport Hovav 2005, henceforth AR). The aspectual theory originates in the classification of events (and the predicates that lexicalize them) according to how the event unfolds over time. The most influential version of this classification is that of Vendler (1967): his four-way division of events into states, activities, achievements, and accomplishments is widely used, despite acknowledgments of its deficiencies (see § 5.3). However, it is unclear whether the aspectual theory actually can contribute to an understanding of the semantic basis of argument realization—choice of subject, object, and oblique form for argument phrases—as well as event lexicalization. Levin and Rappaport Hovav note that "current aspectually driven theories of argument realization typically focus on the relation between choice and morphosyntactic expression of the direct object and notions such as telicity, measure and incremental theme" (AR, 98). While there is much that is valid in this analysis (see § 5.5), it is too limited as a comprehensive theory of argument realization that would account for the realization of other event participants as subjects or as various types of oblique phrases, as Levin and Rappaport Hovav note (AR, 111–112).

The second theory, the causal theory, is not widely found in generative and formal semantic approaches to event lexicalization and argument realization (with one caveat to be discussed in the next paragraph), but is widespread in cognitive semantics, beginning with Talmy (1976, 1985/2000) and followed by DeLancey (1985) and Langacker (1987) as well as myself (Croft 1991, 1993, 1994, 1998a, b, 2009, 2012). The causal approach is characterized by conceptualizing events as a causal chain linking participants in the event in terms of the transmission of force from one participant

W. Croft (✉)
Department of Linguistics, University of New Mexico, MSC 03 2130,
Albuquerque, NM 87131-0001, USA
e-mail: wcroft@unm.edu

R. G. de Almeida, C. Manouilidou (eds.), *Cognitive Science Perspectives on Verb Representation and Processing,* DOI 10.1007/978-3-319-10112-5_5

to another. In addition to providing a model of event lexicalization—predicates lexicalize segments of the causal chain—it also provides a model of argument realization, articulated in greatest detail in Croft (1991, 2012) and described briefly in § 5.2.

Causal structure does play a role in generative approaches to event lexicalization, namely in theories that exploit event decomposition. Early generative analyses decomposed events into subevents such as CAUSE, DO, BECOME, and (result) STATE. These approaches have been elaborated by many researchers, including Pinker (1989), Jackendoff (1990), Rappaport Hovav and Levin (1998), and Van Valin (Foley and Van Valin 1984; Van Valin and LaPolla 1997). These subevents are causal (CAUSE) as well as aspectual (BECOME), while subevents such as DO appear to be both causal and aspectual, involving both agency and process.

The theories cited in the preceding paragraph represent a particular type of event decomposition: one that includes causal, aspectual, and other subevents, and also nests the subevents in a representation such as [CAUSE [x, [BECOME [FLAT y]]], where the CAUSE subevent relates a causer argument to the caused subevent, or [[x ACT] CAUSE [BECOME [FLAT y]]], where the CAUSE predicate links two subevents. The causal theory found in cognitive semantic approaches also decomposes an event into subevents, but purely in terms of transmission of force, and linearly rather than in terms of a nested structure.

The theories proposed for event lexicalization and argument realization, and the evidence put forward to support those theories, suggest that both causal and aspectual structure of events play a role in their lexicalization as predicates and in the syntactic realization of their arguments. The real question, then, is what is the distinct contribution that each makes to lexical and grammatical realization of clauses? In this chapter, I will argue that the contributions of causal and aspectual structure can be most clearly identified by using a representation of event structure that includes both causal and aspectual structure but clearly distinguishes the two.

Section 5.2 reviews the evidence supporting the hypothesis that the causal (force dynamic) structure of events is the primary determinant of argument realization, given a particular lexicalization of an event. Section 5.3 presents a fine-grained analysis of aspectual structure (Croft 2009, 2012), and introduces the category of directed change, which is the aspectual category that appears to play the most important role in understanding event lexicalization. Section 5.4 introduces the combined causal–aspectual representation, and Sect. 5.5 proposes that directed change plays a central role in understanding constraints on event lexicalization.

5.2 Force Dynamics in Event Lexicalization and Argument Realization

Most, if not all, theories of argument realization have the following theoretical constructs (Croft 1998a, pp. 21–23; compare the organization of AR):

(a) Event structure: a representation of the semantic structure of events that is relevant to syntax. This is where event lexicalization enters into theories of argument realization (cf. AR, Chap. 4).
(b) Participant roles: a way of defining participant roles, usually based at least in principle on properties of event structure (cf. AR, Chap. 2).
(c) Ranking of participant roles: some way of ordering participant roles, such as a thematic role hierarchy or force-dynamic ordering, which is used in argument realization (cf. AR, Chap. 6).
(d) Role designation: some way to designate a special status for certain participant roles that generally links them to subject or object grammatical roles, to account for voice and argument structure alternations that cannot be accounted for by the ranking of participant roles in (c) (cf. AR, Chap. 3). Examples of role designation are macro-roles, proto-roles, underlying syntactic relations, and event profiling (see below).
(e) Mapping rules: rules that map the participant roles into grammatical roles such as subject and object, based on properties defined in (a)–(d) (cf. AR, Chap. 5).

AR provides a detailed critical survey of theories of argument realization; a critique of some of the more common theories is found in Croft (1998a). In this section, I review briefly the causal (force dynamic) theory of argument realization found in Croft (1991, 1998a, b, 2009, 2012), the evidence supporting it, and the shortcomings of its representation in my earlier publications (those before 2000).

The event structure that is proposed in the causal theory of argument realization is a linear causal chain defined by the transmission of force from one participant to another. For example, example (1) illustrates the casual chain structure of the event.

(1) *Sue broke the coconut for Greg with a hammer.*

Sue	⟶	hammer	⟶	coconut	- - -→	Greg
SBJ		A.OBL		OBJ		S.OBL

The representation in (1) provides a semantic structure that achieves (a)–(d) in the argument realization model given above. First, the event structure itself is the causal or force-dynamic chain. Causation is defined in broad terms, to include a variety of causal relations. These involve not only physical causation but also an intentional being either initiating an action (volition or agency) or having one's mental state altered as a result of an action (what Talmy (1976) calls affective causation). They also involve not just "billiard-ball" causation but also "letting causation" (as in *I dropped the ball*) and maintaining a static situation (as in *I was holding the ball*), as described in Talmy's (1988/2000) force-dynamic model. (Noncausal relations will be discussed below.)

Participant roles can be defined in terms of the position of the participant in the causal chain. In fact, as many have noted, participant roles defined in absolute terms appear to play little role in argument realization. Instead, the ranking of participant roles is far more significant. In many theories, the ranking of participant roles is independent of event structure: properties of event structure are not used to define the ranking, and participant roles from different kinds of events are lumped together in a single thematic role hierarchy. In the causal theory, the ranking is defined solely within an event, and is defined as the relative ordering of the participants in the event. In particular, subject is antecedent to the object in the causal chain.

The relative position of participants in the causal chain accounts for the high degree of regularity in the mapping of the participants in transitive events to subject and object roles across languages. Where there is variation across languages (and within languages) in the choice of subject versus object, it can be attributed to indeterminacy in the ordering of participants in a causal chain. A clause construes an event as a single linear, asymmetric causal chain, but not all events are of this type. For example, in predicates involving mental states such as *see, know*, and *like*, there is substantial crosslinguistic variation in whether the experiencer or the stimulus is coded as subject or object (Croft 1993, 2012, pp. 233–236). This is illustrated by the well-known English argument realization patterns in (2)–(4):

(2) *I like Beethoven's Seventh Symphony.*
(3) *I am enjoying Beethoven's Seventh Symphony.*
(4) *Beethoven's Seventh Symphony pleases me.*

In mental events such as those expressed in (2)–(4), the force dynamics is bidirectional: the experiencer attends to the stimulus, and the stimulus causes a certain mental state in the experiencer. Hence, the variability. In fact, the variability is limited to examples like (2). Causative predicates that focus on the change of mental state always have the stimulus as subject, as in (4), because they describe the transmission of force from stimulus to experiencer. Activity verbs that describe how the experiencer is attending to the stimulus always have an experiencer subject as in (3), because they describe the transmission of force (in Talmy's broad sense of "force") from experiencer to stimulus. Sentence (2) denotes a state, and hence there is no transmission of force; it is these that are expressed variably across languages (for examples, see Croft 1993).

Role designation is not stipulated in the causal model, but is part of the semantic structure of the event. The solid arrows in example (1) represent the segment of the causal chain that is denoted, or profiled, by the predicate in the clause (in this case, *break*). I use the term "profile" here basically as it is used in cognitive grammar (Langacker 1987): it represents the concept denoted by a word against its semantic frame (Fillmore 1982, 1985), in this case the entire causal chain in example (1). Differences in verbal profile result in differences in argument realization. For example, in the classic locative alternation, different segments of the causal chain are profiled, as examples (5)–(6).

(5) *Jack loaded the furniture on the truck.*

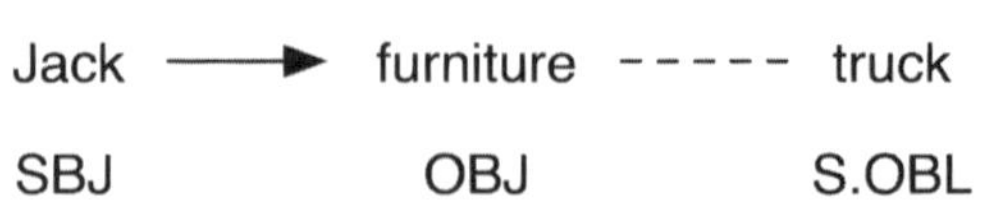

(6) *Jack loaded the truck with the furniture.*

Jack ——▶ furniture ——— truck

SBJ A.OBL OBJ

Examples (5)–(6) illustrate two further properties of the causal structure theory. The caused location event represented in examples (5)–(6) involves a noncausal relation: while the agent causes the change in spatial configuration, the spatial relation of figure (the furniture) to ground (the truck) is not causal. It turns out that, crosslinguistically, the figure is conceptualized or construed as antecedent to the ground in a "causal" chain; the noncausal relation is represented by a line without an arrowhead in examples (5)–(6). The theoretical notion of construal plays a major role in the analysis here. It is a central component of cognitive semantic theories (Croft and Cruse 2004, Chap. 3), but also occurs in generative and formal semantic theories, usually under the name of coercion (see Croft 2012, pp. 84–92, 358–393). I will use the term "construal" here.

The second property is the differentiation of oblique case marking (adposition or case affix) into two types, antecedent and subsequent. Antecedent obliques encode participants antecedent to the object in the causal chain; subsequent obliques encode participants subsequent to the object in the causal chain. Whether a participant is antecedent or subsequent depends of course on which participant in the causal chain is encoded as object, which in turn depends on which segments of the causal chain are profiled.

The division of obliques into antecedent and subsequent is a consequence of the causal theory (or predicted by it, if you prefer). It is supported by extensive crosslinguistic evidence, much of which is presented in Croft (1991, 2012). Case markings are often polysemous, used for multiple semantic roles; but they are generally polysemous only with antecedent roles (as in the case of English *with*) or only with subsequent roles (as with English *for*), as shown in a 40-language typological study (Croft 1991, p. 196). Croft (1991) gives numerous examples of argument structure alternations in a variety of languages in which antecedent and oblique case markings alternate as predicted. Croft (1998a, p. 40) presents developmental evidence that children learning English respect the antecedent/subsequent distinction in their argument realization patterns, even if they use the wrong preposition or an argument structure alternation that does not occur in the adult language.

The profiling of a segment of the causal chain introduces an important new element to the semantic representation of an event. What factors determine what segment of the causal chain is profiled? To some extent, a verbal profile, as reflected by the choice of subject and object, is simply a matter of lexicalization. However, the lexicalization patterns indicate some general principles that appear to govern the choice of the verbal profile. One important principle is that the prototypical simple verb profiles a highly individuated segment of the causal chain, that is, one that is the most "cut off" from the rest of the causal network. Thus, volitional agents are most likely to be subjects—initiators of the profiled segment—because they are construed as autonomous causes. Conversely, completely affected patients are most likely to be objects—end points of the profiled segment—because they are construed as not causing any further change (see Croft 1994). This is just one general principle governing the choice of verbal profile; the rest of this chapter describes other principles governing the choice of verbal profile.

The examples given to illustrate how the causal theory of argument realization represents components (a)–(d) also illustrate (e), the mapping rules that govern the encoding of participants in syntactic roles. The mapping rules that accompany the causal event representation are small in number and simple in formulation (Croft 1998a, p. 24, 2012, p. 207):

1. Subject and object delimit the verbal profile.
2. Subject is antecedent to object in the causal chain (SBJ → OBJ).
3. Antecedent oblique is antecedent to the object in the causal chain; subsequent oblique is subsequent to the object in the causal chain (A.OBL → OBJ → S.OBL).
4. Incorporated arguments are between subject and object in the causal chain (SBJ → INCORP → OBJ).

The crosslinguistic evidence in support of the causal theory of argument linking has withstood the test of time, and causal theories are widely used in cognitive semantics. However, the representation that I presented in earlier publications, which attempted to combine causal structure with aspectual structure, is seriously inadequate (the changes introduced in Croft 1998a do not substantially improve it). Example (7) is the representation for the same sentence as in example (1) in the framework from Croft (1991).

(1) *Sue broke the coconut for Greg with a hammer.*

```
    CAUSE     CAUSE     CHG-ST       STATE      BENEF
  o -----> o -----> o ------> (o) ----- (o) -----> o
  *************break*********************+++for+++
 Sue     hammer    coconut    (coconut) (coconut)   Greg
 Sbj      A.Obl                            Obj      S.Obl
```

This representation suffers from several shortcomings. The arrow notation is applied to processes happening to a single participant (the coconut changing state) as well as to a causal chain (Sue acting on the hammer, the hammer acting on the coconut, etc.). This is in part due to the debt that the representation owes to the traditional event decomposition by McCawley, Gruber and Jackendoff and their successors, which mixes causal and aspectual structure in a single-event decomposition (see § 5.1). Second, the only aspectual distinction that is captured is state versus process—less than Vendler proposed, let alone the additional aspectual types introduced by Vendler's successors (see § 5.3). These shortcomings are the consequence of an unsuccessful integration of aspectual and causal structure in this representation.

The representation is problematic for the causal theory as well. There is reentrant (repeated) representation of participants, e.g., the coconut. The causal theory crucially depends on identifying the verbal profile, yet the representation is ambiguous as to the extent of the verbal profile, since the coconut is represented multiple times in the "causal" chain. Finally, the causal representation itself does not represent events causing other events; it represents only participants acting on other participants. It could be argued that transmission of force is the only relevant model of causation for argument realization; nevertheless, the philosophical position is that causation relates one event to another, and some linguists using event decompositions (e.g., Rappaport Hovav and Levin) represent subevents causing other subevents, not participants acting on other participants.

There is a solution to this representation problem: develop a more fine-grained model of aspectual structure (§ 5.3) and integrate it with a pure causal chain model (§ 5.4). The integrated model is used here to illustrate the role of an aspectual property, directed versus undirected change, in describing certain types of event structure constructions.

5.3 A Two-Dimensional Model of Aspect

The model of aspectual structure presented here is presented in greater detail in Croft (2012, Chaps. 2–4; see also Croft 2009). The aspectual structure of an event describes how the event unfolds over time. This definition implicitly requires two dimensions. The first, of course, is time. The second is what it means to say an event "unfolds." Unfolding characterizes the states and changes of state that take place over the time interval in which the event occurs. These are its phases.

A number of linguists have proposed phasal models of how an event unfolds, that is, a temporal decomposition of the event into discrete phases (see Croft 2009, pp. 149–151, 2012, pp. 45–52). The model presented here is also a phasal model, but unlike most previous proposals, it treats the qualitative states as points on a second dimension, and change as transitions from one state to another on that dimension. Example (8) illustrates the model for the perceptual event of seeing.

(8) *Aspectual contour:* *Achievement profile:* *State profile:*

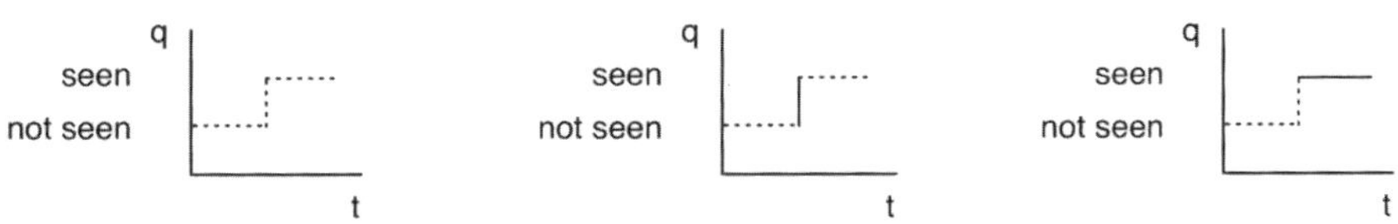

The x axis is the time dimension (t), and the y axis is the qualitative state dimension (q). The aspectual contour in example (8) is how the seeing event unfolds. Seeing has two defined states on q: not seeing something and seeing something. Seeing something is a transitory state, that is, one starts and stops seeing a particular object over one's lifetime. Seeing has at least three phases: not seeing something; the transition from not seeing something to seeing it, which is construed as an instantaneous jump from one state to the other and represented by a vertical line; and seeing that thing. The sequence of phases describes the aspectual contour of the event.

The English verb *see* in a particular usage profiles one phase of the event; a solid line indicates the profiled phase. Hence, the aspectual contour functions as the semantic frame for the profiled phase or phases of the event. The verb *see* may be construed as an achievement (the transition from not seeing something to seeing it) or as a state (the resulting state), as shown in example (8). Part of the challenge in analyzing aspect is the great flexibility of predicates in English to occur in different aspectual construals, without any morphological change in the verb form (Croft 2009, 2012). As will be seen in § 5.4, there are shifts in aspectual construal in argument structure alternations as well.

The two-dimensional *t/q* diagrams allow us to provide distinct representations of all of the aspectual types (or construals) that have been discussed in the aspectual literature, and to make sense of the bewildering variety of aspectual construals. These aspectual types go under different names as several of them have been discovered independently in different analytical traditions (generative, formal semantic, and cognitive semantic). Here I briefly describe the aspectual types and the *t/q* diagrams that represent them. There are ten aspectual types that are lexicalized by simple predicates, and their grouping under the four Vendler aspectual classes provides a framework for analyzing them.

There are three types of states, that is, events in which no change takes place, one of which has two subtypes, as illustrated in example (9):

(9) *The door is open.* *The window is shattered.* *She is French.* *The sun is at its zenith.*

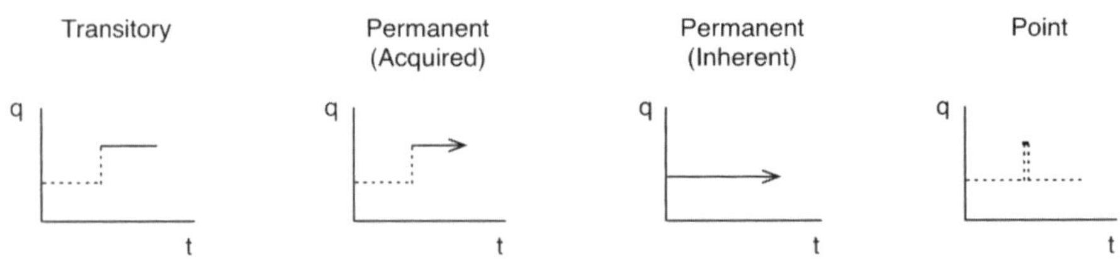

Transitory states are states that are not permanent; they correspond to (stative) stage-level predicates (Carlson 1979, pp. 56–57). Permanent states are represented by an arrow indicating that the state is permanent, i.e., continues for the lifetime of the entity (the time dimension is relative to the lifetime of the entity). Permanent states falls into two subtypes, acquired permanent states (a shattered window will never become whole, but it was whole before) and inherent permanent states (being French is an inherent property of a person, for that person's entire lifetime). Permanent states correspond to absolute states (Comrie 1976, p. 104) and object-level predicates (G. Carlson 1979, pp. 56–57). Finally, a little-known aspectual type are point states (Mittwoch 1988, p. 234), which describe a state that holds for only a point in time; hence, it entails that the point state no longer holds after that point in time, indicated on the *t/q* diagram by the unprofiled phases following the profiled point state.

Corresponding to these three types of states are three types of achievements, that is, instantaneous changes of state that profile the transition phase of the same aspectual contour. They are illustrated in example (10):

(10) *The door opened.* *The window shattered.* *The mouse squeaked.*

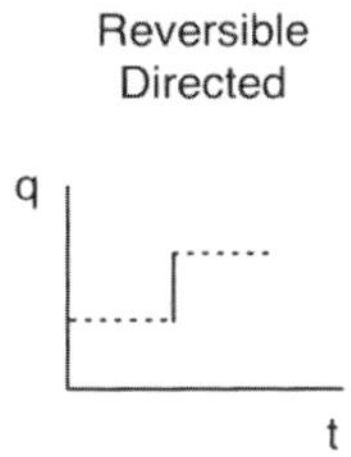

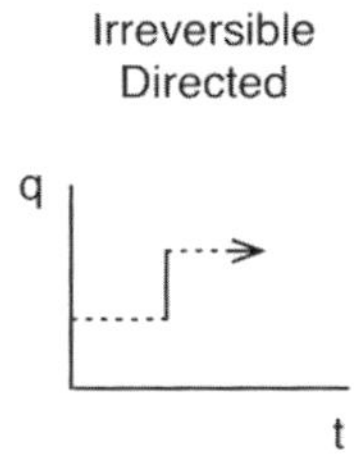

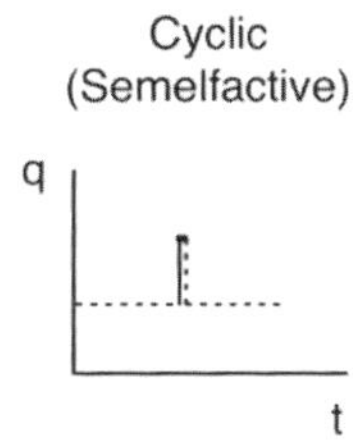

Reversible directed achievements result in transitory, hence reversible, result states; Talmy (1985/2000, p. 68) describes them as resettable verbs. Irreversible directed achievements result in permanent, hence irreversible, result states; Talmy (ibid.) describes them as nonresettable verbs. Cyclic achievements result in point states, which then revert to the rest state. These correspond to semelfactive (Smith 1997, pp. 29–30) or momentaneous (Carlson 1981, p. 39) predicates.

There are two types of activities, that is, durative but unbounded processes, illustrated in example (11):

(11) *The soup cooled.* *The girls chanted.*

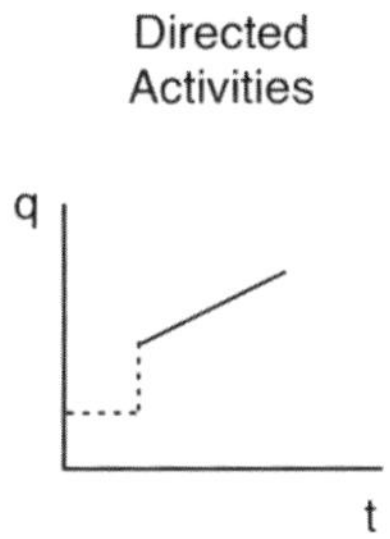

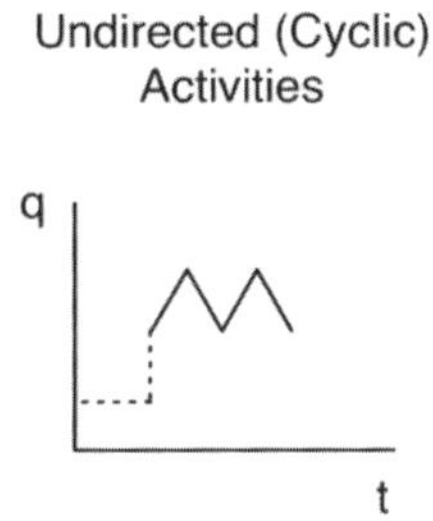

The first is a type of activity that involves an incremental change in a single direction on the q dimension. They are called directed activities here, following Hay et al. (1999, p. 132); they have also been called degree achievements (Dowty 1979, pp. 88–90), dynamic predicates (Carlson 1981, p. 39), and gradient verbs (Talmy 1985/2000, p. 68). The second is an undirected activity, the prototypical activity that Vendler seems to have had in mind. This type of activity essentially involves a process construed as repeated cyclic events of the same kind: talking is repeated sound emissions, dancing is repeated bodily motions, etc. For this reason, it is also described here as a cyclic activity, to emphasize its semantic relationship to cyclic achievements (semelfactives).

Finally, there are two types of performances or bounded processes, which correspond to the two kinds of activities, and are illustrated in example (12):

(12) *I ate an apple pancake.* *Harry repaired the computer.*

Incremental ccomplishments

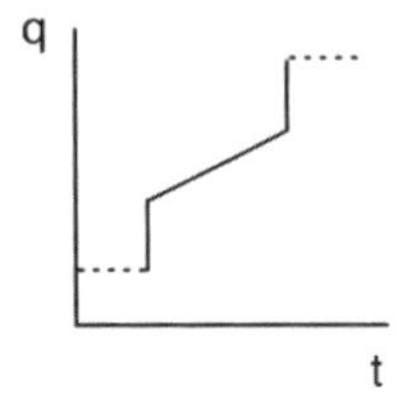

Nonincremental accomplishments

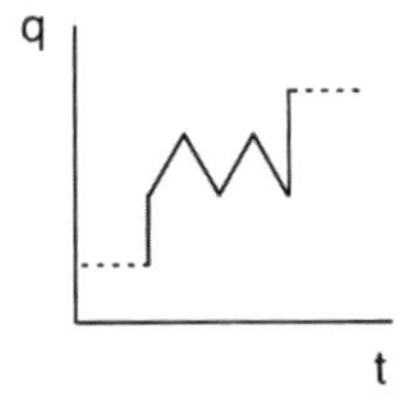

The first type is the prototypical Vendlerian accomplishment: an incremental change leading to a definite resulting state. An incremental accomplishment is temporally bounded, which is represented here, unsurprisingly, by profiling the inception and completion of the event. The resulting state is therefore entailed. (Actually, there are two subtypes of accomplishments, reversible and irreversible, depending on the nature of the result state; but this distinction plays no major role in what follows.)

The second type is called a nonincremental accomplishment (Croft 2012, pp. 62–63; it was called a run-up achievement in Croft 1998, p. 74). This aspectual type was recognized as an alternative construal of some directed achievements, such as *die* in *Help! He's dying!, reach the summit* in *They reached the summit in four hours*, and *fall asleep* as in *She is falling asleep* (Vendler 1967, p. 101, 104; Dowty 1991, p. 137). Rothstein (2004, pp. 98–99) identifies several predicates as having a default construal as what we call nonincremental accomplishments, including *repair X*. A nonincremental accomplishment describes an activity which will or may ultimately lead to a transition to a resulting state, but the activity cannot be described as incrementally accomplishing that resulting state. One can mow half the lawn (an incremental accomplishment), but if one is dying (nonincremental accomplishment), one is still alive until the point at which bodily functions ultimately fail.

It can be seen here that Vendler's examples of activity versus accomplishment conflate the directed/undirected distinction and the unbounded/bounded distinction: Vendler's activities are unbounded and undirected, while his accomplishments are bounded and directed. This may account for the fact that the two distinctions were not clearly separated at first (AR, 95).

The aspectual types illustrated in examples (9)–(12) represent all of the distinct aspectual types that I am aware of in the aspect literature. They also represent all the types with one phase profiled, and all of the bounded types (i.e., with inception and completion phases also profiled; in achievements, the inception and completion phases are one and the same). The only exception is a bounded state. A possible example of a bounded state might be the construal in *The movie starts in five minutes*: it could be argued that nothing is happening in the five-minute period, hence the container adverbial *in five minutes* is bounding a state between the present moment and the start of the movie (I am grateful to Paul Kay for this example). The two-dimensional model thus provides an account of why these are all the aspectual types of simple predicates that have been identified (more complex aspectual contours and profiles can be obtained using complex predicates or certain adverbial expressions; see Croft 2012, pp. 101–110). It also provides a means to represent a variety of semantic relationships among aspectual types.

One of those relationships plays an important role in analyzing the phenomena that will be discussed in § 5.5. Directed achievements, directed activities, incremental accomplishments, and (arguably) nonincremental accomplishment form a coherent class, which will be called directed changes (Croft 2012, pp. 70–77; this category is essentially the same as Beavers' (2008, pp. 250–252) scalar changes; see also Wechsler 2005, pp. 262–268; Rappaport Hovav and Levin 2010, pp. 28–34). In all of these aspectual types, the end point of the profiled phase(s) is higher on the q dimension than the starting point of the profiled phase(s). For all of these types, except nonincremental accomplishment, every point in the profiled phase(s) is higher on the q dimension than every preceding point in the profiled phase(s).

Directed changes cut across the Vendler classification: they include only some achievements (not cyclic achievements) and some activities (not undirected/cyclic activities). Directed changes also cut across the commonly used semantic features that identify aspectual types: directed activities are incremental but unbounded, while directed achievements are bounded but not incremental. Yet there is further evidence that supports the concept of a directed change than just the semantic coherence of the category.

Directed changes include the aspectual type that contain incremental themes as defined by Dowty (1991). Dowty identified three types of incremental themes; Hay et al. (1999) add a fourth (see also references cited in Kennedy and McNally 2005, pp. 362–363). The types are illustrated in (13)–(16):

(13) Mereological incremental themes:
Bill mowed the lawn.

(14) Property incremental themes:
The balloon expanded.

(15) Holistic (path) incremental themes:
a. *They walked across the park.*
b. *He grew into an adult.*

(16) Representation-source themes:
Jane read/scanned War and Peace.

The classic type of incremental theme that Dowty presents can be characterized as mereological (Krifka 1989, Dowty 1991): the incremental progress of the action is manifested in the transformation of incremental parts of an argument (the incremental theme). Hay et al. (1999) identify another type of incremental theme, in which a gradual change in a property of the whole object defines the incremental progress of the action, as in expanding, cooling, etc. Dowty also identifies two other types of incremental themes in which the incremental progress of the action is indirectly associated with a particular argument of the predicate. With holistic themes, the incremental progress is change of location along a path (literal or metaphorical) which is not overtly expressed in the clause; the theme argument is the figure whose change of location is being charted. With representation-source themes, the incremental progress is in the incrementation of the representation of the source (a mental or physical representation); the theme argument is the source whose representation is being created.

In all of these cases, there is a single argument that is associated with the incremental progress of the action, albeit indirectly so in the case of holistic and representation-source themes. Hay et al. analyze the incremental progress of the action as change on a scale that is (directly or indirectly) associated with the theme participant. But the four subtypes are not restricted to gradual directed changes. Directed achievement predicates describe the same four types of directed changes, but as instantaneous transitions rather than incremental progress. We can describe the relevant arguments of directed achievements as transition themes:

(17) Mereological transition themes:
Bill cut the rope.

(18) Property transition themes:
The light turned green.

(19) Transition holistic (path) themes:
a. *They reached the summit.*
b. *She became president of the company.*

(20) Transition representation-source themes:
I saw/photographed Mount Tamalpais.

The convergence in subtypes of themes for incremental themes—bounded (accomplishments) or unbounded (directed activities)—and directed achievements further supports the semantic relevance of the category of directed changes. Moreover, directed changes can be straightforwardly defined in the two-dimensional model as a monotonic function from t to q of the profiled phase(s) of the aspectual contour of the event. (This definition would have to be loosened to include nonincremental accomplishments, where only the initial and final profiled points represent a monotonic function; see § 5.5.) As we will see in § 5.5, after adding the causal dimension to the analysis in § 5.4, the category of directed change is useful for the analysis of constraints on event lexicalization.

5.4 The Three-Dimensional Model: Integrating Causal and Aspectual Structure

The solution I propose for integrating the fine-grained aspectual analysis in § 5.3 with the causal model for argument realization in § 5.2 is simply to add the causal chain as a third dimension to the two-dimensional aspectual representation (Croft 2009, pp. 161–164, 2012, Chaps. 5, 6), as in example (21).

(21) *Jack broke the vase.*

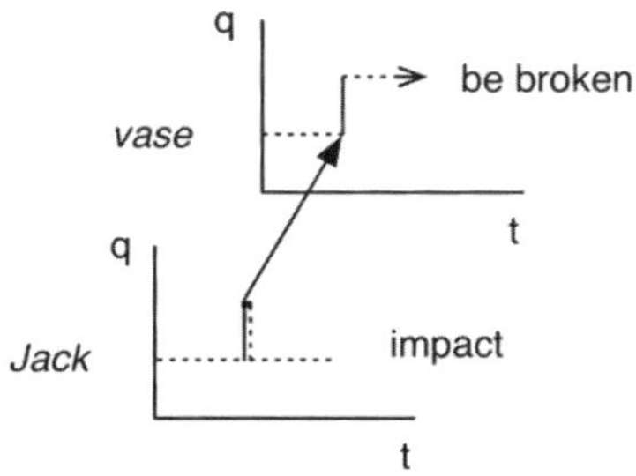

The crucial feature of this representation is that each participant has its own subevent in the causal chain. The subevent is the aspectual profile/contour for that participant's activity in their role in the larger event. Informally, this can be thought of as what each individual participant does or undergoes during the course of the event. Each participant's subevent then stands in a causal relation to the subevent of the next participant in the causal chain (or a noncausal relation, e.g., a spatial relation as in the locative alternation).

Three-dimensional representations are of course difficult to apprehend on a two-dimensional page or screen. Hence, I have adopted the representation in example (22), which more or less collapses the causal and qualitative state dimensions onto the vertical dimension (Croft 2009, pp. 161–162, 2012, pp. 212–213).

(22) *Jack broke the vase.*

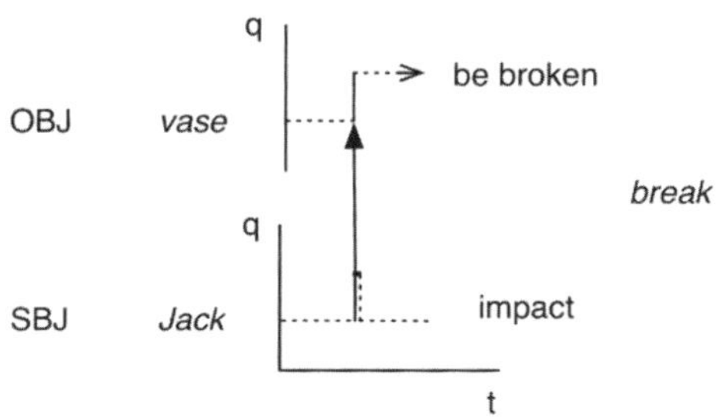

The advantage of this way of reducing the three-dimensional representation onto two dimensions is that the temporal alignment of the subevents is clearly indicated. The qualitative state scales for each participant/subevent are kept separate, in order to remind the viewer that they actually belong on a third dimension.

Example (23) gives the new representation for the sentence in example (1) (Croft 2009, p. 163, 2012, p. 214)

(23) *Sue broke the coconut for Greg with a hammer.*

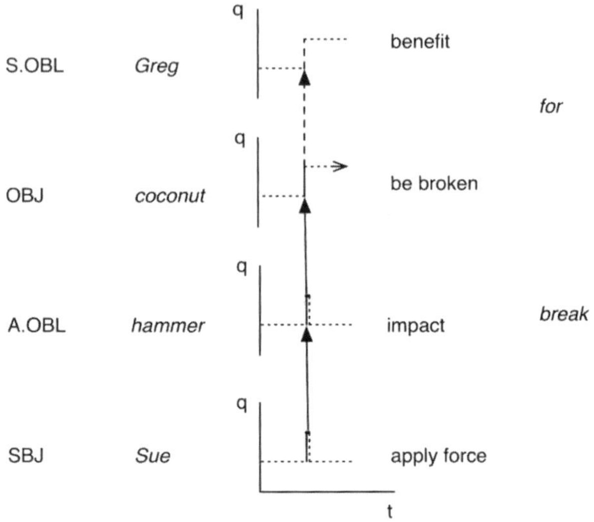

Each participant has its own subevent: Sue applies force to the hammer, the hammer makes impact with the coconut, the coconut undergoes an irreversible change of state, and Greg comes to benefit from the outcome. All of the subevent profiles must be aligned temporally; the entire event is punctual. There is no longer any problem with defining the end point of the verbal profile: the coconut is involved in only one subevent.

Example (24) illustrates how a durative event is represented, namely *Jane read "War and Peace"*.

(24) *Jane read "War and Peace".*

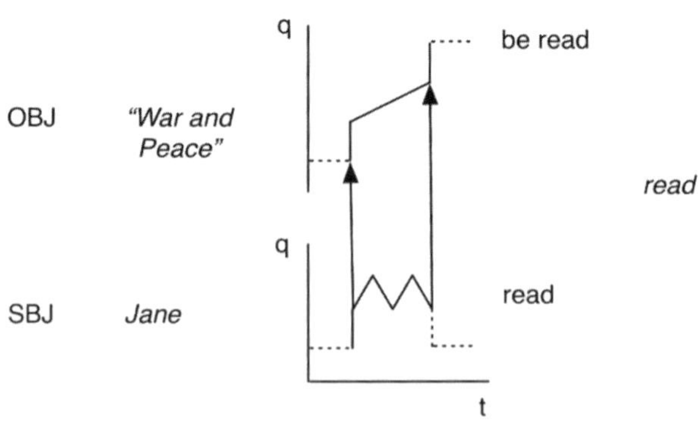

The transmission of force takes place for the profiled temporal phase of the event, but for convenience, it is only represented by the causal arrows at the beginning and the end of the profiled phase.

The event in example (24) contains two subevents: the one involving Jane is an undirected activity and the one involving *War and Peace* is an accomplishment. The overall event profiled by the verb is an accomplishment. In general, if the verb or predicate profile includes a directed change subevent, then the aspectual type of the predicate as a whole is a directed change of the relevant type (Croft 2012, pp. 286–288). There also appears to be another constraint on predicate profiles, at least for simple predicates: only one directed change subevent occurs in a predicate profile. This is the equivalent principle in the model presented here to Tenny's constraint on verbs (i.e., event lexicalizations): "there can be no more than one measuring out for any event described by a verb" (Tenny 1994, p. 11). (In fact, there is one significant exception to this generalization. If a recipient or beneficiary are involved in an event via a directed change, and if there is also a patient or theme undergoing a directed change in the same event, then there are two directed changes in overall event. Not coincidentally, the combination of patient/theme and beneficiary/recipient is the one most likely to be expressed with a double object construction such as the English ditransitive construction (Goldberg 1995), or an obligatory applicative construction, thereby realizing the beneficiary or recipient as a direct object and hence part of the verbal profile; see Croft 2012, pp. 286–288.)

The model presented here has several important advantages over the representation in Croft (1991) as well as the representations proposed by other authors (Croft 2012, pp. 216–217). First, it clearly distinguishes the aspectual and the causal structure of events. Besides the semantic cleanness of this feature, it also allows us to more clearly recognize the distinct contributions that aspectual structure and causal structure make to the grammar of predicates and arguments. Second, it allows us to employ the fine-grained aspectual analysis presented in § 5.3 along with the causal structure analysis presented in § 5.2. Third, it represents both types of causal structure, namely the transmission of force relation—participants acting on other participants—and the standard understanding of causation in terms of events causing other events. This is because of the principle that each participant has its own subevent in the causal structure of the overall event. As noted above, the causal subevents can be informally thought of as what each participant does or undergoes in the overall event. Finally, this model demonstrates that events can be decomposed in three distinct ways: temporally, in terms of temporal phases; qualitatively, in terms of the states defined on the q dimension for each participant's subevent; and causally, in terms of the segments of the causal chain.

The three-dimensional geometrical representation constitutes the semantic structure of the linguistic representation. Outside of that representation are the morphosyntactic structures that are linked to the semantic structure. I use a constructional model of syntax here (Fillmore et al. 1988; Goldberg 1995, 2006; Croft 2001; Croft and Cruse 2004). The leftmost column, with syntactic roles in all capitals, indicates the syntax of the argument structure construction, which is associated with the semantics of the causal chain. The next column to the left of the semantic structure, in italic typeface, indicates the syntax of the argument phrases that instantiate the argument roles of the argument structure construction. They are associated with the individual causal subevents, which describe what each participant does or undergoes in the event. The column to the right of the semantic structure, also in italic typeface, indicates the syntax of the predicate and satellite phrases that instantiate the predicate and satellite (if any) roles in the argument structure construction. They are associated with the participant's subevents that they each profile, including the causal (or noncausal) relations between the subevents that they also profile (indicated by the vertical links in the semantic structure).

A crucial, but incompletely analyzed, element in the semantic structure is the subevents themselves. The diagrams in examples (23) and (24) give suggestive labels for each participant's subevent. Unlike the linguistic forms in italics, these are part of the semantic structure, and are given in roman typeface. In fact, they are only suggestive labels. A proper description would be based on defining all the well-defined states on the qualitative dimension for each participant's subevent. This would be of course an analysis of what has been called the semantic root of a predicate (AR, 71–72).

In the next section of this chapter, I use this representation to provide an analysis of several phenomena that have been associated with a proposed contrast between manner and result predicates. I argue that the crucial semantic distinction between the two types of event structures is the distinction between directed change and undirected change.

5.5 Directed Change in Event Lexicalization

Levin and Rappaport Hovav (1991) address the question of what argument structure alternations are allowed for individual predicates in English. They observe that some predicates allow for many argument structure alternations, using the example of *wipe*.

(25) Transitive: *Kay wiped the counter.*
[surface contact meaning]
(26) Resultative: *Kay wiped the counter clean.*
[result from surface contact]
(27) Removal: *Kay wiped the fingerprints from the counter.*
[remove by means of surface contact]
(28) Application: *Kay wiped the polish onto the table.*
[apply by means of surface contact]

In contrast, other predicates such as *break* and *open* appear to allow very few argument structure alternations (Rappaport Hovav and Levin 1998, pp. 100–103).

Levin and Rappaport Hovav describe verbs of the *wipe* type as manner verbs, and verbs of the *break* type as result verbs. They treat this as a major distinction in event lexicalization patterns, and trace the distinction back to Fillmore (1970). Talmy (1988/2000) makes essentially the same distinction between verb-framing (result) and satellite-framing (manner) realizations of events.

Rappaport Hovav and Levin (1998) propose an explanation in terms of the structural complexity of events (see also AR, 115–17). Manner verbs basically describe simple events, consisting of one subevent in their representation. Further subevents can be combined with a simple manner subevent, leading to the argument structures found in (16)–(18). Result verbs, on the other hand, form complex subevents, including among other things the result state. For the most part, further subevents cannot be combined with a complex result event. As a consequence, result verbs do not allow for as many argument structure alternations.

Rappaport Hovav and Levin's (1998) explanation is based on a highly abstract property of event structure, namely its complexity in their event decomposition model, rather than on specific semantic properties of the event or any of its subevents. It is therefore highly sensitive to the way in which an event is decomposed in a semantic analysis. For example, in the model of event decomposition presented here, virtually every event is complex in at least one of the three dimensions (time, qualitative states, and causal chain). The only type of event that is simple (= consists of only one subevent on all dimensions) is a one-participant inherent permanent state. Thus, we must find a more specific semantic property of events that will distinguish manner and result verbs. That property appears to be directed change (Croft 2012, pp. 337–339).

Manner verbs do not profile a directed change in their most "basic" or lowest valency construal. For example, the lowest valency construal for *wipe* is the transitive construction in example (25), repeated as example (29) (Croft 2012, p. 302):

(29) *Kay wiped the counter.*

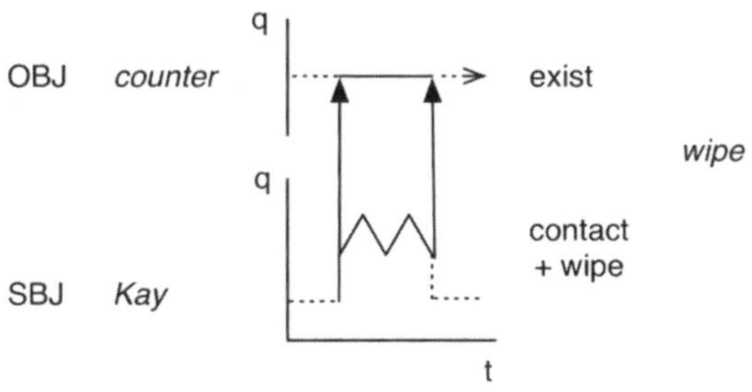

Higher valency constructions add a directed change to the verbal profile. Since there is no directed change in the lowest-valency profile, one can construe different types of directed changes for the manner event. A resultative construction for *wipe* as in (26), repeated as (30), construes the locus of surface contact as possessing a scalar property that is gradually brought about by the manner (Croft 2012, p. 338).

(30) *Kay wiped the counter clean.*

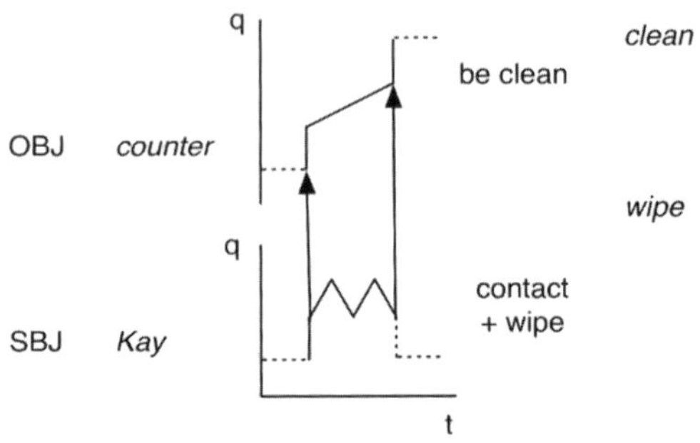

The removal construction in (27), repeated as example (31), construes the entity that is removed as a figure in a directed change of location.

(31) *Kay wiped the fingerprints from the counter.*

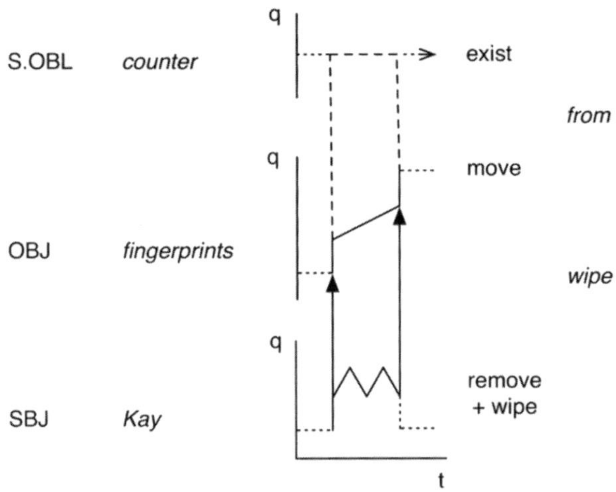

The application construction in (28), repeated as example (32), construes an applied entity as a figure in another kind of directed change of location (Croft 2012, pp. 338–339).

(32) *Kay wiped the polish onto the table.*

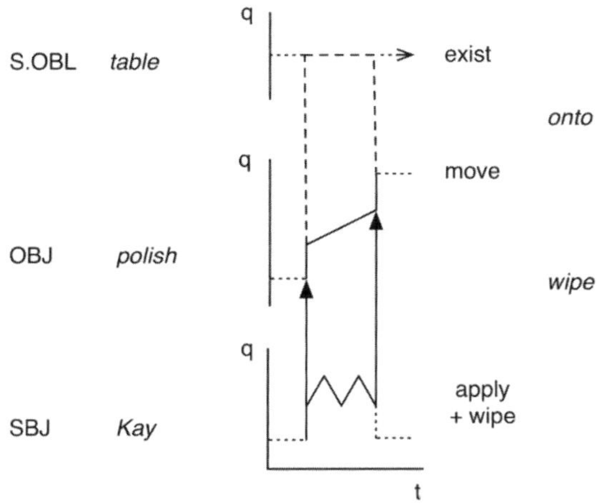

In contrast, a result verb profiles a directed change in its lowest valency form. Hence, it cannot occur in a construction that adds another directed change, because of the constraint allowing only one directed change subevent in a verbal profile.

The suggestive label for the undirected activity subevent in examples (29)–(32) is a combination that incorporates manner (wipe) and a second predicate that describes what is being done in that manner (contact, remove, apply). The manner semantic component is incorporated into a subevent which otherwise appears to describe a result of the manner activity: the wiping appears to cause contact, removal, or application in the examples. In a causal model, one might expect two separate subevents, which in this case would share the same participant.

However, this is not generally the case. For example, while the manner of motion causes the directed motion in (33), the manner of sound emission in (34) does not cause the directed motion, nor does the manner of sound emission in (35) cause the production of the linguistic utterance:

(33) *She swam to the other side.*
(34) *The car screeched around the corner.*
(35) *He growled his answer.*

Thus, it appears that "manner" here really is some sort of accompanying activity, and the verb in each of these argument structure constructions is being construed as (also) being a verb of contact, removal, or application (Croft 2012, pp. 301–302; see also Talmy 1985/2000, pp. 42–47). This characteristic of manner will be addressed again at the end of this section, when we discuss the contrast between manner and result verbs.

The directed change analysis of manner versus result verbs also appears to account for a broad difference between the possible types of resultatives in English and Japanese described by Washio (1997). Washio observes that the Japanese translation equivalents of the English resultatives in (36)–(37) are grammatical, but the Japanese equivalents of the English resultatives in (38)–(39) are ungrammatical:

(36) *I froze the ice cream solid.*
boku-wa aisu kuriimu katikati-ni koorase-ta.

(37) *He wiped the table clean.*
kare-wa teeburu-o kirei-ni hui-ta.

(38) *The horses dragged the logs smooth.*
**uma-ga maruta-o subesube-ni hikizut-ta.*

(39) *They ran the soles of their shoes threadbare.*
**karera-wa kutu-no soko-o boroboro-ni hasit-ta.*

Washio argues that the four examples represent four different semantic classes of predicates, and the semantic differences between them account for the differences in grammaticality between the English and Japanese resultative constructions. I propose a slightly different analysis (Croft 2012, pp. 339–341), which can be made more precise than Washio's thanks to the verbal semantic representation presented in this chapter.

Washio proposes that in the class represented by (36), *freeze X solid*, the verb specifies a change of state, hence the patient undergoes the change of state. The resulting state can therefore be expressed in the resultative construction in either English or Japanese. The alternative analysis here is that the verb in its lowest valency can only be construed as a directed activity. Putting a verb of this class into the resultative construction construes the directed change as an accomplishment; the resultative secondary predicate merely profiles the result state phase.

Washio proposes that in the class represented by (37), *wipe X clean*, the verb specifies that the patient is affected. A change of state is not necessary, but change is specified in a certain direction. The alternative analysis here is that the verb in its lowest valency can be construed as either a directed or an undirected activity. A sentence such as *Kay is wiping the table* can either describe an undirected wiping process or a gradual cleaning of the surface of the table that has not yet reached its end state. When such a verb is put into the resultative construction, it takes on the directed change construal and profiles an accomplishment, with the result phrase profiling the result state phase of the accomplishment.

The next two classes that Washio describes do not have equivalent resultative expressions in Japanese that are grammatical. Washio proposes that in the class represented by (38), *drag logs smooth*, the verb specifies that the patient is affected, but the change of state is not necessary, nor is it specified to occur in a certain direction. The alternative analysis here is that the verb in its lowest valency (*drag logs*) is construed only or chiefly as an undirected activity. There is no (easily) available construal as a directed activity. In English, it is possible to take such a verb and place it into a resultative construction. The outcome is a directed change, or more precisely the marginal member of that category, a nonincremental accomplishment. The reason that the Japanese counterpart is unacceptable is that Japanese does not allow such a radically different construal of the default aspectual type of the lowest valency version of the predicate, or that the Japanese resultative construction cannot denote a nonincremental accomplishment, or both.

Finally, Washio proposes that in the class represented by (39), *run soles threadbare*, the verb specifies neither a patient nor a change of state in any direction. The alternative analysis here is that the verb in its basic, lowest valency form is construed only or chiefly as an intransitive undirected activity. Placing such a verb in a resultative construction adds a participant as well as a resulting state. As with the class exemplified by (38), the event is construed as a nonincremental accomplishment. The reason that the Japanese translation equivalent is ungrammatical is basically the same.

In sum, the analysis presented here captures at least the spirit of Washio's explanation. But it dispenses with the need to make reference to a participant role such as patient, and it makes more precise the aspectual notion that Washio appeals to in the phrase "change of state in a certain direction."

The example of *run soles threadbare* illustrates what is sometimes called the "fake noun phrase (NP)" resultative in English. Rappaport Hovav and Levin (2001) present a semantic analysis of the contrast between a normal resultative on the one hand and a fake NP or fake reflexive resultative on the other. They give attested examples of the same verb occurring in both constructions; (40a–b) gives their examples for *wriggle*:

(40a) *One woman gets up to leave, but Red-Eyes grabs her roughly by the arm and pulls her into his lap.* ***She wriggles free***, *but remains seated obediently beside him.* (The Ottawa Citizen, 30 Nov 1997, p. D10)

(40b) *Mr. Duggan became alarmed about being caught in the door of a lift which was about to begin its descent and* ***wriggled himself free***. (The Irish Times, 2 Dec 1994, p. D11)

They argue that the semantic difference between the two is that in the normal resultative, the two subevents, wriggling and becoming free, unfold together, while in the reflexive resultative, the two subevents do not unfold together:

> A reflexive resultative is required whenever wiggling, wriggling, or kicking is used to bring about a state that is not incrementally brought about by moving in the designated manner since in such instances the events cannot unfold together. (Rappaport Hovav and Levin 2001, p. 778)

Rappaport Hovav and Levin describe unfolding together as temporal dependence (of one subevent on the other), and the lack thereof as temporal independence.

In the model presented in this chapter (see also Croft 2012, pp. 328–332), this is basically the same semantic contrast found between the English resultatives that have Japanese translation equivalents and those that do not. In temporal dependence, as in *She wriggles free* in (40a), repeated as example (41), the relevant participant is involved in only one subevent, which is a directed change.

(41) *She wriggles free.*

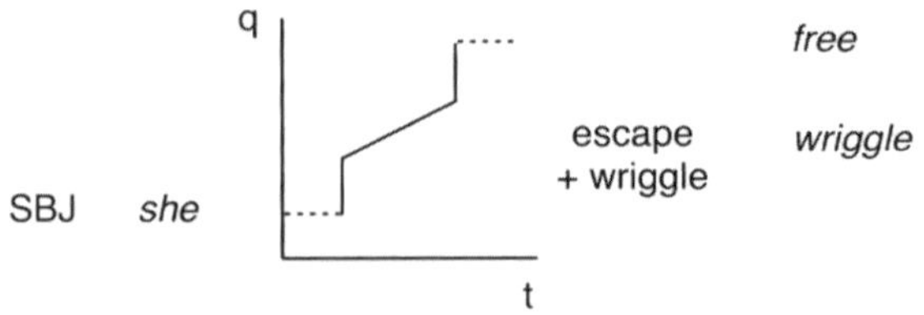

In this analysis, like Rappaport Hovav and Levin's (Rappaport Hovav and Levin 2001, p. 780), there is just one (causal) subevent; the temporal dependence relationship follows by necessity.

The analysis of temporal independence, as in *Mr Duggan wriggled himself free* in (40b), repeated as example (42), is somewhat different from Rappaport Hovav and Levin's.

(42) *Mr Duggan wriggled himself free.*

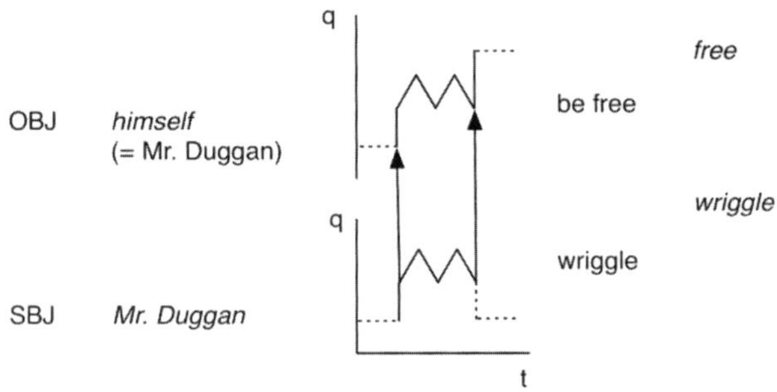

There are two causal subevents in example (42), even though there is only one participant. This is of course manifested grammatically in the fake reflexive. In this respect, the analysis in (42) is the same as the analysis of temporally independent subevents by Rappaport Hovav and Levin (2001, p. 783). Temporal independence is possible when there are two subevents, but another crucial difference is that the aspectual type of the resultative subevent is different from that in example (41). In example (42), the aspectual type of the event is a nonincremental accomplishment, in which there is no directed change until the last instant of the event, when the result state is achieved. As a consequence, these events appear to be temporally independent because the achievement of the result state for the object (Mr. Duggan becoming free) does not occur until the end of the undirected wriggling activity; but that wriggling activity is also nonincremental progression towards the result state. The difference between the two types of English resultatives is basically the same as the difference in resultatives observed by Washio, and in this framework, the analysis is the same. This is not the only constraint governing the occurrence of reflexives. Boas (2003) points out a high degree of idiosyncrasy in resultative predicate choice; Wechsler (1988/2000) argues that the highly conventionalized and specialized reflexives are associated with directed change predicates.

A final manner–result question which the representation presented here may shed light on is the manner–result complementarity hypothesis. Levin and Rappaport Hovav propose that verbs lexicalize (contentful) manner or result, but not both (Levin and Rappaport Hovav 1991, pp. 144–145; Levin and Rappaport Hovav 2008; Rappaport Hovav and Levin 2010). Levin and Rappaport Hovav (2008) argue that some verbs which appear to lexicalize manner and result, such as *cut*, do not lexicalize manner and result simultaneously. Besides semantic intuition, the evidence they offer is that the manner *cut* and the result *cut* occur in different argument structure alternations. For example, manner *cut* occurs in the conative as in the attested example in (27) (from Levin and Rappaport Hovav 2008, ex. 15a, Rappaport Hovav and Levin 2010, p. 36; cf. the transitive *She was cutting the tape*).

(43) *Finally, she got the blade pulled out and started* **cutting at** *the tape on Alex. . .*

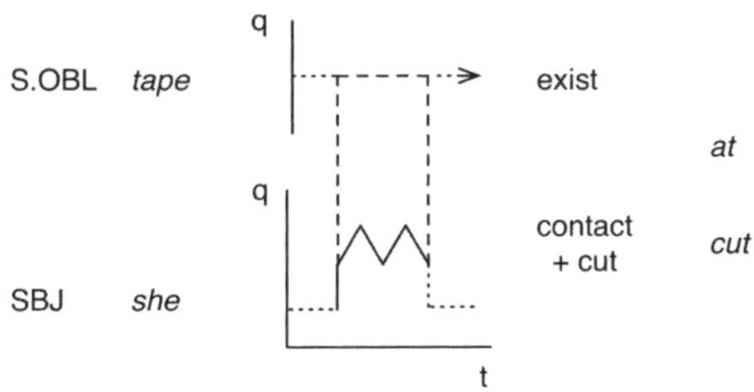

This is the manner *cut*, according to Levin and Rappaport Hovav. As noted above, however, the cutting manner also incorporates the contact event that causally relates the woman and the tape.

On the other hand, result *cut* occurs in the lexical anticausative as in (44) (Levin and Rappaport Hovav 2008, ex. 19a; cf. *The rock cut the rope*):

(44) *. . . the rope* **cut on** *the rock releasing Rod on down the mountain.*

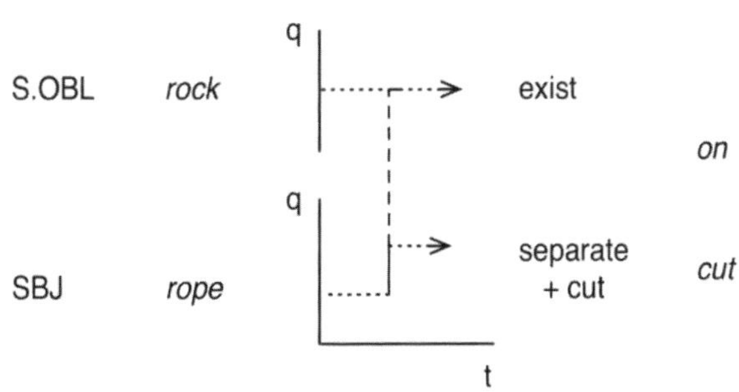

This is the result *cut*, according to Levin. If that is correct, then the label for the rope subevent would be only "separate," not "separate + cut." Does that subevent describe manner as well as result? It seems plausible. The fact of incorporation certainly makes it possible. Goldberg (2010, pp. 46–50) argues that certain motion verbs such as *schuss* and *climb* simultaneously express manner and path/result, since certain types of manners of motion can only be performed in certain paths (e.g., downhill for *schuss*); she also mentions verbs of creation, idea formation, and cooking as verbs that denote both manner and result. Beavers and Koontz-Garboden (2012) argue that verbs of death such as *drown* or *asphyxiate* provide a clear case of result—the victim dies—and clearly different manners in which the result state comes about. So it appears that manner and result may not be complementary.

In fact, it is more complex than this. Manner and result may be combined in lexicalization in (at least) one other way. There are many examples in which a predicate profiles a complex event in which there is an undirected manner subevent and a directed result subevent, as in example (31) (*Kay wiped the fingerprints from the counter*). The overall verbal profile appears to combine manner and result in such events.

What is the difference between manner and result predicates? The analysis of event structure in this chapter suggests an answer. The definition of a result predicate is straightforward: a result predicate contains a directed change subevent in its profile. It is more difficult to define a manner predicate. A manner predicate could be defined as a verbal profile that is dynamic (not a state) but does not contain a directed change subevent in its profile. This definition would preserve complementarity (the definition of manner vs. result offered by Rappaport Hovav and Levin 2010, p. 33, is basically this definition). A manner predicate could also be defined as one that includes an undirected change in its profile—this would allow for complex events with both manner (undirected change) and result (directed change) subevents, as in examples (30)–(32) and (41). However, this definition still does not capture the incorporation of manner in a single subevent, as in example (44). Those subevents are directed changes. But manner is also incorporated in a single subevent with actions such as contact which are not directed changes and yet contrast in manner such as *tap the table, strike the gong, touch the painting*, etc. (see also example (29)). These examples indicate that manner may not be complementary to result (defined as directed change), but instead must be defined in terms of how specifically the qualitative states are characterized on the *q* dimension (so that they potentially contrast with other manner verbs of the same causal–aspectual type). A precise definition of manner will have to await a proper investigation of the qualitative state dimension of event structure.

5.6 Conclusion

The causal theory of event structure demonstrates that there is a simple conceptual basis for the choice of subject, object, and different types of oblique phrases to encode event participants. That work and a fine-grained aspectual analysis show that the aspectual structure of events and the causal structure of events are relevant to the grammar of verbs and clauses, but in different ways (see also the literature surveyed in AR, Chap. 4). This observation implies that the best representation of event structure should represent aspectual and causal structure independently, albeit integrated into a single structure. The three-dimensional geometric representation of event structure presented here provides an efficient, easily visualized model of verbal semantics that can be used to analyze grammatical generalizations in semantic terms. In particular, I have argued that the category of directed change plays a major role in event lexicalization. The most pressing area for further research is in the qualitative dimension, which offers a means to analyze the seemingly unanalyzable verbal root.

References

Beavers, J. (2008). Scalar complexity and the structure of events. In J. Dölling, T. Heyde-Zybatow, & M. Schäfer (Eds.), *Event structures in linguistic form and interpretation* (pp. 245–265). Berlin: Mouton de Gruyter.

Beavers, J. & Koontz-Garboden A. (2012). Manner and result in the roots of verbal meaning. *Linguistic Inquiry, 43,* 331–369.

Boas, H. C. (2003). *A constructional approach to resultatives.* Stanford: Center for the Study of Language and Information.

Carlson, G. N. (1979). Generics and atemporal when. *Linguistics & Philosophy, 3,* 49–98.

Carlson, L. (1981). Aspect and quantification. In P. Tedeschi & A. Zaenen (Eds.), *Tense and aspect: Vol. 14. Syntax and semantics* (pp. 31–64). New York: Academic.

Comrie, B. (1976). *Aspect.* Cambridge: Cambridge University Press.

Croft, W. (1991). *Syntactic categories and grammatical relations: The cognitive organization of information.* Chicago: University of Chicago Press.

Croft, W. (1993). Case marking and the semantics of mental verbs. In J. Pustejovsky (Ed.), *Semantics and the lexicon* (pp. 55–72). Dordrecht: Kluwer.

Croft, W. (1994). Voice: Beyond control and affectedness. In P. Hopper & B. Fox (Eds.), *Voice: Form and function* (pp. 89–117). Amsterdam: John Benjamins.

Croft, W. (1998a). Event structure in argument linking. In M. Butt & W. Geuder (Eds.), *The projection of arguments: Lexical and compositional factors* (pp. 1–43). Stanford: Center for the Study of Language and Information.

Croft, W. (1998b). The structure of events and the structure of language. In M. Tomasello (Ed.), *The new psychology of language: Cognitive and functional approaches to language structure* (pp. 67–92). Mahwah: Lawrence Erlbaum Associates.

Croft, W. (2009). Aspectual and causal structure in event representations. In V. Gathercole (Ed.), *Routes to language development: In honor of Melissa Bowerman* (pp. 139–166). Mahwah: Lawrence Erlbaum Associates.

Croft, W. (2012). *Verbs: Aspect and causal structure.* Oxford: Oxford University Press.

Croft, W, & Cruse, D. A. (2004). *Cognitive linguistics.* Cambridge: Cambridge University Press.

DeLancey, S. (1985). Agentivity and syntax. In W. H. Eilfort, P. D. Kroeber & K. L. Peterson (Eds.), *Papers from the parasession on causatives and agentivity, 21st Regional Meeting, Chicago Linguistic Society* (pp. 1–12).

Dowty, D. (1979). *Word meaning and Montague grammar.* Dordrecht: Reidel.

Dowty, D. (1991). Thematic proto-roles and argument selection. *Language, 67,* 547–619.

Fillmore, C. J. (1970). The grammar of *hitting* and *breaking*. In R. Jacobs & P. Rosenbaum (Eds.), *Readings in English transformational grammar* (pp. 120–133). Waltham: Ginn.

Fillmore, C. J. (1982). Frame semantics. In The Linguistic Society of Korea (Ed.), *Linguistics in the morning calm* (pp. 111–137). Seoul: Hanshin.

Fillmore, C. J. (1985). Frames and the semantics of understanding. *Quaderni di semantica, 6,* 222–54.

Fillmore, C. J., Kay, P., & O'Connor, M. C. (1988). Regularity and idiomaticity in grammatical constructions: The case of *let alone*. *Language, 64,* 501–538.

Foley, W. A., & Van Valin, R. D. Jr. (1984). *Functional syntax and universal grammar*. Cambridge: Cambridge University Press.
Goldberg, A. E. (1995). *Constructions: A construction grammar approach to argument structure*. Chicago: University of Chicago Press.
Goldberg, A. E. (2006). *Constructions at work: The nature of generalization in language*. Oxford: Oxford University Press.
Goldberg, A. E. (2010). Verbs, constructions, and semantic frames. In M. Rappaport Hovav, E. Doron, & I. Sichel (Eds.), *Lexical semantics, syntax, and event structure* (pp. 39–58). Oxford: Oxford University Press.
Hay, J., Kennedy, C., & Levin, B. (1999). Scalar structure underlies telicity in "degree achievements". *Proceedings of SALT, 9,* 127–144.
Jackendoff, R. (1990). *Semantic structures*. Cambridge: MIT Press.
Kennedy, C., & McNally, L. (2005). Scale structure, degree modification, and the semantics of gradable predicates. *Language, 81,* 345–381.
Krifka, M. (1989). Nominal reference, temporal constitution and quantification in event semantics. In R. Bartsch, J. van Benthem, & P. van Emde Boas (Eds.), *Semantics and contextual expression* (pp. 75–115). Dordrecht: Foris.
Langacker, R. W. (1987). *Foundations of Cognitive Grammar, vol. I: Theoretical prerequisites*. Stanford: Stanford University Press.
Levin, B. & Rappaport Hovav, M. (1991). Wiping the slate clean: a lexical semantic exploration. *Cognition, 41,* 123–151.
Levin, B. & Rappaport Hovav, M. (2005). *Argument realization*. Cambridge: Cambridge University Press.
Levin, B. & Rappaport Hovav. M. (2008). Lexicalized manner and result are in complementary distribution. Paper presented at the Twenty-fourth Annual Conference of the Israel Association for Theoretical Linguistics, The Hebrew University of Jerusalem, Jerusalem, Israel.
Mittwoch, A. (1988). Aspects of English aspect: On the interaction of perfect, progressive and durational phrases. *Linguistics & Philosophy, 11,* 203–54.
Pinker, S. (1989). *Learnability and cognition*. Cambridge: MIT Press.
Rappaport Hovav, M. & Levin, B. (1998). Building verb meanings. In M. Butt & W. Geuder (Eds.), *The projection of arguments: Lexical and compositional factors* (pp. 97–134). Stanford: Center for the Study of Language and Information.
Rappaport Hovav, M. & Levin, B. (2001). An event structure account of English resultatives. *Language, 77,* 766–797.
Rappaport Hovav, M. & Levin, B. (2010). Reflections on manner/result complementarity. In M. Rappaport Hovav, E. Doron, & I. Sichel (Eds.), *Lexical semantics, syntax, and event structure* (pp. 21–38). Oxford: Oxford University Press.
Rothstein, S. (2004). *Structuring events*. Oxford: Blackwell.
Smith, C. (1997). *The parameter of aspect* (2nd ed.). Dordrecht: Kluwer.
Talmy, L. (1976). Semantic causative types. In M. Shibatani (Ed.), *The grammar of causative constructions: Vol. 6.Syntax and semantics*, (pp. 43–116). New York: Academic.
Talmy, L. (1985/2000). Lexicalization patterns: Semantic structure in lexical forms. In T. Shopen (Ed.), *Language typology and syntactic description, Vol. 3: Grammatical categories and the lexicon* (pp. 57–149). Cambridge: Cambridge University Press (Revised and expanded version published as 'Lexicalization patterns' In L. Talmy, *Toward a cognitive semantics, vol. II: Typology and process in concept structuring* (pp. 21–146). Cambridge: MIT Press).
Talmy, L. (1988/2000). Force dynamics in language and cognition. *Cognitive Science, 12,* 49–100 (Expanded version published in L. Talmy, *Toward a cognitive semantics, vol. I: concept structuring systems* (pp. 409–470). Cambridge: MIT Press).
Talmy, L. (1991/2000). Path to realization: A typology of event integration. *Buffalo Working Papers in Linguistics* 91-01.147-87 (Expanded version published as 'A typology of event integration'. In L. Talmy, *Toward a cognitive semantics, vol. II: Typology and process in concept structuring* (pp. 213–288). Cambridge: MIT Press).

Tenny, C. L. (1994). *Aspectual roles and the syntax-semantics interface*. Dordrecht: Kluwer.

Van Valin, R. D. Jr., & LaPolla, R. J. (1997). *Syntax: Structure, meaning and function*. Cambridge: Cambridge University Press.

Vendler, Z. (1967). *Verbs and times. Linguistics in philosophy* (pp. 97–121). Ithaca: Cornell University Press (Originally appeared in *The Philosophical Review, 66,* 143–160 [1957]).

Washio, R. (1997). Resultatives, compositionality and language variation. *Journal of East Asian Linguistics, 6,* 1–49.

Wechsler, S. (2005). Resultatives under the 'event-argument homomorphism' model of telicity. In N. Erteschik-Shir & T. Rapoport (Eds.), *The syntax of aspect: Deriving thematic and aspectual interpretation* (pp. 255–273). Oxford: Oxford University Press.

Chapter 6
Neural Processing of Verbal Event Structure: Temporal and Functional Dissociation Between Telic and Atelic Verbs

E. Malaia, J. Gonzalez-Castillo, C. Weber-Fox, T. M. Talavage and R. B. Wilbur

Anything that happens in the world—a storm in the afternoon, a baby starting to crawl, a vase falling on the floor and breaking—is parsed by humans into individual events. This ability—termed *event segmentation*—helps humans analyze, memorize, and compare events that occur around them in order to survive. Individuated events can also be communicated to others in predicative units: sentences. Each well-formed sentence in human languages is constructed around a predicate, typically expressed by a verb. Verbs across languages parse and formulate observable events in a logically restricted fashion (e.g., Son and Cole 2008; Borer 1994; Ritter and Rosen 1998; Davis and Demidarche 2000; Hale and Keyser 1993; Van Valin 2007). Linguists have known for a long time that semantic features of verbs can influence the grammar of the sentence, like the number of arguments, or the typically used tense. The facets

E. Malaia (✉)
University of Texas at Arlington, Arlington, TX, USA
e-mail: malaia@uta.edu

J. Gonzalez-Castillo
Section on Functional Imaging Methods, Laboratory of Brain and Cognition,
National Institute of Mental Health, Building 10, Room 1D80, 10 Center Dr. MSC 1148,
Bethesda, MD 20892-1148, USA
e-mail: javier.gonzalez-castillo@nih.gov

C. Weber-Fox · R. B. Wilbur
Department of Speech, Language, and Hearing Sciences, Purdue University,
Lyles-Porter Hall, 715 Clinic Drive, West Lafayette, IN 47907, USA
e-mail: weberfox@purdue.edu

R. B. Wilbur
Linguistic Program, School of Interdisciplinary Studies,
Purdue Universtiy, West Lafayette, IN, USA
e-mail: wilbur@purdue.edu

T. M. Talavage
School of Electrical and Computer Engineering, Purdue University, Electrical Engineering Building, 465 Northwestern Ave., West Lafayette, IN 47907, USA
e-mail: tmt@purdue.edu

R. G. de Almeida, C. Manouilidou (eds.), *Cognitive Science Perspectives on Verb Representation and Processing,* DOI 10.1007/978-3-319-10112-5_6

of verb meaning which have an effect on the structure of sentences in which they appear are called grammatically relevant semantic features (Pinker 1989).

This decompositional view of verbal meaning, which includes both event and argument structure, has been gaining currency in recent theoretical linguistic and neuroscience research (Kemmerer and Gonzalez-Castillo 2008). The present work describes an effort to extrapolate linguistic theory of event structure into the realm of language processing, in order to understand the neurological mechanisms underlying the difference between different types of verbal events.

6.1 Telicity and Event Structure in Linguistic Theory

Linguistic theory classifies verbs according to whether the event denoted by the verb is seen as having an inherent (telic) end-point (*fall, drop*), or whether the event is considered homogenous, or atelic (*read, worship*). Telicity, while an overtly semantic feature, is a component of verbal-event structure, relating verbal meaning to syntactic frames. Event structure of the verb is closely tied to its argument structure: for example, presence of a resultant state in telic verbs increases the number of obligatory arguments (Ramchand 2008). Interactions between event structure, especially telicity, and grammatical phenomena have been described in many typologically distinct languages, including English, Dutch, Russian, Bengali, Icelandic and Scottish Gaelic, and ASL (for a survey, see Folli and Harley 2006). In theoretical work, Ramchand (2008) has made an attempt to model these interactions in a regular manner, by unifying the predicate's event and argument properties into a cohesive framework (see Fig. 6.1) for representation of event-argument cohesion as related to event types. In Ramchand's model, events can be represented as having three phases: the Initiation phase (InitP), the Process phase (ProcP), and the Result phase (ResP). The participants involved in each phase of the event assume the roles of, respectively, the Initiator, the Undergoer, and the Resultee (note that one argument can be linked to one, two, or all three of these roles).

This system captures the fact that verbal morphology of individual languages can represent individuated elements of event structure, which allows the use of a single verbal root—though tied to different event structures—to yield telic or atelic meanings. The explanatory power of this system on the level of theoretical coverage of existing linguistic data is compelling. However, complete analysis of all known human languages with respect to their linguistic structure is, at the moment, unfeasible. An alternative ground for testing a linguistic theory is empirical evidence from an orthogonal field of language processing.

Behavioral studies have provided early evidence for telicity affecting sentence processing. A word maze[1] study by O'Bryan et al. (2003) has demonstrated that telicity and transitivity independently affect response times to a word maze task in

[1] In a word maze task, the first word of the sentence is followed by a choice of two words, only one of which can be a grammatically correct continuation of the sentence. Once the participants

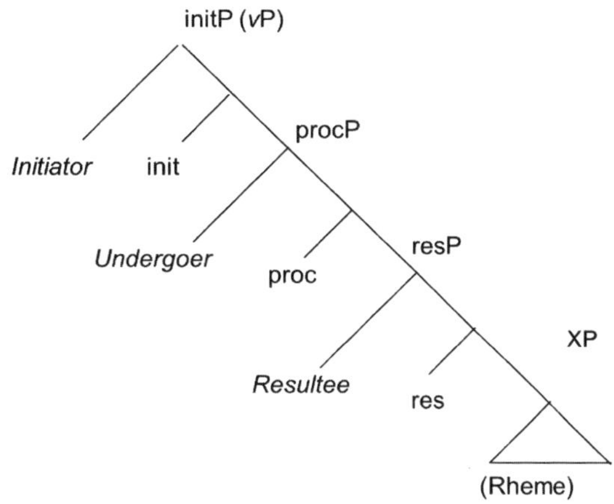

Fig. 6.1 Event structure tree model of syntax–semantics interface, after Ramchand (2008)

Object reduced relative clauses (such as "The actress awakened by the writer left in a hurry"). The experiment demonstrated an advantage to processing of sentences with telic verbs, which was evident in response time to the word "by," and an independent advantage for integration of the second argument in sentences with transitive verbs.

Another behavioral study (Friedmann et al. 2008) used a cross-modal priming technique to compare processing of sentences with intransitive atelic (unergative) and intransitive telic (unaccusative[2]) English verbs. This study has demonstrated an argument priming effect for intransitive telic verbs (non-alternating unaccusatives), but not for intransitive atelics (unergatives). From the processing standpoint, this means that the arguments of telic verbs had to be implicitly understood (or base-generated), before the verbal phrase could be processed.

While behavioral psycholinguistic research points to systematic relationships between the complexity of verbal event structure and expenditure of neural resources required for its processing, neurological correlates of verbal event structure processing—temporal and neuroanatomical—are still under investigation. The time-course of interaction between the semantics of the verb and the sentence structure in online language processing, and the mechanisms responsible for processing of event structure in the cortex are the topics of the empirical studies discussed here. We first consider the fine-grained processing timeline of verbal event structure, as evident from electroencephalography (EEG) studies, and the implications of resource use that this timeline entails; we then turn to the anatomical substrate of event processing in spoken and sign languages, and the evidence it provides for the basic mechanism of event segmentation as implemented in language.

choose the word that can correctly continue the sentence, the choice of two words for the next one is presented, and so on, until the sentence is completed. This task helps measure the typical expectancy of the word given prior context.

[2] Not all unaccusative verbs are obviously telic, however: gradient verbs such as *melt, cool, warm* can denote incomplete events—e.g., "melt somewhat, but not completely."

6.2 The Timeline of Telicity Processing in English Reduced Relative Clauses

The EEG studies that examined the influence of distinctions between the telic and atelic verbs in online sentence processing of English indicate that the difference in resource allocation for processing the two distinct verb types is both early and subtle (Malaia et al. 2008, 2012b, 2013). The design of these studies capitalized on the well-known "garden path" effect, as event-related brain potentials (ERPs) were recorded from native English speakers as they read sentences with reduced relative clauses, in which the main verb was either telic or atelic, e.g., "The actress awakened/worshipped by the writer left in a hurry." The linear (word-by-word) presentation of the sentence results in a drastic processing difference between telic and atelic stimuli at the word "by": in the telic ("awakened") version, "the actress" is processed (or base-generated) as the Patient of the verb, regardless of the fact that the noun is the Subject of the matrix clause. In the atelic version of the sentence ("worshipped"), the alternation of thematic role assignment ("actress" as the Patient, rather than the Agent of worshipping), required for recovery from garden-pathing is more complex, consuming resources of the working memory.

The ERPs in the two conditions—sentences with reduced relative clauses (RRCs) headed by telic and atelic verbs—were compared to each other, and to unreduced relative clause processing (URCs). As earlier behavioral studies of event structure processing (O'Bryan et al. 2003; Friedmann et al. 2008; etc.) reported that telic verbs facilitate interpretation of frame structure alternations in sentences with garden-paths in terms of reaction times, the ERPs to the atelic verbs were interpreted as indexing additional processing demands, previously reported as early negativities in ERP literature[3]. Overall, participants showed significant telicity effects, but the timing of the exact effects differed based on the individual processing resources. ERPs from the group with normal syntactic proficiency first diverged at the second argument, with the atelic condition eliciting larger negativity at the N100, and continuing to the P200 interval. In contrast, ERPs from the high-proficiency group diverged earlier in the sentence, on the preposition "by." This group's ERPs in atelic condition were also characterized by increased negativity relative to the telic condition, which became significant at the P200 interval (200–320 ms), and continued into the later 320–500 ms interval over fronto-central electrode sites.

The difference between the telic and atelic ERP waveforms in the normal proficiency group over the 100–200 ms interval (N100) was similar to that reported for grammatical and ungrammatical sentences requiring phrase structure re-analysis (Yamada and Neville 2007). The frontal and right distribution of this, and the following 200–320 ms component, was similar to the distribution reported by Yamada and Neville (2007), who attributed it to the ongoing processes of syntax-semantics integration. Both investigations converge on the conclusion that previously encountered

[3] All stimuli sentences were completely grammatical, so re-analysis effects typically seen for ungrammatical or semantically incorrect sentences, such as P600 or N400, could not be expected.

semantic information (verbal telicity, for example) may affect the way in which the following syntactic processing is carried out.

The fact that ERPs for telic and atelic conditions in the high-proficiency group differed earlier than in the normal proficiency group is consistent with data in Weber-Fox and Neville (2001) showing that high-proficiency subjects have greater reliance on closed-class words. Another explanation for differential processing found between the groups might lie in likely variations in verbal working memory capacity, which leads to different processing strategies (see also, Newman et al. 2013). Readers with large working memory capacity manage to keep more than one parsing possibility active, and subsequently choose the appropriate interpretation as later sentence information becomes available. ERP studies on verbal working memory reported similar ERP components in verb gapping sentences in English (Kaan et al. 2004) and anaphor resolution in German (Streb et al. 2004).

In general, EEG data provide further evidence that thematic roles defined by the verb can influence parsing decisions (cf. Frazier and Rayner 1982; MacDonald et al. 1994). The timing of this influence may in turn depend on the parsing strategy used by the comprehender; the latter might be the function of his or her linguistic proficiency, and depend on non-linguistic cognitive processes, such as the use of verbal working memory.

6.3 What Does Differential Processing of Grammatically Relevant Semantic Features Suggest for Language Processing and Linguistic Theory?

From the linguistic standpoint, the data on telicity processing are best explained by a combination of event structure and parallel processing theories. According to Ramchand's event structure model, telic verbs alternate between non-causal (intransitive) and causal (transitive) interpretation with the Subject of intransitive verb, or Object of transitive verb occupying the same Undergoer–Resultee thematic roles[4]. An additional argument, when it is introduced in the "by" construction, is added to the existing verbal phrase frame as an external Agent (or causer), but does not necessitate re-assignment of thematic roles to the already-processed argument.

Atelic verbs, on the other hand, initially assign the Agent and Undergoer roles to the first argument. When a new argument is encountered (and the verbal frame

[4] There is still a bit of a controversy regarding whether telicity of the predicate, or affectedness (or quantization) of the object argument is the relevant feature of the predicate that contributes to telicity computation. Ramchand's (2008) model encompasses both affectedness of the object and telicity in a cohesive structure, without suggesting that they are the same thing. In fact, as Ramchand (2008) notes, it is possible to have an affected quantized object in an atelic sentence (he pushed the cart around for hours), and non-quantized object in a telic predicate (they found gold in only 3 years). Importantly, telicity and object quantization tend to correlate in Germanic languages (cf. Ritter & Rosen 1998), but not in Slavic ones (cf. Malaia 2004).

changes from intransitive to transitive), thematic role re-assignment becomes necessary. It is this re-assignment of Agent and Undergoer roles between the subject and the object of the reduced relative clause with an atelic verb which elicits more negative ERPs as compared to simple addition of an extra argument in a vacant thematic role in RRCs with telic verbs.

The linguistic interpretation of the ERP data is consistent with lexically-driven parsing models of sentence processing, which suggest that basic syntactic information available with the verb controls the initial stages of comprehension, but can be quickly modified by the information coming later in the sentence. It is, however, evident that information about a predicate's event structure (and thus telicity) is available for processing at the syntax–semantics interface as soon as argument integration is to take place.

Linguistic ubiquity and processing applicability of event-structural information at the sentence level lead to two important questions related to the role of telicity construct in language development.

Second, how would a mechanism like telicity come to be realized (albeit by different means) in such a vast survey of languages (see Folli and Harley 2006, for review)? One possible explanation suggests that perceptual qualities of events, such as rapid motion in transition scenes, can be a cue to event segmentation (Zacks and Swallow 2007). Perceptual features denoting events could, in time, come to "fossilize" in the language and be coded at the syntax–semantics interface. While demonstrating this on modern spoken language material would be difficult, the study of sign languages is a fruitful testing ground for such hypothesis. Since sign languages are tied to visual modality in both production and perception, they provide the missing link to event-structure building properties of perceived events, by replicating salient perceptual cues to event segmentation during verb sign production. For this reason, sign languages are a great ground to test the hypothesis of telicity representation at the syntax–semantic interface.

6.4 Neural Link Between Processing Event Boundaries and Verb Meaning

The idea that semantic telicity plays a recognizable role in American Sign Language grammar is well-established. Studies have shown that delayed completive aspect only applies to telic stems (Brentari 1998), durative and continuative aspects cannot apply to telic predicates (Wilbur 2003, 2008, 2009); and certain mouth non-manuals are distributed according to predicate telicity type in both Austrian Sign Language and American Sign Language (Schalber 2004). Additionally, motion capture studies (Wilbur and Malaia 2008, Malaia and Wilbur 2012a, b, c; Malaia et al. 2013) demonstrated a kinematic production difference reflecting the semantic distinction between telic and atelic predicates in two unrelated sign languages: the signs representing telic events decelerate to a stop with a 50 % steeper slope than those representing atelic events. The signers, thus, appear to provide perceptual cues to the recipient as

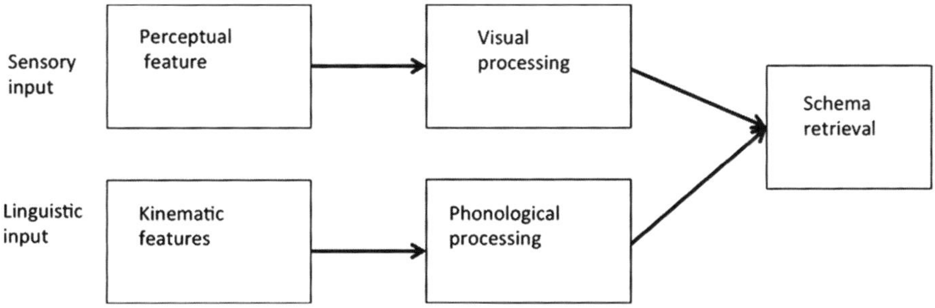

Fig. 6.2 Sensory and sign language processing parallels in event segmentation

to the event structure of the predicate. But are these cues actually received by the comprehender and processed as part of the syntax–semantics interface?

Perceptual research indicates that the manner in which reality is segmented into events affects memory encoding and updating processes (Swallow et al. 2009), and propagates the use of the perceptual features (e.g., object velocity) that relate to the retrieved event schema for future event processing (Kurby and Zacks 2008). Event segmentation theory (EST; Zacks and Swallow 2007) suggests that the information flow from visual cortex is taken apart into significant features identifying event boundaries (with velocity being processed in area MT+, for example). Those features are then used for event schema retrieval from long-term memory, possibly gated by posterior cingulate/precuneus, which is typically activated in contrasts involving event boundary (Zacks et al. 2001).

A similar mechanism appears to be in place for visual processing of event boundary, which is identified by greater deceleration in American Sign Language. The only difference is that the visual features of the linguistic signal are also processed as linguistic features (in case of ASL, phonological). A neuroimaging study (Malaia et al. 2012a) indicated that the contrast between neural activations elicited by telic and atelic ASL verb signs demonstrated activations related to event schema retrieval (posterior cingulate [MNI 18 −54 10]), and syllable weight processing (right STG and cerebellum). These data suggest that the visually expressed boundaries of events in ASL are then mapped to linguistic features of overt hand articulator motion for event schema retrieval from long-term memory (see Fig. 6.2) for the model comparison on sensory and linguistic processing of visual event boundaries.

6.5 Discussion: The Role of Telicity at the Syntax–Semantics Interface in Spoken and Signed Languages

The combined results of neuroimaging and ERP experiments point to early interaction of syntax and semantics in human languages, and suggest that grammatically relevant semantic features of the predicate's event structure, such as telicity, are used for strategic allocation of neural resources during language processing. What follows from empirical evidence in sign and spoken languages is that not only semantics, but also syntax of human languages cross-modally are grounded in what can be construed as biological perception. In other words, the complexity of the interaction between semantics and syntax is not limited to consistent occurrence of certain structures in a specific language, as claimed by constructionist approaches, but rather operates through the complexity of linguistic structures. We suggest that events in the real world are perceived, conceptualized and verbalized in a way which takes advantage of the syntax–semantics interface with the built-in account of real-world events (Malaia and Wilbur 2014; Malaia 2014).

The evidence that the predicates which differ in visual telicity features in ASL differentially engage resources during linguistic processing highlights the theoretical relevance of event structure modeling for language processing. Finally, the combined results of the ERP experiments on English, and neuroimaging experiments in ASL suggest a direction for further research into the biological bases of human languages by identifying the links between language universals and perceptual-level features affecting event segmentation and language processing.

Acknowledgments This work was partially supported by National Science Foundation Grant 0345314 to RBW, NIH DC00524 grant to RBW, NIH EB003990 grant to TMT, and Research Enhancement Program grant from the University of Texas at Arlington to EM.

References

Borer, H. (1994). The projection of arguments. In E. Benedicto & J. Runner (Eds.), *Functional projections(University of Massachusetts occasional papers 17)* (pp. 19–47). Amherst: GSLA, University of Massachusetts.

Brentari, D. (1998). *A prosodic model of sign language phonology*. Cambridge: MIT Press.

Davis, H., & Demidarche, H. (2000). On lexical verb meaning: Evidence from Salish. In C. Tenny & J. Pustejovsky (Eds.), *Events as grammatical objects*. Stanford: CSLI.

Folli, R., & Harley, H. (2006). What language says about the psychology of events. *Trends in Cognitive Sciences, 10*(3), 91–92.

Frazier, L., & Rayner, K. (1982). Making and correcting errors during sentence comprehension: Eye movements in the analysis of structurally ambiguous sentences. *Cognitive Psychology, 14*(2), 178–210.

Friedmann, N., Shapiro, L. P., Taranto, G., & Swinney, D. (2008). The leaf fell (the leaf): The online processing of unaccusatives. *Linguistic Inquiry, 39*(3), 355–377.

Hale, K., & Keyser, S. J. (1993). On argument structure and the lexical expression of syntactic relations. In K. Hale & S. J. Keyser (Eds.), *The view from building 20: A festschrift for Sylvain Bromberger* (pp. 53–109). Cambridge: MIT Press.

Kaan, E., Wijnen, F., & Swaab, T. Y. (2004). Gapping: Electrophysiological evidence for immediate processing of "missing" verbs in sentence comprehension. *Brain and Language, 89*(3), 584–592.

Kemmerer, D., & Gonzalez-Castillo, J. (2008). The two-level theory of verb meaning: An approach to integrating the semantics of action with the mirror neuron system. *Brain and Language, 112*(1), 54–76.

Kurby, C. A., & Zacks, J. M. (2008). Segmentation in the perception and memory of events. *Trends in Cognitive Sciences, 12*(2), 72–79.

MacDonald, M. C., Pearlmutter, N. J., & Seidenberg, M. S. (1994). The lexical nature of syntactic ambiguity resolution. *Psychological Review, 101*(4), 676.

Malaia, E. (2014). It still isn't over: Event boundaries in language and perception. *Language and Linguistics Compass, 8*(3), 89–98.

Malaia, E. (2004). *Event structure and telicity in Russian: An event-based analysis for telicity puzzle in Slavic languages*. Ohio State University Working Papers in Slavic Studies, Vol. 4, (pp. 87–98), Columbus.

Malaia, E., & Wilbur, R. B. (in press) Enhancement of spatial processing in sign language users. In D. R. Montello, K. E. Grossner, & D. G. Janelle (Eds.), *Space in mind: Concepts and ontologies for spatial thinking*. Cambridge: MIT Press.

Malaia, E., & Wilbur, R. B. (2012a). Telicity expression in visual modality. In L. McNally & V. Delmonte (Eds.), *Telicity, change, and state: A cross-categorial view of event structure* (pp. 122–138). Oxford: Oxford University Press.

Malaia, E., & Wilbur, R. B. (2012b). Motion capture signatures of telic and atelic events in ASL predicates. *Language and Speech, 55*(3), 407–421.

Malaia, E., & Wilbur, R. B. (2012c). What sign languages show: Neurobiological bases of visual phonology. In A. M. Di Sciullo (Ed.), *Towards a biolinguistic understanding of grammar: Essays on interfaces* (pp. 265–275). John Benjamins: Amsterdam.

Malaia, E., Wilbur, R. B., & Weber-Fox, C. (2008). ERP evidence for telicity effects on syntactic processing in garden-path sentences. *Brain and Language, 108*(3), 145–158.

Malaia, E., Ranaweera, R., Wilbur, R. B., & Talavage. T. M. (2012a). Event segmentation in a visual language: Neural bases of processing American Sign Language predicates. *NeuroImage, 59*(4), 4094–4101.

Malaia, E., Wilbur, R. B., & Weber-Fox, C. (2012b). Down the garden path in EEG: Telicity effects on thematic role re-assignment in relative clauses with transitive verbs. *Journal of Psycholinguistic Research, 41*(5), 323–345.

Malaia, E., Wilbur, R. B., & Milković, M. (2013a). Kinematic parameters of signed verbs at morpho-phonology interface. *Journal of Speech, Language, and Hearing Research, 56,* 1–12.

Malaia, E., Wilbur, R. B., & Weber-Fox, C. (2013b). Event end-point primes the undergoer argument: A look at neurobiological bases of event structure. In B. Gehrke & B. Arsenijevic (Eds.), *Subatomic semantics of event predicates* (pp. 231–248). Springer: Dordrecht.

Newman, S., Malaia, E., Seo, R., & Hu, C. (2013). The effect of individual differences in working memory capacity on sentence comprehension: An fMRI study. *Brain Topography, 26*(3), 458–467.

O'Bryan, E. L., Folli, R., Harley, H., & Bever, T. G. (2003). *Event structure is accessed immediately during comprehension*. Linguistics Society of America Annual Meeting, Atlanta.

Pinker, S. (1989). *Learnability and cognition*. Cambridge: MIT Press.

Ramchand, G. (2008). *Verb meaning and the lexicon: A first phase syntax*. Cambridge: Cambridge University Press.

Ritter, E., & Rosen, S. (1998). Delimiting events in syntax. In M. Butt & W. Geuder (Eds.), *The projection of arguments* (pp. 135–164). Stanford: CSLI.

Schalber, K. (2004). Phonological visibility of event structure in Austrian Sign Language: A comparison of ASL and ÖGS. Master's thesis, Purdue University.

Son, M., & Cole, P. (2008). An event-based account of *-kan* constructions in standard Indonesian. *Language, 84*(1), 120–160.

Streb, J., Hennighausen, E., & Rösler, F. (2004). Different anaphoric expressions are investigated by event-related brain potentials. *Journal of Psycholinguistic Research, 33*(3), 175–201.

Swallow, K. M., Zacks, J. M., & Abrams, R. A. (2009). Event boundaries in perception affect memory encoding and updating. *Journal of Experimental Psychology: General, 138*(2), 236.

Van Valin, R. (2007). Some universals of verb semantics. In R. Mairal & J. Gil (Eds.), *Linguistic universals* (pp. 155–178). Cambridge: Cambridge University Press.

Weber-Fox, C., & Neville, H. J. (2001). Sensitive periods differentiate processing of open- and closed-class words: An ERP study of bilinguals. *Journal of Speech Language and Hearing Research, 44*(6), 1338–1353.

Wilbur, R. B. (2003). Representations of telicity in ASL. *Chicago Linguistic Society, 39*(1), 354–368.

Wilbur, R. B. (2008). Complex predicates involving events, time and aspect: Is this why sign languages look so similar? In J. Quer (Ed.), *Signs of the time* (pp. 217–250). Hamburg: Signum.

Wilbur, R. B. (2009). Productive reduplication in ASL, a fundamentally monosyllabic language. *Language Sciences, 31,* 325–342.

Wilbur, R., & Malaia, E. (2008). Event visibility hypothesis: Motion capture evidence for overt marking of telicity in ASL. Presentation, Linguistic Society of America meeting, Chicago.

Yamada, Y., & Neville, H. J. (2007). An ERP study of syntactic processing in English and nonsense sentences. *Brain Research, 1130*(1), 167–180.

Zacks, J. M., & Swallow, K. M. (2007). Event segmentation. *Current Directions in Psychological Science, 16*(2), 80–84.

Zacks, J. M., Braver, T. S., Sheridan, M. A., Donaldson, D. I., Snyder, A. Z., Ollinger, J. M., et al. (2001). Human brain activity time-locked to perceptual event boundaries. *Nature Neuroscience, 4*(6), 651–655.

Chapter 7
Argument Structure and Time Reference in Agrammatic Aphasia

Roelien Bastiaanse and Artem Platonov

7.1 Introduction

Aphasia is a language disorder due to brain damage (stroke, traumatic brain injury), usually in the left hemisphere. The nature and severity of the aphasia are dependent on the site and size of the brain lesion. Some patients suffer from problems at the word level, while others encounter difficulties at the sentence level. The focus of this chapter is aphasic patients who suffer from a grammatical disorder, also called "agrammatism" or "Broca's aphasia."

Agrammatism is usually caused by a left frontal brain lesion and is characterized by nonfluent, effortful speech, consisting of mainly content words (nouns, verbs, adjectives), such that free and bound grammatical morphemes are typically omitted and/or substituted. The following sample is an illustration of agrammatic speech, where the patient is asked to tell about her plans for Christmas. (... indicate pauses; *italics*: interviewer)

> *what are you going to do for Christmas?* Eat tasty things... presents Christmas... draw numbers. All get presents... ten guilders... ten guilders each *are you going away?* No... we sold house... our house... new around March... we saving pennies.

What can be seen from this sample is that agrammatic speakers have a preference for base order sentences; complex structures are avoided. Sentences with derived word

R. Bastiaanse (✉)
Center for Language and Cognition Groningen (CLCG), University of Groningen, Postbus 716, 9700 AS Groningen, The Netherlands
e-mail: y.r.m.bastiaanse@rug.nl

A. Platonov
Donders Institute, Radboud University Nijmegen, Montessorilaan 3, 6525 HR Nijmegen, The Netherlands

European Master's of Clinical Linguistics, Universities of Groningen, Groningen, The Netherlands
e-mail: a.platonov@donders.ru.nl

R. G. de Almeida, C. Manouilidou (eds.), *Cognitive Science Perspectives on Verb Representation and Processing*, DOI 10.1007/978-3-319-10112-5_7

order (for example, yes–no questions or passives in English) are rarely produced. This is more visible in languages with a larger variety of word orders, such as Dutch, German, and Turkish, as will be illustrated below.

Although it has often been mentioned that this "telegraphic speech" consists predominantly of nouns, verbs, and adjectives, the category of verbs is vulnerable (e.g., Saffran et al. 1989; Thompson et al. 1995; Bastiaanse et al. 2002). Thompson (2003) showed, for example, that argument complexity plays a role in agrammatic verb production, not only in spontaneous speech but also in action naming: the more complex argument structure is, the more difficult it is for an agrammatic speaker to produce a verb or use it in a sentence. Similarly, the diversity of verbs in spontaneous speech is reduced and the verbs that are used are often not inflected for tense and agreement (Bastiaanse and Jonkers 1998). Basically, the characteristics of the agrammatic speech have been described along three dimensions: word order, argument structure, and verb inflection. The present chapter makes a first attempt to connect these problems, by presenting a model that makes predictions about agrammatic performance with respect to these three dimensions.

7.1.1 Word Order Problems

As can be seen in the above sample, the sentence structures that are produced by this agrammatic speaker are very basic. This is one of the features of speech production in this type of aphasia. It turns out that nonbasic structures, also known as "derived" structures, are harder to produce than base structures. This has been shown for several languages, among which are Dutch and English (e.g., Bastiaanse and Thompson 2003) and Turkish (Yarbay Duman et al. 2007, 2008). In Dutch and German, for example, so-called subject–object–verb (SOV) languages, the base position of the verb is clause-final position as can be seen in (1a–b). In the matrix clause, the finite verb is in the second position (1c). According to theoretical linguistics, the finite verb in the matrix clause is in "derived" position; "i" denotes the original position of the verb, which is co-indexed with its antecedent.

(1a)	de jongen	die	een boek	*leest*
	the boy	who	a book	*reads*
(1b)	de jongen	wil	een boek	*lezen*
	the boy	wants to	a book	*read*
(1c)	de jongen	*leest*$_i$	een boek	i
	the boy	*reads*	a book	

Bastiaanse et al. (2002, 2003) show that the "object–finite verb" string in embedded clauses, such as (1a) are significantly easier to produce than the "finite verb–object" strings in matrix clauses for Dutch agrammatic speakers. Similarly, English "yes–no" questions, in which the auxiliary is in derived position, are more difficult to produce for English agrammatic speakers than comparable declarative sentences ("is$_i$ the student i helping the biker?" is more difficult than "the student is helping the biker"; Bastiaanse and Thompson 2003). These problems with derived order have not only been shown for verbs in derived position but also for sentences with scrambled objects in Dutch (Bastiaanse et al. 2003), German (Burchert et al. 2008), and Turkish (Yarbay Duman et al. 2007), and for sentences in which the theme is in subject position, so-called unaccusative constructions in Dutch (Bastiaanse and Van Zonneveld 2005), English (Lee and Thompson 2004), and Russian (Dragoy and Bastiaanse 2009).

On the basis of these findings, Bastiaanse and Van Zonneveld (2005) formulated the Derived Order Problem Hypothesis (DOP-H) that predicts that sentences in the derived order are more difficult for agrammatic speakers to produce and comprehend than sentences in the base order. For the present chapter, these findings from Bastiaanse and Van Zonneveld (2005) are relevant and are briefly summarized.

7.1.1.1 Verbs with Alternating Transitivity

Bastiaanse and Van Zonneveld (2005) did a study on the production of sentences with verbs with alternating transitivity (see Levin 1993). These are verbs that have both a transitive and an unaccusative reading, such as "to break" ("the woman breaks a glass" vs. "the glass breaks"), and "to burn" ("the man is burning the book" vs. "the book is burning"). In the unaccusative reading, the theme ("glass," "book") is no longer in its base—that is the object—position, but is a so-called derived subject. This implies a derived structure ("the book$_i$ is burning i") and, hence, the DOP-H predicts that the unaccusative reading will be more difficult for agrammatic speakers to produce. Bastiaanse and Van Zonneveld (2005) developed a test to elicit both structures from agrammatic speakers. An example of an item is given in Fig. 7.1.

Sentences like "the book is burning" turned out to be significantly more difficult to produce than sentences like "the man is burning the book," which supports the DOP-H.

7.1.2 Argument Structure Problems

Another linguistic property that influences verb and sentence production in agrammatic aphasia is "argument structure." Although processing of argument structure seems to be spared (Shapiro and Levine 1990), the production of the full range of verb–argument structures is compromised, as was noticed first in the narrative speech of English agrammatic speakers (Thompson et al. 1997) and later confirmed for Dutch (Bastiaanse and Jonkers 1998). Agrammatic speakers have a preference for simple verb–argument structures, that is, they overuse intransitive structures, have a reduced proportion of transitive structures, and avoid verbs with three arguments

Fig. 7.1 Example of a transitive (*left; target "the man is burning the book"*) and an unaccusative (*right; target "the book is burning"*) item of the test for verbs with alternating transitivity

(e.g., "to give") and with sentential arguments (e.g., "believe that..."). In a later study, Thompson et al. (1997) demonstrated that this reduction of argument structure complexity does not only affect the use of verbs in narrative speech but also influences verb retrieval in an action-naming test. That is, agrammatic speakers have more problems retrieving verbs with a complex argument structure than a simple argument structure. Not only the simple number of arguments is crucial here but also the number of *possible* argument structures of a verb plays a role. For example, "to knit" can be used both with and without a theme and is, therefore, more difficult to produce than "to fix" which has only one possible argument structure. Further research showed that the complexity of the "internal" structure of the verb and its argument(s) also affects verb production. Simple unergative verbs, like "to sleep," are easier to retrieve than unaccusative verbs, like "to fall," in which, as mentioned in the previous section, the theme is in subject position.

These findings have been captured under the Argument Structure Complexity Hypothesis (ASCH) as formulated in Thompson (2003): the more complex the verb–argument structure, the more difficult it is for an agrammatic speaker to retrieve the verb. Notice that the DOP-H and the ASCH are complementary, although there is some overlap. While the DOP-H is a hypothesis about word order, the ASCH is about argument structure; they both predict that unaccusative verbs will be difficult for agrammatic speakers: the first, because the word order is derived; the second, because of the property of unaccusatives to have the theme in the subject position.

7.1.3 Verb Inflection Problems

As shown in the above example, agrammatic speakers have a tendency to omit finite verbs. In languages like Dutch and English, there is an overuse of infinitives, gerunds, and participles (usually without a finite auxiliary) compared to normal speech. It has been argued that this is due to (1) the position of the finite verb in the syntactic tree (Hagiwara 1995; Friedmann and Grodzinsky 1997), (2) an underspecification of the interpretable features of tense (Wenzlaff and Clahsen 2004, 2005) or of tense and agreement (Burchert et al. 2005), and (3) a problem with the implementation of morphological rules (Lee et al. 2008).

None of these theories alone can account for all the data. The theories that relate the problems to tense or to a morphological interpretation problem fail to explain the verb inflection problems that arise in one sentence position but not in another. In Dutch and German, as mentioned above, the base position of the verb is at the end of the clause, but in a matrix clause, the finite verb is in second position. Agrammatic speakers of these languages have more problems with finite verbs in derived position than in base position (Bastiaanse et al. 2002; Rausch et al. 2005). The theories that assume that the position in the tree is crucial cannot explain the selective problems with time reference with both finite and nonfinite verbs. In the next section, these problems are discussed.

7.1.3.1 Time Reference Problems

Recent findings on agrammatic verb inflection in Dutch, as reported by Bastiaanse (2008), cannot be explained by the DOP-H nor by any of the theories that postulate a general tense problem. In this study, a sentence completion paradigm was used to elicit finite and nonfinite verbs in base position. The sentences referred either to the past or to the present. Surprisingly, the results showed that (1) production of finite verbs referring to the past was significantly more impaired than production of finite verbs referring to the present; (2) this difference between reference to past and present was also observed for the production of nonfinite verbs, that is, participles were more difficult than infinitives. In an additional study on Turkish, Yarbay Duman and Bastiaanse (2009) showed a similar difference between production of finite verb in past and future tense: past tense was significantly more difficult than future tense. Bastiaanse et al. (2011) and Bastiaanse (2013) showed that the time reference problem is not restricted to tense and holds for all verb forms that refer to the past, including those with perfect aspect, even in combination with present tense (Dragoy and Bastiaanse 2013; Bos and Bastiaanse 2014). Bastiaanse et al. (2011) and Bastiaanse (2013) argue that this is due to the fact that reference to the past requires discourse linking (Zagona 2003), and discourse linking is hard for individuals with agrammatic aphasia (Avrutin 2006). This was coined the PAst DIscourse LInking Hypothesis (PADILIH). These findings are not entirely incompatible with the theories of Wenzlaff and Clahsen (2004, 2005) and Burchert et al. (2005) that this effect has to do with the interpretable features (or, in terms of Burchert et al., with "sentence external relationships"); however, these theories seem to be too restrictive: the findings of Bastiaanse (2013) and Yarbay Duman and Bastiaanse (2009) suggest that the problems are (a) not restricted to tensed verbs but extend to nonfinite participles as well and (b) most severe for reference to the past.

In all, several theories have been formulated on the underlying disorder as a cause for the verb production deficits of agrammatic speakers. It seems as though word order, verb argument structure, and time reference each play a role. The question is how these three concepts are related.

7.2 The Aspect Assignment Model (AAM)

Verbs can be classified along several dimensions. The dimensions which are important here are (1) argument structure/ transitivity and (2) telicity. *Argument structure* refers to the thematic roles that belong to the verb (e.g., intransitive verbs ("to run") have no internal argument, transitive verbs have one internal argument ("to read") and ditransitive verbs ("to give") have two internal arguments) and the rules that are needed to use these verbs and their arguments in a sentence (e.g., the theme of an unaccusative verb ("to fall") is in subject position). *Telicity* has to do with the fact that certain actions result in a change of state, whereas others do not. Telic verbs imply a certain endpoint, whereas atelic verbs do not. For example, "to break" implies a change of state and "to run" does not. Telic verbs include both accomplishments and achievements in Dowty–Vendler terms (Dowty 1979; Vendler 1967). Here, the term *atelic* will be used to refer to verbs that signify events without such an endpoint, including activities, semelfactives and states.

Verbs also have the ability, at least in most languages, to express the relation of the event to past, present and future—through Tense—and to whether the action has been finished or not—through Aspect. In short, the relevant concepts here are (1) argument structure, (2) telicity, and (3) time reference (Tense and Aspect).

These three characteristics are related, as shown by both preferences of normal speakers and data from language acquisition. With respect to argument structure and telicity, it was first noted by Perlmutter (1978) that intransitive telic verbs (e.g., "to arrive") are usually unaccusatives. Similarly, intransitive atelic verbs tend to be unergatives (e.g., "to chirp"). The relation between telicity and Tense was shown by Torrence and Hyams (2004): English-speaking children tend to use past Tense with telic verbs and present Tense with atelic verbs. Also, there is a close relation between children's early use of telic verbs and perfective Aspect in the past Tense on the one hand, and between atelic verbs and imperfective Aspect in the present Tense on the other. These latter relations have been reported for many languages, such as English (Shirai and Andersen 1995), Russian (Stoll 1998; Gagarina 2000) and German (Behrens 1993).

In Table 7.1, the relations between argument structure, telicity and time reference that are relevant for the present study are given.

Table 7.1 Preferences for the combinations of argument structure, telicity, and time reference (tense and aspect)

Argument structure	Telicity	Tense	Aspect
Transitive	Telic	Past	Perfect
Intransitive	Atelic	Present	Imperfective

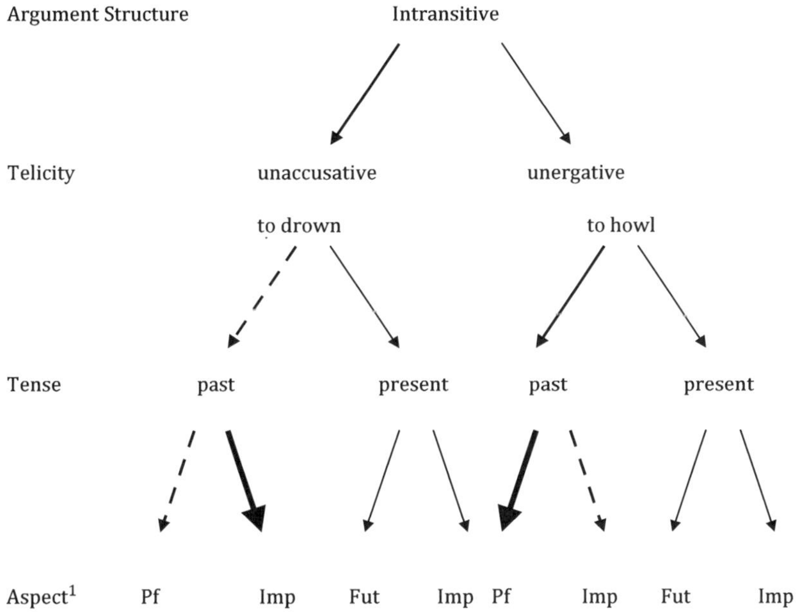

Fig. 7.2 The relevant part of the Aspect Assignment Model. *Bold lines* mean that this variant is in conflict with speakers' preferences; *bold broken lines* mean that although at the higher level the preference is in conflict, there is a preference at the present level; *extra bold lines* mean that there is a conflict at more than one level. *Pf* perfect, *Imp* imperfect; since in Russian, there is only imperfect present, perfect future tense has been used here

These preferences have been used by Platonov (2007) to build the *Aspect Assignment Model* (AAM). In this model, information on argument structure and time reference, concepts that have been shown to influence agrammatic production, have been combined. For reasons of clarity in Fig. 7.2, only that part of the model that is relevant for the present study is given. The complete model is given in Appendix 1.

If it is assumed that what is preferred by normal speakers is relatively well preserved in agrammatic aphasia and what is marked for normal speaker is difficult for agrammatic speakers, then this model makes predictions on agrammatic behavior. For example, considering the preferences mentioned in Table 7.1 and taking into account Perlmutter's theory that intransitive telic verbs tend to be unaccusatives and intransitive atelic verbs tend to be unergatives, the model makes the following predictions for unaccusative and unergative verbs (see Fig. 7.2):

- Unergatives prefer present tense, so past tense is a conflict (represented by a bold line in Fig. 7.2)
- Unergatives prefer imperfect aspect (broken bold line = relatively easy), so perfect aspect is a conflict (extra bold line = very difficult)
- Unaccusatives are in conflict with the preference of an intransitive verb being unergative (bold line)
- Unaccusatives prefer past tense (broken bold line)
- Unaccusatives prefer perfect aspect (broken bold line), so imperfect aspect is conflict (extra bold line)

In sum, this model—taking not only argument structure but also time reference into account—predicts that for agrammatic speakers, unergatives will be easier than unaccusatives in the past tense, imperfect aspect; the opposite pattern is expected for sentences in past tense, perfect aspect. Hence, for the imperfect aspect condition, the same pattern should be observed as reported by Bastiaanse and Van Zonneveld (2005) and Lee and Thompson (2004). For the perfect aspect condition, the opposite pattern is predicted by Platonov's model. This hypothesis has been tested using an experiment that elicited the relevant sentence structures. The experiment was performed in Russian, where both perfective and imperfective aspects are expressed through the finite verb.

7.3 Methods

7.3.1 Participants

Twelve agrammatic speakers (nine male, three female) were tested. The mean age was 43.2 years. They had been diagnosed as suffering from efferent motor aphasia in Lurian terms (Luria 1973), which is equivalent to Broca's aphasia. The aphasia type was established by the analysis of spontaneous speech, which was clearly telegraphic, and confirmed by the language assessment of the speech pathologist and the neuropsychologist. None of the patients suffered from apraxia of speech (range 22–70 years). Eight subjects were aphasic due to a single stroke in the left hemisphere, one subject had two strokes, and three subjects' aphasia resulted from traumatic brain injury caused by a car accident. All subjects were at least 4 months post onset.

Twelve non-brain-damaged speakers served as controls (mean age 45.6; six male, six female). The control subjects performed faultlessly on the test and, therefore, their data will further be ignored.

Fig. 7.3 Two examples of the stimuli used in the study. The first two pictures (*on the left*) illustrate the verb *to drown* in both perfective and imperfective conditions. The second set of pictures (*on the right*) was used to illustrate the verb *to howl* in perfective and imperfective conditions

7.3.2 *Materials*

The subjects were given two pictures in which a different person, animal, or object was involved in the same action (see Fig. 7.3). A sentence-prompting paradigm was used. Instructions accompanying the task were read aloud by the experimenter at the beginning of a trial, and then at the beginning of the experimental task: "I will tell you the sentence describing what is going on in the first picture. I want you to tell me, using a similar sentence, what is happening in the other picture."

The test started with two practice trials on nontest stimuli that were repeated until it was clear that the subject understood the task. There were four verb form conditions: unaccusative perfective, unaccusative imperfective, unergative perfective, and unergative imperfective. There were 14 sentences in each of the four conditions. Examples of the four conditions are:

Unaccusatives (two conditions)

For both conditions, the introduction sentence is:

Eto turist, a eto kot
This is a tourist and this is the cat
Condition 1 (unaccusative, perfect):
Experimenter: Tourist utonul v reke
[Patient: "Kot utonul v reke"]
Experimenter: The tourist drowned (Past, Pf) in the river
[Patient: "The cat drowned (Past, Pf)"] in the river

Condition 2 (unaccusative, imperfect):
Experimenter: Tourist tonul v reke
[Patient: "Kot tonul v reke"]
Experimenter: The tourist drowned (Past, Imp) in the river
[Patient: "The cat drowned (Past, Imp)"] in the river

Unergatives (two conditions)

For both conditions, the introduction sentence is:
Eto pes, a eto volk
This is a dog and this is the wolf
Condition 1 (unergative, perfect):
Experimenter: *Pios zavil na lunu*
[Patient: "Volk zavil na lunu"]
Experimenter: The dog howled (Past, Pf) at the moon
[Patient: "The wolf howled (Past, Pf) at the moon"]
Condition 2 (unergative, imperfect)
Experimenter: *Pios vil na lunu*
[Patient: "Volk vil na lunu"]
Experimenter: The dog howled (Past, Imp) at the moon
[Patient: "The wolf howled (Past, Imp) at the moon"]

Every picture was used twice, once in perfective and once in imperfective aspect. The sentences were mixed and presented in random order (though the actual order was the same for each subject).

7.3.3 Scoring

The answers of the participants were scored by a native Russian speaker. Self-corrections were allowed and the final answer was scored. No time limits were imposed. When failing to produce a verb, a subject was prompted to try it once again. No feedback was provided during the test.

Both quantitative and qualitative analyses were done. For the quantitative analysis, the number of correct/incorrect responses was counted. Responses were considered to be correct when a verb with the proper aspect was produced. For the qualitative comparisons, an error analysis was performed. Errors were classified post hoc, based on the most frequent errors made by the agrammatic speakers during the test. These included (a) aspect substitutions: verbs incorrectly inflected for aspect (i.e., production of perfect instead of imperfect and vice versa), (b) verb omissions, (c) tense errors, and (d) others (agreement and unspecified errors).

Table 7.2 Mean number (and percentages) of errors (max = 14) made by agrammatic speakers

Verb form	Mean (%)
Per construction (max = 14)	
Unaccusative perfective	1.83 (13.1)
Unaccusative imperfective	4.50 (32.1)
Unergative perfective	4.58 (32.7)
Unergative imperfective	1.08 (7.7)
Per argument structure (max = 28)	
Unaccusative	6.33 (22.6)
Unergative	5.67 (20.6)
Per aspect (max = 28)	
Perfective	6.42 (22.9)
Imperfective	5.58 (19.9)

7.4 Results

7.4.1 Quantitative Analysis

In Table 7.2, the group results are given.

Since the data was not normally distributed, statistical testing was done nonparametrically with the Wilcoxon signed-rank test. The *Aspect Assignment Model* of Platonov (2007) predicted that:

1. Unergatives will be easier in past tense, imperfect aspect than in past tense, perfect aspect ($z = -2.921, p = 0.003$).
2. Unaccusatives will show the opposite pattern: easier in past tense, perfect aspect than in past tense, imperfect aspect ($z = -2.83, p = 0.005$).

As can be seen from the statistic comparisons, the data supported both predictions. Contrary to the predictions of both the DOP-H and the ASCH, there is no overall difference between the unergatives and unaccusatives ($z = -0.302$, $p = 0.763$). Also, overall, perfect aspect is not more difficult than imperfect aspect ($z = -0.397$, $p = 0.692$).

7.4.2 Qualitative Analysis

The majority of errors were substitutions of one aspect for another (64.29 % of all errors). The second largest category was "verb omissions," leading to ungrammatical sentences (24.02 % of all errors). Tense errors constituted relatively small group (9.74 %). The remaining errors (1.95 %) consisted of two agreement errors and one unspecified error. In Table 7.3, the numbers of errors per experimental condition are given.

Table 7.3 Number of errors made per experimental condition

	Unaccusative perfective	Unaccusative imperfective	Unergative perfective	Unergative imperfective	Total
Aspect substitution	11	40	36	2	89
Verb omission	8	12	14	3	37
Tense substitution	1	1	5	8	15
Others	2	1	0	0	3
Total	22	54	55	13	144

Both aspect substitutions and verb omissions occur significantly more often in the unaccusative imperfective and the unergative perfective conditions than in the two other conditions (aspect substitutions: $\chi^2 = 43.89$, $p = 0.0001$; verb omissions: $\chi^2 = 5.19$, $p = 0.0227$). Tense substitutions and other errors are distributed equally over the conditions (tense substitutions: Fisher's exact, $p > 0.05$).

7.4.3 Summary of the Results

The results support the predictions made by the *Aspect Assignment Model*: the combination of arguments structure, telicity (in this case, unaccusatives vs. unergatives), tense, and aspect determines agrammatic performance. This is not only shown by the number of correct sentences produced but also by the error pattern. If agrammatic speakers make errors, these are predominantly (1) production a verb in the aspect form that is preferred for the argument structure and (2) verb omissions.

Discussion

Contrary to the DOP-H's and the ASCH's predictions and to all other theories that focus on only one aspect of agrammatic speech, no dissociation between unaccusative and unergative verbs was found. An explanation for this is that Lee and Thompson's (2004) and Bastiaanse and Van Zonneveld's (2005) studies did not take preferences for tense and aspect into account. When this is done, as in the present study, then the *Aspect Assignment Model* offers a better description of the agrammatic performance. Of course, the present study only tested a small part of the *Aspect Assignment Model* and in only one language. "Aspect" is a very complex notion that is not expressed similarly in every language. For example, the difference between perfect and imperfect is not the same in Dutch and English. Actually, roughly speaking, it is reversed: where English uses perfect aspect, Dutch uses imperfect and vice versa. Another difference between Russian (the language used in the study) on the one hand and

Dutch and English on the other is that perfect aspect is expressed through the finite verb in Russian but with a periphrastic form in Dutch and English ("has written"). It is, therefore, probable that the *Aspect Assignment Model* should be adjusted per language.

As mentioned in the Introduction, different theories on the nature of the grammatical impairment resulting in agrammatic speech refer to difference characteristics: the word order problem, the problem with complex verb–argument structures, and the problem with inflected verbs, more specifically with time reference through verbs. The question was whether these problems are related. The *Aspect Assignment Model* demonstrated that the difficulties with argument structure and time reference are related. However, the model makes many more predictions that still need to be tested.

It is not exactly clear how the model can be related to the obvious word order problems that agrammatic speakers have (as shown for Dutch matrix clauses and object scrambling in several languages; see Introduction), that are now captured under the DOP H. For now, however, the *Aspect Assignment Model* seems to be new approach to the argument structure and time reference problems in agrammatic aphasia.

Acknowledgment The authors wish to thank Prof. Shklovsky V.M., Director of the Federal Center of Speech Pathology and Neurorehabilitation in Moscow, Russia, and Dr. Malukova N.G., Head of the Department of Psychology and the therapist working in the center for their willingness to select the patients and their help with patients' diagnostics. We are also very grateful to Allison Smith for her comments on an earlier version.

References

Avrutin, S. (2006). Weak syntax. In Y. Grodzinsky & K. Amunts (Eds.), *Broca's region* (pp. 49–62). Oxford: Oxford University Press.

Bastiaanse, R. (2008). Production of verbs in base position by Dutch agrammatic speakers: Inflection versus finiteness. *Journal of Neurolinguistics, 21,* 104–119.

Bastiaanse, R. (2013). Why reference to the past is hard for agrammatic speakers. *Clinical Linguistics and Phonetics, 27,* 244–263.

Bastiaanse, R., & Jonkers, R. (1998). Verb retrieval in action naming and spontaneous speech in agrammatic and anomic aphasia. *Aphasiology, 12,* 951–969.

Bastiaanse, R., & Thompson, C. K. (2003). Verb and auxiliary movement in agrammatic Broca's aphasia. *Brain and Language, 84,* 286–305.

Bastiaanse, R., & Van Zonneveld, R. (2005). Sentence production with verbs of alternating transitivity in Broca's agrammatic aphasia. *Journal of Neurolinguistics, 18,* 57–66.

Bastiaanse, R., Hugen, J., Kos, M., & Van Zonneveld, R. (2002). Lexical, morphological and syntactic aspects of verb production in Dutch agrammatic aphasics. *Brain and Language, 80,* 142–159.

Bastiaanse, R., Koekkoek, J., & Van Zonneveld, R. (2003). Object scrambling in Dutch Broca's aphasia. *Brain and Language, 86,* 287–299.

Bastiaanse, R., Bamyeci, E., Hsu, C., Lee, J., Yarbay Duman, T., & Thompson, C. K. (2011). Time reference in agrammatic aphasia: A cross-linguistic study. *Journal of Neurolinguistics, 24,* 673.

Behrens, H. (1993). The relationship between conceptual and linguistic development: The early encoding of past reference by German children. *Chicago Linguistic Society, 29,* 63–75.

Bos, L. S., & Bastiaanse, R. (2014). Time reference decoupled from tense in agrammatic and fluent aphasia. *Aphasiology, 28,* 533–553.

Burchert, F., Swoboda-Moll, M., & De Bleser, R. (2005). Tense and agreement dissociations in German agrammatic speakers: Underspecification vs. hierarchy. *Brain and Language, 94,* 188–199.

Burchert, F., Meißner, N., & De Bleser, R. (2008). Production of non-canonical sentences in agrammatic aphasia: Limits in representation or rule application? *Brain and Language, 104,* 170–179.

Dowty, D. (1979). *Word meaning and Montague grammar. The semantics of verbs and times in generative semantics and in Montague's PTQ: Synthese language library*. Dordrecht: Reidel.

Dragoy, O., & Bastiaanse, R. (2009). Verb production and word order in Russian agrammatic speakers. *Aphasiology, 24,* 28–55.

Dragoy, O., & Bastiaanse, R. (2013). Aspect of time: Time reference and aspect in Russian aphasic speakers. *Journal of Neurolinguistics, 26,* 113–128.

Friedmann, N., & Grodzinsky, Y. (1997). Tense and agreement in agrammatic production: Pruning the syntactic tree. *Brain and Language, 56,* 397–425.

Gagarina, N. (2000). The acquisition of aspectuality by Russian children: The early stages. *ZAS Papers in Linguistics, 15,* 232–246.

Hagiwara, H. (1995). The breakdown of functional categories and the economy derivation. *Brain and Language, 50,* 92–126.

Lee, M., & Thompson, C. K. (2004). Agrammatic aphasic production and comprehension of unaccusative verbs in sentence contexts. *Journal of Neurolinguistics, 17,* 315–330.

Lee, J., Milman, L. H., & Thompson, C. K. (2008). Functional category production in English agrammatism. *Aphasiology, 22,* 893–905.

Levin, B. (1993). *English verb classes and alternations: A preliminary investigation*. Chicago: University of Chicago Press.

Luria, A. R. (1973). *Outline of the neuropsychological examination* (in Russian). Moscow: Moscow University Press.

Perlmutter, D. M. (1978). *Impersonal passives and the unaccusativity hypothesis*. In Proceedings of the Fourth Annual Meeting of the Berkeley Linguistic Society, Berkeley Linguistic Society, University of California at Berkeley, pp. 157-189.

Platonov, A. (2007). *Testing the incremental model of aspect assignment in Russian Broca's Aphasia*. Thesis, University of Groningen, University of Potsdam: European Master's for Clinical Linguistics.

Rausch, P., Burchert, F., & De Bleser, R. (2005). Parallels in the breakdown of CP and DP-internal movement processes in agrammatism. A case study. *Brain and Language, 95,* 129–130.

Saffran, E. M., Berndt, R. S., & Schwartz, M. F. (1989). The quantitative analysis of agrammatic production: Procedure and data. *Brain and Language, 37,* 440–479.

Shapiro, L. P., & Levine, B. A. (1990). Verb processing during sentence comprehension in aphasia. *Brain and Language, 38,* 21–47.

Shirai, Y., & Andersen, R. W. (1995). The acquisition of tense-aspect morphology: A prototype account. *Language, 71,* 743–762.

Stoll, S. (1998). The role of Aktionsart in the acquisition of Russian aspect. *First Language, 18,* 351–378.

Thompson, C. K. (2003). Unaccusative verb production in agrammatic aphasia: The argument structure complexity hypothesis. *Journal of Neurolinguistics, 16,* 151–167.

Thompson, C. K., Shapiro, L. P., & Schendel, L. (1995). Analysis of verbs and verb-argument structure. A method for quantification of aphasic language production. *Clinical Aphasiology, 23,* 121–140.

Thompson, C. K., Lange, K. L., Schneider, S. L., & Shapiro, L. P. (1997). Agrammatic and non-brain-damages subjects: Verb and verb argument structure production. *Aphasiology, 11,* 473–490.

Torrence, H., & Hyams, N. (2004). On the role of aspect in determining finiteness and temporal interpretation in early grammar. In J. van Kampen & S. Baauw (Eds.), *Proceedings of GALA 2003: Vol 2. Generative approaches to language aquisition* (pp. 481–491). Utrecht: LOT.

Vendler, Z. (1967). *Verbs and times. Linguistics in philosophy*. Ithaca: Cornell University Press.

Wenzlaff, M., & Clahsen, H. (2004). Tense and agreement in German agrammatism. *Brain and Language, 89,* 57–68.

Wenzlaff, M., & Clahsen, H. (2005). Finiteness and verb-second in German agrammatism. *Brain and Language, 92,* 33–44.

Yarbay Duman, T., & Bastiaanse, R. (2009). Time reference through verb inflection in Turkish agrammatic aphasia. *Brain and Language, 108,* 30–39.

Yarbay-Duman, T., Aygen, G., Özgirgin, N., & Bastiaanse, R. (2007). Object scrambling and finiteness in Turkish agrammatic production. *Journal of Neurolinguistics, 20,* 306–331.

Yarbay Duman, T., Aygen, G., & Bastiaanse, R. (2008). The production of Turkish relative clauses in agrammatism: Verb inflection and constituent order. *Brain and Language, 105,* 149–160.

Zagona, K. (2003). Tense and anaphora: Is there a tense-specific theory of co-reference? In A. Barrs (Ed.), *Anaphora. A reference guide* (pp. 140–171). Oxford: Blackwell.

Chapter 8
Building Aspectual Interpretations Online

E. Matthew Husband and Linnaea Stockall

8.1 Introduction

The classification of linguistic events in terms of whether they last for any length of time and whether they have an inherent end point has a long history (going back at least to Aristotle, see Rosen (1999) for discussion) and continues to excite considerable interest today in both linguistic (Borer 2005b; MacDonald 2008, 2010; Ramchand 2008; Rothstein 2004, 2008; Thompson 2006) and psycholinguistic (Bott 2008; Frazier et al. 2006; Paczynski et al. 2014; Pickering 2006; Piñango et al. 1999; Piñango et al. 2006; Piñango and Deo 2012; Proctor et al. 2004; Todorova et al. 2000) research. Much of this recent work has focused on the linguistic properties of telicity, including which properties of a sentence determine terminative and durative interpretations, and whether these interpretations are constructed early during sentence comprehension.[1]

We would like to thank Alan Beretta, for bringing us together to work on this project a rather long time ago – without him this work would never have happened. We also thank David Adger, Daniel Harbour, Sarah VanWagenen and colleagues and audiences too numerous to mention for feedback at various stages of the development of this work.

[1] The terminology of aspect is notoriously complicated and some clarification for this chapter is in order. The term *aspect* here will refer to what is called lexical aspect, situation aspect, inner aspect, or aktionsarten in other work. Grammatical aspect (also called viewpoint aspect or outer aspect)

E. M. Husband (✉)
St. Hugh's College, University of Oxford,
St. Margaret's Rd., Oxford OX2 6LE, UK
e-mail: matthew.husband@ling-phil.ox.ac.uk

L. Stockall
Department of Linguistics, Queen Mary University of London,
Mile End Road, London E1 4NS, UK
e-mail: l.stockall@qmul.ac.uk

R. G. de Almeida, C. Manouilidou (eds.), *Cognitive Science Perspectives on Verb Representation and Processing*, DOI 10.1007/978-3-319-10112-5_8

Event classification systems, of which the most well known and widely adopted comes from Vendler (1957), have attempted to establish the range of possible event types and demonstrate how these classes interact with other aspects of the linguistic system. Dowty (1979), for instance, showed that the classes proposed in Vendler (1957) can be distinguished on the basis of adverbial modification tests, among others. The test most relevant to our current concerns is *in X time*. This test for telicity rests on the acceptability of end-point modification; for instance a terminative event can be modified with *in an hour* with the interpretation that the event took an hour to finish. An event is said to be terminative, or telic, if it allows this end-point modification, and durative, or atelic, if it does not. Research on aspect has also focused on the linguistic primitives which contribute to the building of aspectual interpretations.

The telicity of some events appears to be entirely determined by the lexical semantics of the verb itself. *Explode* and *find*, for instance, are inherently terminative (1), allowing end-point modification, whereas *sleep* and *fly* are inherently durative (2), blocking end-point modification. This observation has led researchers to refer to this phenomenon as lexical aspect.

(1) Inherently terminative events
  a. The bomb exploded in ten minutes.
  b. John found beer in ten minutes.
  c. John found the beer in ten minutes.
(2) Inherently durative events
  a. John slept #in 8 hours.[2]
  b. John flew planes #in 8 hours.
  c. John flew the plane #in 8 hours.

For these verbs, other elements of the sentence, including changes in the properties of the internal argument, do not affect the event interpretation. However, Verkuyl (1972) demonstrated that for many events, telicity often depends on the count or mass syntax of the internal argument, and not on lexical properties of the verb alone (3).

(3) Internal argument dependent events
  a. John drank beer #in ten minutes.
  b. John drank the beer in ten minutes.
  c. John built planes #in eight hours.
  d. John built the plane in eight hours.

will be explicitly mentioned when necessary. We will focus in particular on the dimension of telicity (whether an even has a natural end or not) and use *durative* (atelic) and *terminative* (telic) to refer to the two aspectual interpretations of interest here.

[2] Sentence judgments are as follows: a * indicates ungrammaticality, a # indicates an unavailable reading. Often in this chapter, # will indicate that the event modified by in X time cannot receive an end-point interpretation.

For verbs like *drink* and *build*, the terminative interpretation depends on properties of the internal argument: count internal arguments give rise to terminative interpretations while mass internal arguments give rise to durative interpretations[3]. Based on these observations, Verkuyl (1972) argued that 'lexical' aspect is actually a VP phenomenon since the VP is the first point where verbal and nominal sources of aspectual information can combine together. Subsequent research has supported this phrasal-level understanding of aspectual interpretation, although studies on 'achievement' verbs (such as *explode*, *find*, and *notice*) have continued to argue for their inherently terminative nature (Borer 2005b; Mittwoch 1991; Smith 1997). Taken together, a calculus of telicity emerges based on both verbal and nominal properties.

This summary suggests that the presence of any kind of terminative element, including an inherently terminative verb or a count internal argument, leads to a terminative, or telic, interpretation. The absence of both of these elements leads to a durative, or atelic, interpretation.

In § 1.1, we review both semantic (§ 1.1.1) and syntactic (§ 1.1.2) accounts for how the properties of the internal argument (and other elements in the clause) can affect aspectual interpretations. Critically, we point out that the two approaches make different assumptions about whether telic or atelic events are more structurally complex. We typically expect that additional complexity is associated with additional processing costs (Fodor and Garrett 1967)[4]. In § 2, we review the previous psycholinguistic and neurolinguistic literature on the processing of inner aspect. We show that the basic assumptions about complexity and processing cost are supported, and that there is clear and compelling evidence that aspectual interpretations are generated and evaluated rapidly. However, the bulk of this existing research focuses on intransitive, inherently telic predicates, and thus does not allow us to see how verbal and nominal elements are put together to generate aspectual interpretations, or whether aspectual interpretation is truly a VP level phenomenon.

In § 3, we discuss a series of experiments that directly investigate aspectual compositionality. We find clear evidence for the individual contributions of both lexical verbs and internal argument DPs to aspectual interpretation, and, find moreover that these interpretations only seem to be generated at the level of the VP (not as properties of individual lexical verbs). In some experiments, we find that atelic interpretations are significantly more costly to process than telic, consistent with semantic models of

---

[3] Both mass nouns like *beer* and bare plurals like *planes* are known for allowing durative interpretations. Bare plurals are argued to allow 'aspectual leaks' because of the cumulativity of their denotation Verkuyl (1987). A syntactic hallmark of these two phrases, and of mass interpretation in general, is their lack of determiner in languages which allow bare nominals.

[4] Note that Fodor and Garrett failed to find a correspondence between the number of syntactic transformations assumed in Chomsky (1965), and the dependent measures in their experiments (e.g. number of errors in a paraphrase task). This may or may not be compelling evidence against, say, the theory of the passive transformation in Chomsky (1965), but it is certainly not a compelling reason to assume there should be no relationship between representational and derivational linguistic complexity and processing costs. See Phillips (1996, Chap. 5) for further discussion.

telicity (§ 1.1.1). In other experiments, we find hints of the inverse—greater processing costs for telic predicates, consistent with syntactic accounts (§ 1.1.2)—though these are merely suggestive.

We conclude by summarising what we think our results mean for how aspectual interpretations are generated in real time, and the implications these results have for our more general understanding of potential limits on the incrementality of parsing and syntactic constraints on the domains for semantic interpretation.

8.1.1 Theories of Event Complexity

While current approaches to aspect generally agree that aspectual interpretation requires composition, precisely which verbal and nominal properties contribute to telicity and how these elements are composed to yield aspectual interpretations have been the object of considerable debate. Of the many questions that have come out of this work, we are particularly interested in the representational complexity of events here. Claims about representational complexity have largely depended on the framework adopted by the theory in question since the representations which syntactic theory and semantic theory assume are not necessarily the same.

Differences in the kind of formal approach to events that is adopted have led to different analyses of the source of telicity and different conclusions concerning the representational complexity of events. In semantic theories, the properties of the model-theoretic interpretation of events are central to understanding the interpretation of telicity. Syntactic theories focus on the functional primitives needed to account for syntactic structures which are linked to different aspectual interpretations.

8.1.1.1 The Semantics of Aspect

Semantic theories of aspect have focused on the ontological commitments necessary to capture the differences between telic and atelic interpretations. Bach (1986) represents an early approach which connected algebraic structures familiar in the analysis of count and mass individuals to telic and atelic events. Enriching the domain of events in a way similar to Link's (1983) enrichment of the domain of individuals has allowed researchers to understand how nominal properties like the count/mass distinction in the domain of individuals can affect events. This work has also lead researchers to define notions such as *homogeneity* over the model structures of individuals and events which capture the differences between count and mass interpretations and durative and terminative interpretations, and link up with other important properties, such as the sub-interval property of durative events (Bennett and Partee 1978).

A major concern has been to understand how telicity is related to the properties of these event structures. Using Bach's enriched event domain, Krifka (1992, 1998) and Verkuyl (1993) formulated formal properties of predicates to capture the telic/atelic distinction. Krifka presented the first compositional semantic theory of telicity in a series of influential papers which made two important insights. First, he proposed a mapping between the individual denoted by the internal argument of the verb and the event denoted by the verb itself. This mapping captured the intuition behind Tenny's (1987, 1994) measuring out of an event by its internal argument, creating a relationship between the part structure of the individual and the part structure of the event. He also proposed to capture the homogeneity of atelic events through a *cumulative* property of their resulting event structure. The event structure of telic events in turn was argued to be *quantized*. Krifka's approach formalized these properties of event structure with the following definitions.

(4) *Cumulative:* $\exists x,y\ [P(x) \wedge P(y) \wedge \neg x = y] \wedge \forall x,y\ [P(x) \wedge P(y) \rightarrow P(x \oplus y)]$
P is cumulative if there is an x and y (x distinct from y) with property P such that the sum of x and y also have property P.

(5) *Quantized:* $\forall x,y\ [P(x) \wedge P(y) \rightarrow \neg y <_P x]$
P is quantized if for all x and y with property P, y is not a proper part of x.
(Krifka 1998)

Consider the atelic event *John drank beer* which is argued to be cumulative. According to Krifka's definition of cumulativity, if two distinct events can be described as *John drank beer*, then their sum should also be able to be described as *John drank beer*. For instance, suppose that we describe an event from 5 to 6 pm as *John drank beer*, and an event from 6 to 7 pm as *John drank beer*. Then we can describe their sum from 5 to 7 pm as *John drank beer*.

Now consider the telic event *John drank a beer* which is argued to be quantized. According to Krifka's definition of quantization, if an event can be described as *John drank a beer*, then there should be no proper part of this event that can also be described as *John drank a beer*. So, if we describe the event from 5 to 6 pm as *John drank a beer*, and we look at a proper part of this event, say the part from 5 to 5:30 pm, this proper part cannot be described as *John drank a beer*, which is the case since John is likely only about halfway through drinking his beer given this proper part.

Although these definitions have been shown to be too restrictive and to fail to capture certain kinds of telic and atelic predicates (Borer 2005b), they represent an important advance in understanding the model-theoretic representation of events. By assuming that events have structure, one can formulate a relationship between the structure of an event and the aspectual interpretation of that event.

Refinements of Krifka's initial properties have lead to a deeper understanding of telicity and the semantic distinctions between predicates. Kiparsky (1998) emphasized other important properties such as *divisiveness* as a requirement of atelic predicates. This property captured those predicates which are cumulative but have telic interpretations, such as *eat more than two apples*. Borer (2005b), noting that other examples of telic interpretations, like *read fewer than three books* and *fill the room with smoke* were problematic for earlier approaches, modified Krifka's notion of quantization and made further refinements by proposing that while homogeneous predicates (6) are syntactically simple, they carry both cumulative and divisive requirements for their model-theoretic interpretation, leading to the revised formulations in (7) and (8), and the addition of (9).

(6) *Homogeneous:* P is homogeneous if P is cumulative and divisive.
(7) *Cumulative:* $\forall x,y\ [P(x) \wedge P(y) \rightarrow P(x \cup y)]$
P is cumulative if for all x and y with property P, the union of x and y have property P.
(8) *Divisive:* $\forall x\ [P(x) \rightarrow \exists y\ [P(y) \wedge y < x] \wedge \forall x,y\ [P(x) \wedge P(y) \wedge y < x \rightarrow P(x,y)]$
P is divisive if for all x with property P there is a y with property P that is part of x and for all x and y with property P such that y is a part of x, the subtraction of x and y also has property P.
(9) *Quantity:* P is quantity iff P is not homogeneous.

(Borer 2005b)

Consider again the atelic event *John drank beer* which is argued to be homogeneous and therefore both cumulative and divisive. To be cumulative, the union of two events described as *John drank beer* should also be an event of *John drank beer*. So given that we describe an event from 5 to 6 pm as *John drank beer* and an event from 6 to 7 pm as *John drank beer*, we are happy to describe the event from 5 to 7 pm as *John drank beer*. To be divisive, an event described as *John drank beer* should have a proper part that is also described as *John drank beer*, and if we take out that proper part, we should be able to describe the remainder as *John drank beer*. So given that we again describe an event from 5 to 6 pm as *John drank beer*, we can find a proper part that is also described as *John drank beer*, say the part from 5 to 5:30 pm, and if we look at the remainder, the part from 5:30 to 6 pm, that part can also be described as *John drank beer*. In this way, atelic events are homogeneous in their model-theoretic interpretation.

Returning to the telic event *John drank a beer*, we should find that it denotes a quantity event. To be quantity, an event must not be homogeneous; that is, it must fail to be either cumulative or divisive. In this particular case, the telic event *John drank a beer* fails to be both. Concerning the failure of cumulativity, if we take two events described as *John drank a beer*, one from 5 to 6 pm and one from 6 to 7 pm, we cannot describe their union as *John drank a beer*. Instead, we would describe such an event as *John drank two beers*. Concerning the failure of divisiveness, given an event from 5 to 6 pm described as *John drank a beer*, we will be unable to find a proper part within that event that is also described as *John drank a beer* because any such proper part can only be described as *John drank some part of a beer*. In this way, telic events are quantity in their model-theoretic interpretation.

The notion that there is something more complicated underlying the interpretation of atelic events pervades inquiry into the semantics of aspect. Since the representations of atelic predicates require the ability to see inside their temporal interval and make reference to their sub-events, the model-theoretic objects needed to capture the homogeneity of atelic events are complex in a way that those needed for the representation of telic events are not. While a telic event like *drink a beer* has no sub-events which are also *drink a beer* events and therefore is atomic, an atelic event like *drink beer* has sub-events which are also *drink beer* events. The event models which semantic interpretation builds must encode this kind of distinction. As such, while the event model representation of a telic event does not encode sub-events, the

Fig. 8.1 A semantic model for a homogenous event

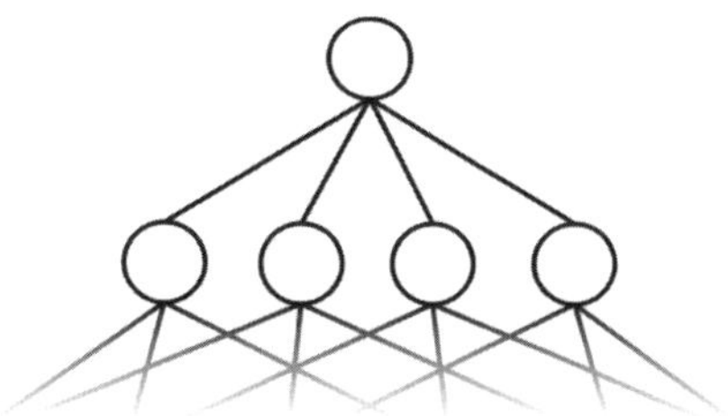

event model representation of an atelic event encodes an often unbounded number of sub-events as schematized in Fig. 8.1 (Bach 1986; Link 1998). By structuring the domain of events, atelic events can be modelled through part structures which have properties like homogeneity.

8.1.1.2 The Syntax of Aspect

In research on the syntax of aspect, the role of the internal argument has been a central concern. Tenny (1987, 1994) represents an early attempt at understanding the syntactic consequences of telicity with respect to the use of the internal argument. Her aspectual-interface hypothesis, which claims that thematic structure and syntactic argument structure are governed by aspectual properties, links the syntactic position of an argument to the argument's event role[5]. She further proposed that internal arguments in some sense 'measure out' or delimit an event, a notion that has continued to resonate in the semantics of aspect. Further research uncovered syntactic phenomena that were tightly related to aspectual interpretations. Dowty (1991), for instance, noted a systematic relationship between unergative/unaccusative diagnostics and telicity; agentive, atelic sentences are always unergative, while those that are non-agentive and telic are always unaccusative. Since unergatives and unaccusatives are diagnosed in part based on the presence of an internal argument, these studies further linked the role of the internal argument to telicity. This work has led researchers to explore telicity in terms of core syntactic features and functional syntactic structure.

Expanding the functional lexicon to include aspectual heads responsible for licensing aspectual interpretations has produced several interesting theories of aspect in recent years. These functional heads have been important in understanding the relationship between the internal argument and aspectual interpretation by formalizing the tight link between argument realization and telicity. Borer (1994, 1998, 2005b) and Ramchand (1997, 2008) exemplify theories that introduced the idea of aspectual phrases (AspP) in the functional syntax. These proposals link verbal arguments to

[5] The issue of how exactly syntactic argument structure relates to event/thematic role is too complicated and controversial to engage within this chapter. Bacrach et al. (2014) review the range of approaches and current state of the debate. For our purposes it suffices that the internal argument of a non-psych transitive verb is typically interpreted as a theme or patient argument

Table 8.1 The calculus of telicity

Verb Type	NP Type	Telicity	VP
Unspecified	Non-count	Durative	Drink beer
Unspecified	Count	Terminative	Drink the beer
Bounded	Non-count	Terminative	Find beer
Bounded	Count	Terminative	Find the beer

their event participant roles by appearing in these aspectual projections at LF, the syntactic representation which receives semantic interpretation. The syntactic mechanisms used to project arguments into these aspectual phrases were often driven by syntax-internal factors such as case assignment, suggesting that aspectual interpretations are triggered by the syntax for reasons unrelated to aspectual interpretation itself. Schmitt (1996), van Hout (1996, 2000), and Ritter and Rosen (1998) made this notion explicit by claiming AgrOP, a phrase initially conceived as the locus of structural case assignment to the internal argument, as the domain for the interpretation of telicity. In these theories, the internal argument is required to move to the specifier of an AspP (AgrOP) dominating the VP. In doing so, it creates the right syntactic configuration for a telic predicate at LF.

Research focused on the syntax of aspect has led to a number of interesting conclusions. Syntactic theories of telicity all argue that the structure of telic predicates is more complex than the structure of atelic predicates. Depending on the particular implementation, the aspectual phrase of atelic predicates remains unfilled by a delimiting argument or may not be projected in the syntax at all. On the other hand, every syntactic theory of telicity requires the aspectual phrase of a telic predicate to be present and licensed either by a verb's event semantics or by a delimitating argument, typically a count noun phrase. In these theories, telic predicates are overall more complex syntactic objects than atelic predicates.

In addition, the syntax of aspect gives us a compositional way to derive the calculus of telicity (Table 8.1). Making a few simplifying assumptions along the guidelines established by research in the syntax of aspect, the derivation of AspP can derive the calculus of events in Table 8.1 in two stages. Once the verb and its direct object have formed the VP, Asp is grammatically licensed and its derivation is triggered. In the first stage, Asp merges with the VP and the verb is allowed to assign an event semantics to Asp if the verb has one to assign (only terminative verbs in (Table 8.1) have event semantics; Fig. 8.2a). In the second stage, the direct object NP moves to Asp for case assignment. If the verb has not already assigned Asp an event semantics, the NP assigns an event semantics to Asp depending on whether the NP has a determiner or not (Fig. 8.2b). NPs with determiners assign terminative event semantics whereas NPs without determiners assign durative event semantics. This derivation of telicity demonstrates how both the verb and internal argument contribute compositionally to the telicity of a sentence.

In the sections that follow, we review previous and new evidence from single word and sentence processing experiments that investigate how comprehenders make use of verbal and nominal information to generate aspectual interpretations online.

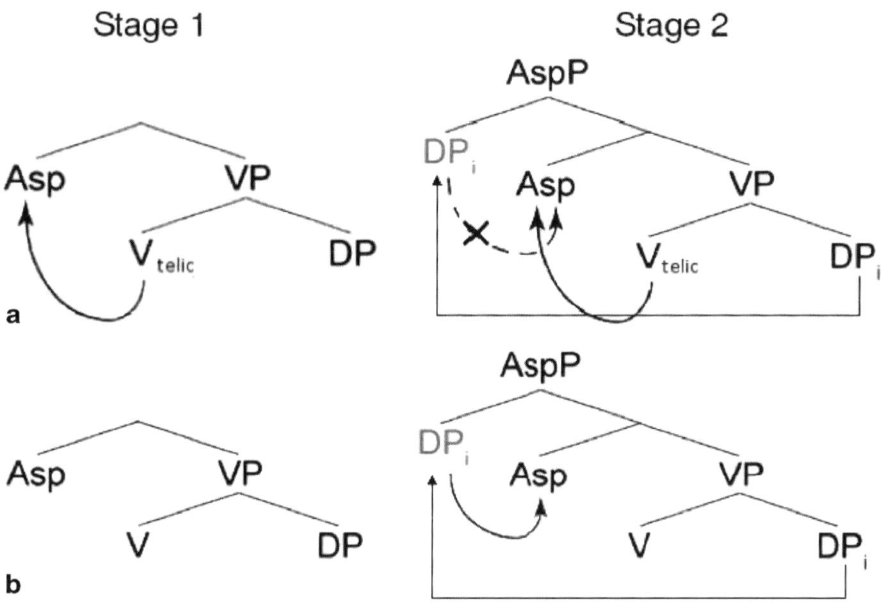

Fig. 8.2 Syntactic structures demonstrating the derivation of the calculus of events

8.2 Online Event Interpretations

The representations discussed above involve several factors important for aspectual interpretation which have implications for the processing of aspect. We identify three critical questions:

1. What is the domain over which the parser interprets aspect?
2. What factors guide the parser in aspectual interpretation online?
3. Does the parser immediately commit to an aspectual interpretation?

Different researchers have adopted different positions on these questions. Much processing research takes the verb to be the initial domain for aspectual processing even though linguistic theory argues that the whole VP is required. A highly incremental parser may in fact use verbal information immediately in guiding interpretative commitments. Indeed, the assumption implicit in much of the research on semantic processing since at least Marslen-Wilson and Tyler (1980) assumes that access to lexical semantic information is extremely rapid and that interpretation begins immediately as each word is processed. See, for instance Amsel et al. (2013) who find access to conceptual information within 160 ms of encountering a written word, or Müller and Hagoort (2006) who argue that semantic information may even be processed before syntactic information (contra Friederici (2002) and related work that argues that syntactic processing precedes semantic). However, Frazier (1999) suggests interpretation may be more complicated and constrained. Frazier (1999) argues for a principle of immediate partial interpretation, which contrasts with the commonly-held assumption that all interpretation is immediate and complete.

(10) Immediate partial interpretation (Frazier 1999)
Perceivers must choose between grammatically incompatible meanings of a word or constituent immediately, by the end of the word or constituent, unless this conflicts with the dictates of the grammar.

Applying this principle requires a clear understanding of what kinds of meanings are grammatically incompatible and what kinds of underspecification the grammar permits. We assume that different model-theoretic events count as distinct and incompatible interpretations. Homogeneous event models are distinct from non-homogeneous event models and give rise to incompatible aspectual interpretations (§ 1.1.1). As such, the parser is not allowed to underspecify telicity after completion of the VP, because telic and atelic event models, and likewise their resulting aspectual interpretations, are incompatible.

We also assume, following Dickey (2001), that distinct LFs cannot be underspecified. The LF of a telic predicate contains a licensed AspP while the LF of an atelic predicate lacks an AspP (§ 1.1.2). Since the grammar requires a complete VP to trigger the derivation of AspP, this proposal further argues for the parser to immediately commit to an aspectual interpretation upon completion of the VP constituent, but not before as this would violate the dictates of the grammar.

Given the above assumptions, immediate partial interpretation provides us with explicit predictions concerning the processing of aspect. First, since aspect is a VP phenomenon, processing related to telicity is predicted to occur only upon completion of the VP constituent since earlier commitment to an aspectual interpretation would be in violation of the grammar of aspect. Even in the face of a verb with unambiguous event semantics, the parser should delay commitment to an aspectual interpretation until it has processed the full VP. Second, the parser should commit to an aspectual interpretation when an event requires a homogeneous or non-homogeneous model for its interpretation. The point at which this decision is anchored is also the VP constituent.

Previous experimental research provides some partial answers to our three questions concerning the domain, factors, and time course of the processing of aspect. Below, we review some of the important findings which have contributed to our current understanding of aspectual processing.

8.2.1 Costs Associated with Accessing Verbal Lexical Semantics

Some recent work has focused on the contribution of lexical properties of the verb to aspectual interpretation. Gennari and Poeppel (2003) examined stative verbs, such as *know*, and eventive verbs, such as *build*, in both self-paced reading and lexical decision experiments. Based on linguistic theories of the lexical semantic representation of verbs, eventive verbs are considered to be more complex than stative verbs

(Dowty 1979; Parsons 1990; Rappaport Hovav and Levin 1998; Vendler 1957), and thus should result in longer processing times due to the recovery of their extra structure. In both lexical decision and self-paced reading studies, Gennari and Poeppel (2003) found that eventive verbs incurred longer processing times compared to stative verbs. These results provide general support for the assumption that representational semantic complexity affects processing times, and specific support for their claim that 'processing verb meanings involves activating properties of the event structure' (Gennari and Poeppel 2003, p. 89). Interestingly, Gennari and Poeppel find similar response time costs when verbs are read in sentential context and when they are read in isolation, suggesting that the stative/eventive distinction (or, possibly, as Gennari and Poeppel propose, the distinction between causal and non-causal event representations) is encoded in the lexicon, not only in the syntax as most constructivist models of grammar would assume (Hale and Keyser 1993; Borer 2005a, b; Ramchand 2008, inter alia)[6].

Focusing on events only, McKoon and Macfarland (2002) examined the processing of externally caused events, such as *break* (John broke the glass.), compared to internally caused events, such as *bloom* (*John bloomed the flower). Externally caused events are predicted to incur a processing cost compared to internally caused events due to the number of event participants: two in externally caused events, one in internally caused events. Using comprehension and production measures, McKoon and Macfarland (2002) found that the more complex externally caused events took longer to process than the less complex internally caused events, independent of the actual number of arguments in the sentence. This provides further evidence that event structure plays an active role in verbal processing.

These studies together suggest that the event semantics of the verb carries immediate consequences for processing. What remains to be shown is whether the processing cost associated with verb event semantics is due to retrieval of semantic information from the lexicon, as assumed by Gennari and Poeppel (2003) and McKoon and MacFarland (2002), or if the costs are associated with generating an event interpretation. Aspectual interpretation offers a chance to compare these alternatives, since both the verb and the internal argument contribute critical information to the interpretation. However, to date, most research on processing aspectual interpretations has focused on using adverbial temporal modification of obligatorily telic predicates, rather than on the initial generation of terminative or unbounded interpretations.

[6] Liina Pylkkänen (p.c.) noted that a closer inspection of the materials used in Gennari and Poeppel (2003) revealed that many of the eventive verbs were in fact also achievement verbs, and therefore telic. Since stative verbs are thought to be generally atelic, it is unclear whether their effect is truly driven by a stative/eventive distinction or by a distinction in telicity.

8.2.2 Using Adverbial Modification

Considerable research has been devoted to investigating the processing mechanisms associated with combining temporal modifiers (*for years*, *until dawn*) and verb phrases with different aspectual interpretations. Brennan and Pylkkänen (2008); Paczynski et al. (2014); Piñango et al. (1999, 2006) and Todorova et al. (2000) all find costs associated with reading sentences such as (11a), in which a durative modifier (*for months*) is combined with a terminative event, as compared to the same terminative event modified by a modifier with neutral aspect (*last month*) (11b).

(11) a. The bunny hopped until dawn. [coercion]
b. The bunny slept until dawn. [control]

Resolving the mismatch between the durative requirements of the adverbial and the terminative event supplied by the verb phrase requires that the event be either iterated into a multiple event interpretation, such that the bunny hopped repeatedly all night long, or somehow stretched beyond its usual temporal extent, such that a single event of hopping can be construed as taking the whole night to complete. Whichever solution is arrived at, successful interpretation of sentences like (11a) requires extra work above and beyond the normal processing required to understand sentences like (11b). This extra work is often called coercion (Pustejovsky 1995; Jackendoff 1997).

Piñango et al. (1999) investigated aspectual coercion using cross-modal lexical decision. Participants heard sentences containing verbs like *jump* or *sneeze*, which denote highly punctual events and adverbial modifiers like *until* that denote extended spans of time. 250 ms after the the offset of the adverbial, marked here by the ∧, a letter string appeared on the screen, and participants made a lexical decision to this letter string. Piñango et al. found that lexical decisions were significantly slower in the coercion sentences (12b) than in the control (12a).

(12) a. The insect glided effortlessly until ∧ it reached ... [Control]
b. The insect hopped effortlessly until ∧ it reached ... [Coercion]

Brennan and Pylkknen (2008) reversed the order of the key elements in the Piñango et al. (1999) study such that the temporal adverbial preceded the verb, in order to more tightly assess the time course in which aspectual interpretations are constructed. They found immediate processing costs associated with sentences like (13a) relative to (13b).

(13) a. Throughout the day the student sneezed in the back of the ... [Coercion]
b. After twenty minutes the student sneezed in the back of the ... [Control]

Using a self-paced reading task[7], they found that punctual verbs like *sneezed* were read more slowly in aspectual coercion sentences compared to controls, again confirming the processing cost associated with aspectual coercion. They also collected magnetoencephalography (MEG) responses to reading these types of sentences to

[7] We return to the issue of experimental task in § 8.2.3

probe for neural correlates of online aspectual interpretation. The punctual verbs in the aspectual coercion condition evoked greater activity than the same verbs in the control condition in two time regions. The first, between 350–380 ms after the presentation of the verb, was localized to frontal and temporal regions, and the second, between 450–460 ms, originated in anterior midline frontal regions. The 350 ms response, called the M350, is a response component associated with the activation of stored lexical semantic representations (Pylkkänen and Marantz 2003; Stockall and Marantz 2006; Harris et al. 2008), and has been argued to be an early component of the well known N400 response typically associated with costs associated with mismatches between sentence meaning and real world conceptual knowledge (Pylkkänen and Marantz 2003; Lau et al. 2008). The later anterior midline frontal response, or AMF, is a response associated with complement coercion (Pylkkänen and McElree 2007; Pylkkänen et al. 2009)[8]

Paczynski et al. (2014) complement these behavioral and MEG studies with an event-related potentials (ERP) study, which also investigated the processing associated with combining a durative adverbial with a punctual predicate. Like Brennan and Pylkkänen, Paczynski et al. (2014) used materials with the temporal adverbials at the beginning of the sentence so as to be able to find processing effects time locked to the onset of the verb. Paczynski and colleagues crossed adverbial type (punctual vs. durative) with verb type (punctual vs. durative), and also embedded their critical test sentences inside a three sentence sequence, to attempt to create a more naturalistic context for event interpretation. Critical items are in (14)[9].

(14) a. After several minutes, the cat pounced on the rubber mouse. [PA-PV]
b. For several minutes, the cat pounced on the rubber mouse. [DA-PV]
c. After several minutes, the cat prowled about the backyard. [PA-DV]
d. For several minutes, the cat prowled about the backyard. [DA-DV]

[8] Like aspectual coercion which results from a mismatch between the aspectual requirements of an adverbial modifier and the aspectual properties of the VP, complement coercion involves a mismatch. Verbs like *begin*, *start*, *try*, etc. typically take clausal complements that denote events, as in (i.b). When these verbs are combined with simple nominal direct objects (i.a), Pustejovsky (1995) argues that comprehenders must coerce the nominal (*the book*) into some kind of event to resolve the mismatch between the verb's eventive selectional requirements and the direct object's non-eventive properties.

(i) a. John began the book.
b. John began to read the book.

A number of studies report reading time and related measures showing that, like aspectual coercion, complement coercion is behaviorally costly (McElree et al. 2001, 2006; Pickering et al. 2005; Traxler et al. 2002; Traxler et al. 2005; Pylkkänen and McElree 2007; Husband et al. 2011). Brennan and Pyllkänen interpret the finding of an AMF response for aspectual coercion in their 2008 study, and for complement coercion in Pylkkänen and McElree (2007); Pylkkänen et al. (2009), as evidence that both phenomena share similar processing mechanisms. See Pylkkänen et al. (2009) for further discussion.

[9] Note that we are simplifying Paczynski et al.'s design slightly for the purposes of highlighting the comparison with the previous literature. We discuss their additional manipulations in § 8.2.3.

Paczynski et al. (2014) find a sustained, late (post 500 ms) negativity for punctual verbs following durative adverbial contexts, as compared to durative verbs in the same contexts. Paczynski and colleagues conclude 'even when simply reading for comprehension in a word-by-word fashion, participants can make a commitment to aspectual interpretation within a few hundred milliseconds after it becomes syntactically licensed'. (Paczynski et al. 2014, p. 8–9).

It's well established, then, that at least for punctual verbs like *jump,* which are always interpreted as having a definite, necessary end point, terminative aspectual interpretations are generated rapidly and automatically during processing. Todorova et al. (2000) address the question of whether aspectual interpretations are also rapidly generated when they involve contributions from both the verb and internal argument. When a verb like *send*, which is usually an accomplishment verb in Vendlerian terms (Vendler 1957), is combined with an internal argument with unbounded semantics, such as *letters*, the resulting event is one with no necessary or specific end point: *John sent letters* denotes an unbounded event in which John sends some unknown number of letters over the space of some unspecified span of time. By contrast *John sent a letter*, with a count internal argument, denotes a single, telic, event of letter sending. Thus when a phrase like *John sent letters* is combined with a durative adverbial, there is no mismatch, and no coercion should be necessary. But *John sent a letter* combined with the same adverbial should trigger aspectual coercion, just as above.

Todorova et al. compare processing costs associated with modifying telic and atelic verb phrases by temporal adverbials that either require a durative event, or are neutral with respect to event duration, as in (15).

(15) a. Even though Howard sent a large check to his daughter for many years, she refused to accept his money. [bounded/durative]
b. Even though Howard sent large checks to his daughter for many years, she refused to accept his money. [unbounded/durative]
c. Even though Howard sent a large check to his daughter last year, she refused to accept his money. [bounded/neutral]
d. Even though Howard sent large checks to his daughter last year, she refused to accept his money. [unbounded/neutral]

Using a self-paced stop-making-sense reading paradigm in which participants evaluated the sensicality of the sentences they were reading, region by region, Todorova et al. (2000) found significantly greater reading times and greater sensicality rejections for the adverbial region in the mismatch condition (15a) as compared to the other conditions.

However, using a self-paced reading design, Proctor et al. (2004) fail to find similar immediate effects of compositional aspectual interpretation. Proctor and colleagues manipulated several factors known to affect telicity, including verb semantics (telic or atelic), the noun semantics of the internal argument (mass or count), and the aspectual requirements of adverbial modification (for eight minutes (atelic)/in eight mintues (telic)) in a self-paced reading study. They also probed subjects' aspectual interpretations offline with a post-sentential comprehension question which asked whether the specified action had been completed halfway through the time given by the adverbial modifier. Below is a subset of their items demonstrating the first two manipulations, with '/' indicating the units used for the self-paced reading design.

(16) a. Leslie consumed/Polar Purity's/ice water/with zeal/ . . . [atelic/mass]
b. Leslie consumed/Polar Purity's/ice cube/with zeal/ . . . [atelic/count]
c. Leslie monitored/Polar Purity's/ice water/with zeal/ . . . [telic/mass]
d. Leslie monitored/Polar Purity's/ice cube/with zeal/ . . . [telic/count]

Each of these sentences was continued with the adverbial modifier region. Proctor et al. found an online effect of aspectual interpretation on the adverbial modifier region; processing slowed when the adverbial modifier conflicted with the telicity of the VP. In addition, they found an offline effect of telicity in comprehension responses. A sentence with an atelic verb, mass noun, and atelic modifier was more likely to be interpreted as atelic (eliciting 'yes' to a question about the sub-interval of the event) than a telic verb, count noun, and telic adverbial modifier. They did not find any processing differences in the nominal region (ice water vs. ice cube) which is the first point at which a telic or atelic interpretation for the verbal event could be generated. Proctor et al conclude that the parser commits to an aspectual interpretation online, but somewhat more slowly than might be expected given assumptions concerning incrementality.

While these results are interesting in that they provide further support for the online interpretation of telicity and its offline consequences in comprehension, some caution must be taken concerning the items used. The reported example item makes use of a possessive noun phrase before the internal argument. Possessive NPs act as determiners and trigger a telic interpretation as shown in (17).

(17) a. Leslie drank ice water #in eight minutes.
b. Leslie drank the ice water in eight minutes.
c. Leslie drank John's ice water in eight minutes.

As a result, (16a) above does not in fact involve an internal argument with mass semantics, and thus is not predicted to contrast with (16b).

Also, the distinction between mass and count nouns may not trigger atelic and telic interpretations respectively. Instead, the syntactic properties of count and mass interpretations, driven by the presence or absence of a determiner seem to be necessary to trigger telicity. Note that singular count nouns require a determiner (18a) and trigger a telic interpretation (18b) while plural count nouns without a determiner still trigger an atelic interpretation (18c).

(18) a. *Leslie ate ice cube in eight minutes.
b. Leslie ate an ice cube in eight minutes.
c. Leslie ate ice cubes #in eight minutes.

The failure to find significant early effects may have resulted from these factors.

Overall, then, research using adverbial modifiers to probe aspectual interpretations during sentence processing has shown that the parser commits to a telic or atelic interpretation of a VP rapidly online. Processing costs increase when the telicity of the VP mismatches the aspectual requirements of an adverbial modifier.

8.2.3 An Apparent Counterexample

Pickering et al. (2006) used self-paced reading and eye-movement measures to investigate aspectual coercion and the time course within which aspectual interpretations are generated. They based their experiments on the materials and manipulations of Piñango et al. (1999) and Todorova et al. (2000).

Pickering et al.'s first two experiments used the same materials as Piñango et al. (1999) with one additional manipulation: the relative order of the verb and the durative adverbial were manipulated, resulting in materials as in (19).

(19) a. The insect glided effortlessly until it reached the far end of the garden. It was in a hurry to return to its nest. [unfronted/unbounded]
b. The insect hopped effortlessly until it reached the far end of the garden. It was in a hurry to return to its nest. [unfronted/bounded]
c. Until it reached the far end of the garden, the insect glided effortlessly under the moonlight. It was in a hurry to return to its nest. [fronted/unbounded]
d. Until it reached the far end of the garden, the insect hopped effortlessly under the moonlight. It was in a hurry to return to its nest.[fronted/bounded]

In two separate experiments, one using a self-paced reading paradigm, the second recording eye-movement measures, Pickering et al. (2006) failed to find any effect of aspectual mismatch between the bounded verb and durative adverbial.

Given that Pickering et al. themselves suggest that Piñango et al.'s (1999) stimuli may not have been sufficiently well-controlled to reliably demonstrate effects (p. 14), and that Brennan and Pylkkänen (2008) and Paczynski et al. (2014) successfully find rapid, significant costs for durative adverb + punctual verb sentences in both self-paced reading times and evoked neural activation, it is likely that the failure to replicate Piñango et al. (1999)'s effects is due to specific problems with the experiment.

Pickering et al. (2006) also raise the issue of whether aspectual coercion effects, and thus evidence for rapid aspectual interpretation, are dependent on the kind of task participants are asked to do. Since they find no effects with self-paced reading and eye-movement paradigms that allow participants to read sentences in a relatively natural way that does not specifically require them to commit to a complete event interpretation, they speculate that the effects previously reported by Piñango et al. (1999) and Todorova et al. (2000) may be task effects. However, in the interim, Brennan and Pylkkänen (2008) and Paczynski et al. (2014) show that this can not be the case—aspectual coercion effects are clearly automatic and rapid.

This leaves us with Pickering et al.'s experiments 3 and 4, which aim to partially replicate Todorova et al. (2000). Instead of using the durative modifiers used by Piñango et al. (1999), Piñango et al. (2006), Todorova et al. (2000), and Brennan & Pylkkänen (2008), Pickering et al. (2006) used adverbials composed of a universal quantifier and a temporal interval, as in (20). Pickering et al also made minor modifications to the basic sentence structure of Todorova et al's materials, by removing the initial complementizer (*although, even though*) and adding additional spillover material immediately after the critical adverb phrase.

(20) a. Howard sent a large check to his daughter every year but as usual, she refused to accept his money. [Singular object/frequency adverb (SF)]
b. Howard sent large checks to his daughter every year but as usual, she refused to accept his money. [Plural object/frequency adverb (PF)]
c. Howard sent a large check to his daughter last year but as usual, she refused to accept his money. [Singular object/neutral adverb (SN)]
d. Howard sent large checks to his daughter last year but as usual, she refused to accept his money. [Plural object/neutral adverb (PN)]

Pickering and colleagues were motivated to use the adverbials in (20a), which they call *frequency adverbials*, by a concern raised by Todorova et al. (2000) that durative adverbials could trigger a range of different event extension interpretations when combined with telic predicates. In most cases, the most natural way to resolve the mismatch would be to interpret the telic event as occurring iteratively, throughout the time span specified by the durative adverbial (*John sent a letter for years* = *John sent letter after letter for years*). However, it is also possible to resolve the mismatch by stretching the duration of a single event beyond its natural time span (*John sent a letter for years* = *It took years for John to send a letter*). These different solutions to the mismatch, which could vary from sentence to sentence and participant to participant, could plausibly introduce a source of noise into the data that Pickering et al. sought to avoid. Frequency adverbials do not allow the extended event interpretation, and thus might seem like a better tool to probe for online event interpretations. Pickering et al. argue in motivating their choice of frequency adverbials that 'The strongest test of whether an aspectual clash causes difficulty involves items with frequency modifiers (e.g. every year) because they avoid the potential ambiguity introduced by durative modifiers (e.g. for many years)' (Pickering et al. 2006, p. 145).

And indeed, the resulting interpretation in (20a) is very similar to that in (15a): Howard engaged in multiple events of letter sending, distributed over years. However, Pickering et al. failed to find any significant costs associated with this interpretation as compared to a control condition with a modifier such as last year (20c), or a durative VP such as *send checks* (20d).

Pickering et al. conclude that the failure to find processing costs suggests that aspectual interpretations are not computed incrementally, but may instead be underspecified. This result conflicts with the previous and subsequent findings of highly incremental aspectual interpretation. However, we argue that the lack of a cost for (20a) is actually expected given the semantics of universal event quantification (Rothstein 1995).

8.2.4 Universal Event Quantification

Rothstein (1995) considers sentences such as (21) in which a universally quantified temporal adverbial imposes a matching relationship on two events.

(21) I met a friend every time I went to the bakery.

Following Boolos (1981), she notes that sentences of this type have the general structure in (22), and that sentences with this structure are truth-conditionally equivalent to 'There are at least as many Bs as As': every event of going to the bakery must be matched with a unique event of meeting a friend for sentence (21) to be true.

(22) For every A, there is a B.

Rothstein (1995) argues that not only do such sentences impose a matching requirement, such that every A event has a corresponding B event, but she also argues that each B event must be a distinct B event, and that this distinctness requirement is determined by the grammar, not by pragmatics. She points to the contrast between the adverbial use of *every time* in (23a) and the nominal use of *every time* in (23b).

(23) a. I regretted it every time I had dinner with John.
 b. I regret every time I had dinner with John.

The sentence in (23a) requires that there be a unique regretting event for each 'having dinner with John' event, while (23b) makes no such requirement. If pragmatic inference were responsible for the uniqueness interpretation in (23a) it is hard to see why the same inference would not be generated for (23b). Furthermore, this requirement is not cancellable, as would be expected if it were the result of pragmatic inference, as illustrated by the contrast in (24) (Rothstein's 24 & 25).

(24) a. Every girl (and there were many of them) saw a movie last night. In fact they all saw Aladdin.
 b. #Every time the bell rang last night (and it rang many times) Mary opened the door. In fact Mary only opened the door once.

Rothstein notes that (24b) is incoherent, since it asserts the existence of many bell ringings, and at least as many door openings as bell ringings, yet also asserts only a single door opening event. In contrast, the implication of multiple movies set up by *every girl saw* in (24a) is easily canceleable. Therefore, the requirement that each bell ringing event be associated with a door opening event in (24b) must be the direct result of grammatical computation, not merely an inference.

In order to account for these facts about the interpretation of sentences involving universal quantification over a temporal interval, Rothstein proposes that phrases such as *every time the bell rang last night* are actually prepositional phrases, headed by a null preposition *M* which is a function from the set of events e' onto the set of events e, such that every e' maps to a distinct e. Nothing crucially hinges on the function *M* being a preposition, rather than some other functional head—what is crucial is that Rothstein's analysis provides an account of universal event quantification over telic events that does not involve any mismatch or coercion.

Rothstein's analysis of matching between events can be extended to matching between events and time span nominals like *year*, *day* or *minute* if we allow M to also establish pairwise matching between events and temporal intervals as in (25).

(25) a. Howard wrote a letter every year.
 b. [Howard [wrote a letter]] [M [every year]]
 c. Every interval of a year has a corresponding event of Howard writing a letter.

Thus the iterated event interpretation that results in sentences with universally quantified temporal modifiers is, we argue, the result of straightforward semantic composition + the generation of an implicature, and not as the result of a mismatch between the requirements of an adverbial and the properties of a verb phrase. We have in mind, here, the kind of neo-Gricean conversational implicature that leads us to interpret *I marked some of the assignments* as *I did not mark all of the assignments*. When comprehenders hear *Howard wrote a letter every year*, they assume that the speaker included *every year* for a reason (Gricean maxims of relevance and informativity (Grice 1975)), and that if they could have truthfully uttered *Howard wrote a letter* they would have done so, so the hearer can reasonably conclude that the speaker has multiple years in mind. Substantial discussion of the semantics, pragmatics and processing of implicatures is beyond the scope of the current chapter. See, among others, Chemla and Singh (in press); Bott and Noveck (2004); Bott et al. (2012); Grodner et al. (2010); Hartshorne and Snedeker (submitted); Huang and Snedeker (2009); Papafragou and Musolino (2003) and Sikos et al. (2013) for recent relevant work on the topic. What is critical to note is that while generating and processing conversational implicatures is generally found to be costly, there is no reason to expect that generating the implicature that *every year* implies a plurality of years should be more difficult in a sentence containing a telic VP than in one containing an atelic VP. Sentences with universal quantification over temporal intervals may be more costly to fully process than those with adverbials that do not license scalar implicatures, but we do not expect a difference between sentences in which an adverbial like *every year* is matched with a terminative event and those where it is matched with an unbounded event. And, indeed, Pickering et al. (2006) find no such differences[10].

Thus, there are no true counter examples to the otherwise robust evidence that commitment to aspectual interpretations happens quite quickly during online sentence processing. The results we have discussed are all consistent with the predictions

---

[10] Todorova et al. (2000) report an Experiment 2, which includes frequency adverbials (the source of Pickering et al's materials). Using the same self-paced stop-making-sense reading task as their Experiment 1 (above), they found that the cost for processing a frequency adverbial (*every years*) following a telic VP (*sent letters*) were significantly greater than for the same adverbial following an atelic VP (*sent a letter*), but that the magnitude and duration of the effect were much smaller for the frequency adverbials than for the durative (*for years*).

Paczynski et al. (2014) also consider frequency adverbials. In addition to the punctual vs. durative adverbial manipulation (14), they also included frequency adverbials as in (26).

(26) a. Several times, the cat pounced on the rubber mouse. [FA-PV]
b. Several times, the cat prowled about the backyard. [FA-DV]

Critically, Paczynski et al used adverbials such as *several times*, which assert the existence of multiple, repeated, events, rather than the universally quantified, morphologically singular adverbials used by Todorova and Pickering and their colleagues. Paczynski et al. were interested in distinguishing between possible explanations for what exactly it is about the durative adverbial + telic event sentences that triggers processing costs (the initial aspectual mismatch? the repeated event interpretation solution?), and the frequency adverbials they chose to use were appropriate for their purposes, but not so useful for ours.

from linguistic theory that the VP is the domain of event and aspect interpretation. However, all of these studies have relied on indirect processing consequences of aspectual interpretation. We present data from four experiments that explore the initial processing of the verb and VP themselves to further test the VP domain of aspectual interpretation hypothesis.

8.3 Direct Measurements of VP Composition

Stockall and Husband (2014) reports the results of two self-paced reading experiments designed to investigate the time course in which aspectual interpretations are initially constructed and the relative costs of terminative vs. durative interpretations. In this 2014 chapter we focused on the question of how different types of verbs are lexically specified for aspectual features, but the first experiment also speaks critically to the issue of the syntactic domain of aspectual interpretation. We summarize the results of this experiment below.

8.3.1 Exp 1: Self-Paced Reading Experiment

We investigated verbs such as *watch*, *run*, *host*, *drink*, *build*, etc. that are lexically unspecified for telicity. When they are combined with direct objects with count semantics, the resulting event has a specific end-point (27b and 27d), but when they are combined with mass direct objects, the resulting event is unbounded, with no necessary end-point (27a and 27c).

(27) a. John drank beer #in ten minutes.
b. John drank the beer in ten minutes.
c. John built planes #in eight hours.
d. John built the plane in eight hours.

We conducted a self-paced reading study specifically investigating the individual contributions of the verb and the direct object to aspectual interpretations. We constructed materials which paralleled the calculus of events in Table 8.1 . Sentences varied in verb class (terminative or unspecified) and NP determiner (definite [count] or null [non-count]) as in (28).

(28) a. The expert physicist lost the files on the formation of black holes. [terminative/definite] (telic)
b. The expert physicist lost files on the formation of black holes. [terminative/null] (telic)
c. The expert physicist read the files on the formation of black holes. [unspecified/definite] (telic)
d. The expert physicist read files on the formation of black holes. [unspecified/null] (atelic)

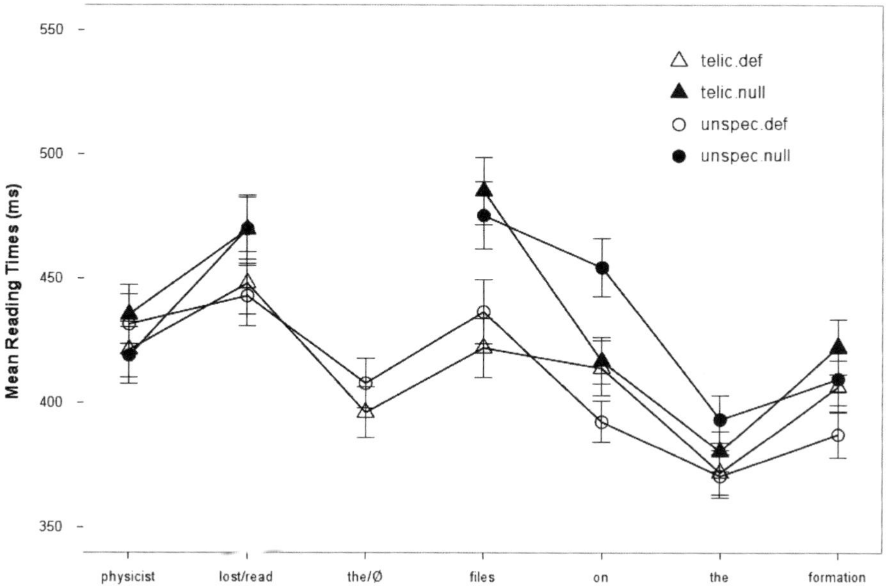

Fig. 8.3 Word by word reading times from Stockall and Husband (2014)

Verbs were classified into bounded or unspecified using Dowty's (1979) tests and rated for acceptability with *in X time* modification. Verbs which were judged acceptable with end point modification were classified as terminative, while verbs which were judged acceptable with *in X time* modification when they took count direct objects, but unacceptable with non-count objects were classified as unspecified. Sixty native English speaking participants performed a self-paced word-by-word moving window reading task. After each sentence, participants were asked to rate the acceptability of sentences on a 1 to 5 scale (5 = good).

We found a main effect of NP determiner on the noun position ($z = 4.97$, $p < 0.0001$)[11], and an interaction between verb class and NP determiner at the noun+1 position ($z = 3.15$, $p = 0.002$). Planned comparisons between means on the noun+1 position revealed significant differences between unspecified definite and unspecified null conditions (28c vs. 28d) and terminative null and unspecified null conditions (28c vs. 28d). Word by word reading times are shown in Fig. 8.3.

The initial main effect of determiner type on noun reading time is not surprising. The overt definite determiner plausibly provides a highly reliable cue that the upcoming word is likely to be a noun, while no such cues are available for nouns without

[11] All statistics reported in this chapter were generated using linear mixed effects modeling techniques, with subjects and items as random factors (Baayen et al. 2008). Model comparison techniques were employed, and the statistics we report are for the best fitting model that converged (Barr et al. 2013). More details about the model and its parameters are in the chapter.

a determiner. Crucially, this initial main effect is very short lived. By the time the next word is encountered, the terminative null sentences no longer evoked different responses from the terminative definite and unspecified definite sentences, while the slowdown for the unspecified null sentence persisted. This is of interest to aspectual interpretation since the unspecified null condition is the only condition which triggers an atelic interpretation; all other conditions trigger a telic interpretation.

This sustained slowdown in the durative event sentences relative to terminative event sentence suggests that properties of both the direct object and the verb are used immediately in aspectual interpretation. Unspecified verbs combined with non-count objects (*build planes*) are associated with significant reading time slow downs as compared to the same verb with a count object (*build the planes*), or the same non-count object with a terminative verb (*crash planes*). To our knowledge, this is the first evidence of a processing correlate directly associated with aspectual composition specifically and compositional semantics more generally that does not rely on coercion due to a mismatch effect (Brennan and Pylkkänen 2008; Piñango et al. 1999; Paczynski et al. 2014; Todorova et al. 2000) or a 'noncompositional' structure (McElree et al. 2001, 2006; Pylkknen and McElree 2007). Interestingly, no significant main effects of verb class were revealed during online sentence processing. This could be evidence that telicity is only computed for full VPs, but it could also be that the lexical verb effects are smaller than the compositional effects and are simply lost in the background in the self-paced reading study. We tested this second possibility in a lexical decision experiment.

8.3.2 Exp 2: Lexical Decision Experiment A

This experiment used the infinitival versions of the verbs in experiment 1 in a single word lexical decision experiment designed to see whether verbs lexically specified for terminative aspect would evoke different processing responses from verbs underspecified for aspect when they were processed in isolation with no sentential context.

Materials consisted of the 48 inherently telic verbs and 48 unspecified verbs from Stockall and Husband (2014), and 96 length-matched pronounceable nonwords from the ARC Nonword Database (Rastle et al. 2002). The verbs were matched for length, lexical frequency, orthographic, and phonological neighborhood density, mean bigram frequency and number of syllables. Fillers consisted of 218 words and 218 nonwords from a separate study.

Sixty native English-speaking students from Michigan State University participated. Each participant saw all stimuli. Following Gennari and Poeppel (2003), words were presented with varying inter-trial times (500–1500 ms) at the center of the screen. Before each stimulus item, a fixation cross was presented for 500 ms. Participants indicated their response via keyboard button press.

Although just such a lexical decision paradigm was successful in uncovering different processing responses to stative vs. eventive verbs (Gennari and Poeppel 2003),

Table 8.2 Reaction times for lexical decision Exp A

Verb Type	M	SE
Unspecified	610	9.0
Telic	614	9.0

and has been argued to be more sensitive than self-paced reading to effects of verbal semantics (Balota 1994), we found no differences in either error rate ($t = 0.999$, $p = 0.323$) or response time ($t = 1.323$, $p = 0.191$) between terminative and underspecified verbs (Table 8.2) . This result suggests that initial lexical processing of terminative verbs and unspecified verbs does not differ, consistent with the findings in the verb region from the self-paced reading experiment. Further evidence supportive of this finding that aspectual interpretations are not generated/parsed at the level of the lexical verb comes from Paczynski et al. (2014), discussed above, who find no differences in evoked response potentials between punctual verbs (*pounce*) and durative/activity verbs (*prowl*).

We have so far found, then, that durative events take longer to process than terminative events. The direction of this effect is predicted by the model-theoretic approaches to aspect outlined in (§ 1.1.1). Durative events are argued to be representationally more complex than terminative events. The observed processing cost in the self-paced reading experiment is thus support for this kind of analysis of event semantics. This effect does not seem to be the direct result of lexical semantic differences between verbs, but instead arises from the composition of the event information supplied by the verb with the event information supplied by the internal argument.

These results also point to differences between terminative and unspecified verb classes. While there were no significant differences between terminative and unspecified verbs in lexical decision, we do see clear differences between them in the interaction of verbal semantics with internal argument semantics in sentence reading. Given both these effects, the lack of any effect in lexical decision is somewhat surprising. One possibility is that infinitival verbs do not project a VP when processed in isolation and aspectual differences require that full VP. Another possibility is that lexical decision is only sensitive to the semantic properties of aspect and not sensitive to the syntactic properties of aspect present in verbs. These two options were tested in two further experiments.

8.3.3 Evidence for Syntactic Complexity?

The above experiments demonstrate effects of semantic complexity due to aspectual interpretation of the VP. Self-paced reading shows semantic complexity effects upon completion of the VP but neither lexical decision measures nor self-paced reading show effects of complexity for verbs themselves. Also, no evidence has been found for the syntactic complexity predicted by the syntax of aspect. In no case have terminative verbs been shown to be more difficult to process compared to verbs with no event specified semantic properties.

Table 8.3 Reaction times for lexical decision Exp B

VERB TYPE	Verb form	M	SE
Unspecified	Infinitival	694.77	259.77
Unspecified	Past	711.14	254.65
Telic	Infinitival	668.37	226.34
Telic	Past	709.73	248.38

Two further experiments tested the syntactic event properties of terminative and unspecified verbs. First, a lexical decision task was constructed to test for effects of the syntactic presence of AspP. According to syntactic theories of aspect, AspP is projected between the verb and the functional projection for tense, TP. To parse a tense morpheme, the parser projects a TP, but, in doing so, the parser may be required to also project AspP following the hierarchy of functional projections (Borer 2005b). If the presence of tense on verbs requires the projection of AspP as well as TP, we may expect to see processing costs associated with its projection when processing terminative verbs which assign their event properties to AspP, compared to unspecified verbs which have no event properties to assign to AspP and may not project AspP at all.

8.3.3.1 Exp 3: Lexical Decision Experiment B

Forty subjects were run in a lexical decision task which manipulated both verb class (terminative vs. unspecified) and tense (past vs. infinitive tense) using the same procedure, verbs, and filler items as above. We found a main effect for tense ($t = -1.04, p < 0.001$)[12], with past tense forms triggering longer reaction times than the infinitive forms, which may simply reflect the length difference between the two conditions. However, we found no effect of verb class, and no interaction. While no statistically significant differences were found, the difference in processing costs associated with adding tense to a terminative verb was 24.99 ms more than that of adding tense to an unspecified verb (terminative: 41.36 ms; unspecified: 16.37 ms) (Table 8.3). Though non-significant, this effect trends in the expected direction, suggesting that a more sensitive measure may find significant differences between our verb classes.

A second study further tested for the syntactic complexity of aspect using MEG in a sentence reading paradigm. While lexical decision and other behavioral studies have failed to find significant differences between our verb classes, MEG may be more sensitive to the early effects of syntactic projection of AspP.

[12] As above in Experiment 1, we employed linear mixed modeling and model comparison techniques and report the statistics of the best fitting model.

Table 8.4 Mean sensor activity by quadrant in 200–400 ms window

VERB TYPE	LEFT ANTERIOR	POSTERIOR	RIGHT ANTERIOR	POSTERIOR
Unspecified	0.88	2.56	– 2.58	– 1.16
Telic	2.28	1.68	– 1.01	– 3.57

8.3.3.2 Exp 4: MEG Experiment

Twelve subjects were run in a word-by-word sentence comprehension study using the sentences from Experiment 1 above. Sentences were presented one word at a time, with each word centred on the screen. The duration of each word was determined by its length. One character words were presented for 280 ms, and each additional character increased the stimulus duration by $16.\overline{6}$[13]. Analyses were time locked to the onset of the verb. An analysis of the average sensor activity was done over four quadrants using several time windows which encompassed known MEG components in visual word recognition studies, including one from 270–400 ms, the time window associated with the M350 response (Pylkkänen and Marantz 2003). Analysis of the 200–400 ms time window revealed a significant interaction of lexical verb type by anteriority ($t = 2.662$, $p = 0.0078$), with a trending effect of telic verbs eliciting more positive activity in the anterior hemisphere ($t = 1.678$, $p = 0.0933$) and more negative activity in the posterior hemisphere ($t = -1.814$, $p = 0.0697$). However, within the left hemisphere, no significant difference in lexical telicity was observed for either the anterior ($t = 1.156$, $p = 0.2476$) or the posterior ($t = -1.000$, $p = 0.3173$) quadrant, and within the right hemisphere, no significant difference in lexical telicity was observed for either the anterior ($t = 1.233$, $p = 0.2176$) or posterior ($t = -1.533$, $p = 0.1253$) quadrant (Table 8.4).

Overall then, these two experiments provide no compelling processing correlate for the increased syntactic complexity hypothesized for telic representations. The non-significant trends we observe for increased processing costs associated with syntactic complexity (RT slow downs and increased MEG activation) are consistent with those models, andk do suggest that further experimentation to find more reliable effects would be worthwhile.

8.4 Conclusions

Half a century of research on the linguistics of event interpretations has established a general consensus concerning the syntactic and semantic representation of telicity. Aspectual interpretation involves a complex system with information distributed over several sentential constituents. To license telicity, these constituents compose together at the structural level of the VP. Interpretation of aspect then proceeds by

[13] Equivalent to one screen refresh on a 60 Hz monitor.

denoting event models with different semantic properties. These syntactic and semantic representations have been used in this chapter to generate predictions concerning the processing of aspect, to which we now turn.

8.4.1 What is the Domain Over Which the Parser Interprets Aspect?

Concerning the processing domain of aspect, our results argue that the domain of aspectual interpretation is the VP constituent. Evidence from our self-paced reading results and the literature on aspectual coercion both point to the role of the VP as the relevant domain for aspectual interpretation. Studies showing earlier effects of event semantic processing (Gennari and Poeppel 2003; McKoon and Macfarland 2002) may be attributable to recovery of features which drive the syntactic parse rather than to commitment of the parser to an aspectual interpretation. This domain also has theoretical weight in linguistic theory, suggesting that the grammar places strong constraints on the decision points at which the parser makes a commitment to interpretation. This is not unlike other interpretative decisions in which the grammar constrains the timing and range of possible decisions the parser must make (Frazier 1999; Dickey 2001).

8.4.2 What Factors Guide the Parser in Aspectual Interpretation Online?

Having surveyed the ingredients of aspectual interpretation and their use online, we affirm the importance of both a verb's event semantics and the count/mass syntax of the internal argument. Both elements play an important role in determining telicity and both are shown here to have consequences for online processing. Much of aspectual processing relies especially on the verb's event semantics, though as demonstrated in several studies, the properties of the internal argument are rapidly composed with those of the verb to yield aspectual interpretations.

8.4.3 Does the Parser Immediately Commit to an Aspectual Interpretation?

In response to our final question about the time course of aspectual interpretation, experimental evidence points to a parser that makes commitments to aspectual interpretation immediately upon parsing the aspectual domain, i.e. the VP. Several studies have demonstrated immediate effects of aspectual interpretation in sentence processing, and evidence for delayed or underspecified aspectual interpretation are likely due to experimental confounds. We also note that the verb constituent alone appears to be unable to license an aspectual interpretation. A highly incremental parser which performs immediate full interpretation should commit to an aspectual interpretation upon encountering an unambiguous verb. However, our evidence suggests that the parser delays commitment to aspectual interpretation until it has parsed the full VP in accord with the grammar, as per the principle of immediate partial interpretation (Frazier 1999).

The results of these studies suggest a two stage model of aspectual processing. In the first stage, verbal and nominal properties license the construction of the VP and project AspP if needed due to either verbal event features or nominal properties. Verbal semantic features alone cannot trigger the projection of AspP, though the hierarchy of functional projections requires that in the presence of tense a verb with event semantics must project AspP, mapping to an event model awaits parsing of a full VP. In the second stage, the parser commits to an aspectual interpretation based on the syntactic structure arrived at through the first stage and constructs an event model with the right structure. If AspP was projected either for reasons of a verb's event semantics, or because of the presence of a count NP, a non-homogeneous model is constructed. If AspP was not projected because the verb had no event semantics and its internal argument was not a count NP, a homogeneous model in constructed. Taken together, this research provides detailed evidence concerning the processing of aspect specifically, and the processing of compositional structures more generally. These results also have important consequences for theories of sentence processing. They continue to argue for incremental commitment to aspectual interpretation, placing the commitment point for telicity at the VP, which is the first point when all the information needed to construct an aspectual interpretation has been provided to the system (i.e. both the verb and the internal argument).

References

Amsel, B. D., Urbach, T. P., & Kutas, M. (2013). Alive and grasping: Stable and rapid semantic access to an object category but not object graspability. *Neuroimage, 77,* 1–13.

Baayen, R., Davidson, D., & Bates, D. (2008). Mixed-effects modeling with crossed random effects for subjects and items. *Journal of Memory and Language, 59*(4), 390–412.

Bach, E. (1986). The algebra of events. *Linguistics and Philosophy, 9,* 5–16.

Bacrach, A., Roy, I., & Stockall, L. (2014). Introduction: Argumenting the structure. In A. Bacrach, I. Roy, & L. Stockall (Eds.), *Structuring the argument* (pp. 1–22). John Benjamins.

Balota, D. A. (1994). Visual word recognition: The journey from features to meaning. In M. A. Gernsbacher (Ed.) *Handbook of psycholinguistics* (pp. 303–358). San Diego: Academic.

Barr, D. J., Levy, R., Scheepers, C., & Tily, H. J. (2013). Random effects structure for confirmatory hypothesis testing: Keep it maximal. *Journal of Memory and Language, 68*(3), 255–278.

Bennett, M., & Partee, B. H. (1978). *Toward the logic of tense and aspect in English.* Bloomington: Indiana University Linguistics Club.

Boolos, G. (1981). For every A there is a B. *Linguistic Inquiry, 12*(3), 465–467.

Borer, H. (1994). The projection of arguments. In E. Benedicto & J. Runner (Eds.) *Occasional papers in linguistics* (Vol. 17, pp. 19–47). University of Massachusetts: Amherst.

Borer, H. (1998). Deriving passive without theta roles. In S. Lapointe., P. Farrell, & D. Brentari (Eds.) *Morphology and its relation to phonology and syntax* (pp. 60–99). Stanford: CSLI.
Borer, H. (2005a). *In name only*. Oxford: Oxford University Press.
Borer, H. (2005b). *The normal course of events*. Oxford: Oxford University Press.
Bott, O. (2008). Doing it again and again may be difficult, but it depends on what you are doing. In *Proceedings of the 27th West Coast Conference on Formal Linguistics* (p. 63). Cascadilla Proceedings Project.
Bott, L., & Noveck, I. A. (2004). Some utterances are underinformative: The onset and time course of scalar inferences. *Journal of Memory and Language, 51*(3), 437–457.
Bott, L., Bailey, T. M., & Grodner, D. (2012). Distinguishing speed from accuracy in scalar implicatures. *Journal of Memory and Language, 66*(1), 123–142.
Brennan, J., & Pylkkänen, L. (2008). Processing events: Behavioral and neuromagnetic correlates of aspectual coercion. *Brain and Language, 106,* 132–143.
Chemla, E., & Singh, R. (in press). *Remarks on the experimental turn in the study of scalar implicature*. Language and Linguistics Compass.
Chemla, E., & Singh, R. (2014). Remarks on the experimental turn in the study of scalar implicature. *Language and Linguistics Compass*.
Chomsky, N. (1965). *Aspects of the theory of syntax*. Massachusetts: MIT.
Dickey M. W. (2001). *The processing of tense*. Dordrecht: Kluwer.
Dowty, D. (1979). *Word meaning and Montague grammar: The semantics of verbs and times in generative semantics and in Montague's PTQ*. Berlin: Springer.
Dowty, D. (1991). Thematic proto-roles and argument selection. *Language, 67*(3), 547–619.
Fodor, J. A., & Garrett, M. (1967). Some syntactic determinants of sentential complexity. *Perception & Psychophysics, 2*(7), 289–296.
Frazier, L. (1999). *On sentence interpretation*. Dordrecht: Kluwer.
Frazier, L., Carminati, M., Cook, A., Majewski, H., & Rayner, K. (2006). Semantic evaluation of syntactic structure: Evidence from eye movements. *Cognition, 99*(2), 53–62.
Friederici, A. D. (2002). Towards a neural basis of auditory sentence processing. *Trends in cognitive sciences, 6*(2), 78–84.
Gennari, S., & Poeppel, D. (2003). Processing correlates of lexical semantic complexity. *Cognition, 89*(1), 27–41.
Grice, H. P. (1975). Logic and conversation. In P. Cole & J. J. Morgan (Eds.), *Syntax and semantics* (Vol. 3, pp. 41–58). New York: Academic.
Grodner, D. J., Klein, N. M., Carbary, K. M., & Tanenhaus, M. K. (2010). Some, and possibly all, scalar inferences are not delayed: Evidence for immediate pragmatic enrichment. *Cognition, 116*(1), 42–55.
Hale, K., & Keyser, S. (1993). On argument structure and the lexical expression of syntactic relations. In K. Hale & S. Keyser (Eds.), *The view from building 20* (pp. 53–109). The MIT Press.
Harris, J., Pylkkänen, L., McElree, B., & Frisson, S. (2008). The cost of question concealment: Eye-tracking and meg evidence. *Brain and Language, 107*(1), 44–61.
Hartshorne, J. K., & Snedeker, J. (submitted). The speed of inference: Evidence against rapid use of context in calculation of scalar implicatures.
Huang, Y. T., & Snedeker, J. (2009). Online interpretation of scalar quantifiers: Insight into the semantics–pragmatics interface. *Cognitive Psychology, 58*(3), 376–415.
Husband, E. M., Kelly, L. A., & Zhu, D. C. (2011). Using complement coercion to understand the neural basis of semantic composition: Evidence from an fmri study. *Journal of cognitive neuroscience, 23*(11), 3254–3266.
Jackendoff, R. (1997). *The architecture of the language faculty*. Cambridge: The MIT Press.
Kiparsky, P. (1998). Partitive case and aspect. In M. Butt & W. Geuder (Eds.), *The Projection of Arguments: Lexical and Compositional Factors*, (pp. 265–307). Stanford: CSLI.

Krifka, M. (1992). Thematic relations as links between nominal reference and temporal constitution. In I. Sag & A. Szabolcsi (Eds.), *Lexical Matters* (pp. 29–53). Cambridge: Cambridge University Press.
Krifka, M. (1998). The origins of telicity. In S. Rothstein (Ed.) *Events and grammar* (pp. 197–235). Dordrecht: Kluwer Academic Publishers.
Lau, E. F., Phillips, C., & Poeppel, D. (2008). A cortical network for semantics:(de) constructing the n400. *Nature Reviews Neuroscience, 9*(12), 920–933.
Link, G. (1983). *The logical analysis of plurals and mass terms: A lattice-algebraic approach. Meaning, Use and Interpretation of Language* (pp. 303–323). Berlin: De Gruyter.
MacDonald, J. E. (2008). Domain of aspectual interpretation. *Linguistic Inquiry, 39*(1), 128–147.
MacDonald, J. (2010). The aspectual influence of the noun: (a)telicity, (a)symmetry, incrementality and universality. *Language and Linguistics Compass, 4*(9), 831–845.
Marslen-Wilson, W., & Tyler, L. (1980). The temporal structure of spoken language understanding. *Cognition, 8*(1), 1–71.
McElree, B., Traxler, M., Pickering, M., Seely, R., & Jackendoff, R. (2001). Reading time evidence for enriched composition. *Cognition, 78*(1), 17–25.
McElree, B., Pylkkänen, L., Pickering, M., & Traxler, M. (2006). A time course analysis of enriched composition. *Psychonomic Bulletin & Review , 13*(1), 53–59.
McKoon, G., & Macfarland, T. (2002). Event templates in the lexical representations of verbs. *Cognitive Psychology, 45*(1), 1–44.
Mittwoch, A. (1991). In defense of Vendler's achievements. *Belgian Journal of Linguistics, 6,* 71–84.
Müller, O., & Hagoort, P. (2006). Access to lexical information in language comprehension: Semantics before syntax. *Journal of cognitive neuroscience, 18*(1), 84–96.
Paczynski, M., Jackendoff, R., & Kuperberg, G. (2014). When events change their nature: The neurocognitive mechanisms underlying aspectual coercion. *Journal of Cognitive Neuroscience, 26*(9), 1905–1917.
Papafragou, A., & Musolino, J. (2003). Scalar implicatures: Experiments at the semantics–pragmatics interface. *Cognition, 86*(3), 253–282.
Parsons, T. (1990). *Events in the semantics of English: A study in subatomic semantics*. Cambridge: MIT Press.
Phillips, C. (1996). *Order and structure*. Ph.D. thesis, Massachusetts Institute of Technology.
Pickering, M., McElree, B., & Traxler, M. (2005). The difficulty of coercion: A response to de almeida. *Brain and Language, 93*(1), 1–9.
Pickering, M., McElree, B., Frisson, S., Chen, L., & Traxler, M. (2006). Underspecification and aspectual coercion. *Discourse Processes, 42*(2), 131–155.
Piñango, M., & Deo, A. (2012). Aspectual verbs and the "coercion" effect. In *Proceedings of Semantics and Linguistic Theory*, 22.
Piñango, M., Zurif, E., & Jackendoff, R. (1999). Real-time processing implications of enriched composition at the syntax–semantics interface. *Journal of Psycholinguistic Research, 28*(4), 395–414.
Piñango, M., Winnick, A., Ullah, R., & Zurif, E. (2006). Time-course of semantic composition: The case of aspectual coercion. *Journal of Psycholinguistic Research, 35*(3), 233–244.
Proctor, A. S., Dickey, M. W., & Rips, L. J. (2004). The time-course and cost of telicity inferences. In *Proceeding of the 26th Annual Meeting of the Cognitive Science Society*, Chicago.
Pustejovsky, J. (1995). *The generative Lexicon*. Cambridge: The MIT Press.
Pylkkänen, L., & Marantz, A. (2003). Tracking the time course of word recognition with MEG. *Trends in Cognitive Sciences, 7*(5), 187–189.
Pylkkänen, L., & McElree, B. (2007). An meg study of silent meaning. *Journal of Cognitive Neuroscience, 19*(11), 1905–1921.
Pylkkänen, L., Martin, A., McElree, B., & Smart, A. (2009). The anterior midline field: Coercion or decision making? *Brain and language, 108*(3), 184–190.

Ramchand, G. (1997). *Aspect and Predication: The Semantics of Argument Structure*. Oxford: Oxford University Press.

Ramchand, G. (2008). *Verb meaning and the lexicon*. Cambridge: Cambridge University Press.

Rappaport Hovav, M., & Levin, B. (1998). Building verb meanings. In M. Butt & W. Geuder (Eds.), The projection of arguments: Lexical and compositional factors (pp. 97–134). Stanford: CSLI Publications.

Rastle, K., Harrington, J., & Coltheart, M. (2002). 358,534 nonwords: The arc nonword database. *The Quarterly Journal of Experimental Psychology Section A, 55*(4), 1339–1362.

Ritter, E., & Rosen, S. (1998). Delimiting events in syntax. In M. Butt & W. Geuder (Eds.), *The projection of arguments: Lexical and compositional factors* (pp. 135–164). Stanford: CSLI Publications.

Rosen, S. T. (1999). The syntactic representation of linguistic events. *Glot International, 4*(2), 3–11.

Rothstein, S. (1995). Adverbial quantification over events. *Natural Language Semantics, 3*(1), 1–31.

Rothstein, S. D. (2004). *Structuring events: A study in the semantics of lexical aspect*. Malden: Blackwell Publishing.

Rothstein, S. (2008). *Structuring events: A study in the semantics of aspect* (Vol. 5). New York: Wiley.

Schmitt, C. (1996). *Aspect and the Syntax of Noun Phrases*. Ph.D. thesis, University of Maryland.

Sikos, L., Tomlinson, S., Traut, H., & Grodner, D. (2013). *Incremental computation of scalar implicatures: An ERP study*. Poster presented at the 26th annual CUNY Sentence Processing Conference (Columbia, SC).

Smith, C. S. (1997). *The Parameter of Aspect*. Dordrecht: Springer.

Stockall, L., & Husband, E.M. (2014). Processing (the) events: Lexical and structural ingredients of inner aspect. In C. T. Schutze & L. Stockall (Eds.), Connectedness: Papers by and for Sarah VanWagenen. UCLA Working Papers in Linguistics, 18 (pp. 275–291).

Stockall, L., & Marantz, A. (2006). A single route, full decomposition model of morphological complexity: MEG evidence. *The Mental Lexicon, 1*(1), 85–123.

Tenny, C. L. (1987). *Grammaticalizing Aspect and Affectedness*. Ph.D. thesis, Massachusetts Institute of Technology.

Tenny, C. L. (1994). *Aspectual Roles and the Syntax-Semantics Interface*. Berlin: Springer.

Thompson, E. (2006). The structure of bounded events. *Linguistic Inquiry, 37*(2), 211–228.

Todorova, M., Straub, K., Badecker, W., & Frank, R. (2000). Aspectual coercion and the online computation of sentential aspect. *Proceedings of the 22nd Annual Conference of the Cognitive Science Society* (pp. 3–8).

Traxler, M., Pickering, M., & McElree, B. (2002). Coercion in sentence processing: Evidence from eye-movements and self-paced reading. *Journal of Memory and Language, 47*(4), 530–547.

Traxler, M., McElree, B., Williams, R., & Pickering, M. (2005). Context effects in coercion: Evidence from eye movements. *Journal of Memory and Language, 53*(1), 1–25.

van Hout, A. (1996). *Event semantics of verb frame alternations: A case study of Dutch and its acquisition*. Ph.D. thesis, Tilburg University.

van Hout, A. (2000). Projection Based on Event Structure. *Amsterdam studies in the theory and history of linguistic science. Series IV, Current issues in linguistic theory,* (197), 403–428.

Vendler, Z. (1957). Verbs and times. *Philosophical Review, 66*(2), 143–160.

Verkuyl, H. J. (1972). *The Compositional Nature of the Aspects* (Vol. 15 of Foundations of Language Supplementary Series). Dordrecht: Reidel.

Verkuyl, H. J. (1987). Nondurative closure of events. In J. Groenendijk, M. Stokhof, & F. Veltman (Eds.), *Studies in discourse representation theory and the theory of generalized quantifiers* (pp. 87–113). Amsterdam: Institute for Language, Logic and Computation.

Verkuyl, H. J. (1993). *A theory of aspectuality: The interaction between temporal and atemporal structure*. Cambridge University Press.

Part IV
Meaning and Structure: Representation and Processing

Chapter 9
Visual and Motor Features of the Meanings of Action Verbs: A Cognitive Neuroscience Perspective

David Kemmerer

9.1 Introduction

The linguistics literature contains several different theories about the nature of verbs. During the past few decades, however, there has been an increasing convergence toward the view that verb meanings have two separate levels of semantic structure (Levin and Rappaport Hovav 2005, 2011). One level of meaning reflects the uniqueness of every verb and has been dubbed the "root" or "constant" (Levin and Rappaport Hovav 1998) because it captures idiosyncratic semantic features that (a) distinguish each verb in a given class from all the others, (b) are often concrete and fine-grained in content, and (c) do not interface with grammar. The other level of meaning consists of a more austere representation that is referred to variously as the "event structure template" (Rappaport Hovav and Levin 1998), the "thematic core" (Pinker 1989), or the "logical structure" (Van Valin 2005), and that is (a) common to all the verbs in a given class, (b) composed primarily of schematic predicates and variables for arguments, and (c) relevant to the grammatical behavior, especially the syntactic argument structure possibilities, of all the verbs in a given class.

Consider, for example, so-called running verbs (Levin 1993, pp. 265–267), which encode different ways in which animate entities, typically humans, locomote (e.g., *run, walk, strut, sashay, trudge,* etc.). All of the verbs in this class share the same simple template—roughly [x ACT < *MANNER OF LOCOMOTION* >]—but they differ with respect to the unique roots that instantiate the "manner" component of that template. In other words, the template represents a schematic or skeletal event structure in which an agent performs a generic action of the locomotion type, with

D. Kemmerer (✉)
Department of Speech, Language, and Hearing Sciences,
Purdue University, 715 Clinic Drive, West Lafayette, IN 47907-2038, USA
Tel.: (765) 494-3826
e-mail: kemmerer@purdue.edu

Department of Psychological Sciences, Purdue University,
West Lafayette,IN, USA

R. G. de Almeida, C. Manouilidou (eds.), *Cognitive Science Perspectives on Verb Representation and Processing,* DOI 10.1007/978-3-319-10112-5_9

the details left unspecified. The multifarious verbs that collectively comprise the running class share this semantic framework, but the unique root of each one fleshes out the "manner" variable in an idiosyncratic way by specifying prototypical values, or ranges of values, for a complex array of conceptual parameters, including visual pattern, motor pattern, rate, effort, social/emotional significance, etc. Thus, as Pinker (2007, p. 83) puts it, "basic conceptual distinctions assemble themselves into a scaffolding of meaning [at the template level], which has hooks here and there on which to hang images, sounds, emotions, mental movies, and the other contents of consciousness [at the root level]."

It is notable that although some universals have been reported in the domain of verb meaning (Van Valin 2006; Croft 2012), a great deal of typological variation has also been documented, especially at the root level (Levinson and Wilkins 2006; Majid et al. 2008). To continue with the example of Running verbs, the English verbs *walk* and *run* encode two of the most biomechanically stable kinds of human gait, and comparable verbs are usually, if not always, found in other languages (Malt et al. 2008). However, the roughly 6000 languages of the world differ substantially in how far beyond these two central "attractors" they go in lexicalizing the complex semantic field of locomotion. English has a relatively rich inventory of Running verbs containing approximately 125 members (Levin 1993, pp. 265–267). It can be divided into multiple subclasses, including verbs of rapid locomotion (e.g., *jog, run, sprint*), leisurely locomotion (e.g., *amble, stroll, mosey*), furtive locomotion (e.g., *sneak, tiptoe, sidle*), awkward locomotion (e.g., *stagger, stumble, lurch*), and so on (Slobin 2000). The root meanings of many of these verbs are so specialized, however, that it is hard to find equivalents in other languages (Slobin 2006; Filipovic 2007). For instance, Spanish *escabullirse* does not distinguish between *glide, slide, slip,* and *slither,* and French *bondir* does not distinguish between *jump, leap, bound, spring,* and *skip* (Slobin 2000). Even English and German, which are closely related, differ somewhat in this domain, since German has no exact matches for English *scamper, scurry, scuttle,* and *scramble,* and English has no exact matches for German *stapfen, stiefen, trampeln,* and *stampfen,* which designate subtly different kinds of firm, heavy walking (Snell-Hornby 1983). Additional typological diversity in the lexicalization of locomotion can be found if one casts a wider net and takes into account languages that routinely employ constructions involving serial verbs (Aikhenvald and Dixon 2006) or coverbs (Wilson 1999; McGregor 2002; Schultze-Berndt 2006). Such extensive crosslinguistic variation is far from trivial, as shown in an especially striking way by recent forensic linguistic analyses of eyewitness testimony (Filipovic 2009).

This chapter focuses on the idiosyncratic root-level semantic features of action verbs—not just Running verbs, but many other classes as well, such as verbs of Hitting, Cutting, Putting, Throwing, etc. (Levin 1993). The main goal is to show how recent developments in cognitive neuroscience have begun to illuminate the representational character of these aspects of verb meaning. Given that root features are typically concrete, it is not surprising that many of the neuroscientific studies that have explored them have been motivated by an increasingly popular theory in the mind/brain sciences that is often called the Embodied Cognition Framework (for reviews, see Barsalou 2008; Binder and Desai 2011; Kiefer and Pulvermüller 2012;

Meteyard et al. 2012; Hauk and Tschentscher 2013). In short, this approach maintains that much of our conceptual knowledge is grounded in modality-specific input/output systems, such that understanding the meanings of many words involves rapidly re-activating some of the sensory, motor, and affective patterns that occurred when the referents were directly experienced. These simulations are modality-specific in format, but they are not necessarily conscious, and their specific content may be modulated by contextual factors.

In what follows, I concentrate on two specific hypotheses that derive from the Embodied Cognition Framework and that involve the neural substrates of two root-level semantic components of action verbs, namely visual features and motor features:

- *Hypothesis 1: The visual–motion features of action verbs depend on the left posterolateral temporal cortex.*
- *Hypothesis 2: The motor features of action verbs depend on the left premotor and primary motor cortex.*

The functional–anatomical details of these hypotheses are elaborated in Sects. 9.2 and 9.3. What matters most here at the outset is the following point. According to the two hypotheses, when one comprehends an action verb like *walk,* one mentally replays, at least in a sketchy way, what it is like to both see and perform the designated type of action, using cortical structures that partially overlap some of the same regions that are engaged during the visual perception and motor execution of that action. As shown below, a number of studies have begun to test various predictions that follow from these hypotheses. Many of the findings are supportive, but others are more challenging and hence generate new questions for future research. My aim is simply to provide a concise survey of this rapidly growing field of investigation.

9.2 Hypothesis 1: The Visual–Motion Features of Action Verbs Depend on the Left Posterolateral Temporal Cortex

It is well established that within the hierarchy of cortical structures comprising the human visual system, area MT+ (also known as V5 or hOc5) is the first region that is specialized primarily for motion processing. As depicted in Fig. 9.1, it is most commonly located in the vicinity of the anterior occipital and lateral occipital sulci (Malikovic et al. 2007). It receives input from multiple earlier visual areas (Nassi and Callaway 2006), and it partially overlaps the extrastriate body area (EBA), which responds preferentially to the sight of human bodies and body parts (Peelen and Downing 2007). MT+ projects forward to the posterior superior temporal sulcus and the posterior middle temporal gyrus—areas that process the complex visual–motion patterns of both living and nonliving things (Beauchamp and Martin 2007; Jastorff and Orban 2009; Saygin 2012) and that are referred to collectively here as the posterolateral temporal cortex (PLTC), as shown in Fig. 9.1.

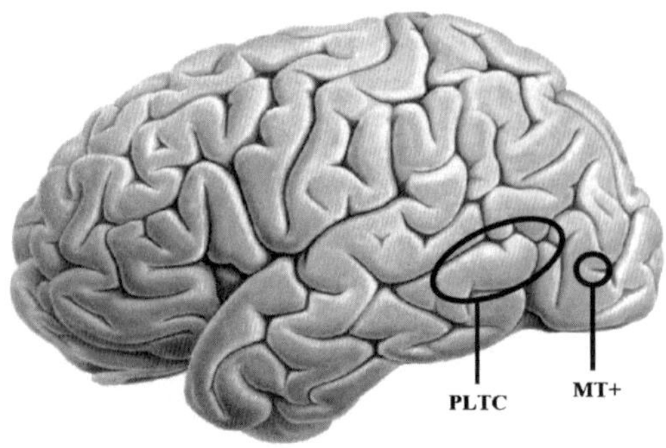

Fig. 9.1 Approximate locations of MT+ and the posterolateral temporal cortex (*PLTC*) in the left hemisphere of the human brain

Hypothesis 1 proposes that the root-level visual–motion features of action verbs—e.g., the idiosyncratic appearances of the different kinds of dynamic body movements encoded by *leap, march,* and *skip*—depend primarily on the left PLTC. This hypothesis makes the following predictions: (1) when action verbs are processed, this area should be engaged; (2) the engagement should be fast and relatively automatic; and (3) it should be functionally relevant to understanding the visual–motion features of the verbs. Below, I discuss evidence bearing on each of these predictions.

9.2.1 Activation Patterns

In keeping with the first prediction, a number of studies employing either positron emission tomography (PET) or functional magnetic resonance imaging (fMRI) have shown that the PLTC is significantly engaged, more strongly in the left than the right hemisphere, when people process the meanings of action verbs, relative to when they perform various baseline tasks (for a review, see Gennari 2012, and for representative studies see Damasio et al. 2001; Kable et al. 2002, 2005; Noppeney et al. 2005; Tranel et al. 2005; Kemmerer et al. 2008; Pirog Revill et al. 2008; Pulvermüller et al. 2009; Lin et al. 2011).

To take a specific example, Kemmerer et al. (2008) used fMRI to investigate the neural substrates of the following five classes of verbs, as defined by Levin (1993): Running verbs (e.g., *run, jog, sprint*), Hitting verbs (e.g., *hit, poke, jab*), Cutting verbs (e.g., *cut, slice, hack*), Speaking verbs (e.g., *yell, shout, sing*), and Change of State verbs (e.g., *shatter, smash, snap*). The main task involved making fine-grained discriminations among triads of verbs within each class—for instance, determining that *limp* is more like *trudge* than *stroll*—and the baseline task involved making similarity judgments about strings of meaningless characters in a peculiar font called Wingdings. Relative to the baseline condition, the five verb classes elicited widely distributed patterns of brain activity that differed from each other in many theoretically interesting ways. For present purposes, however, what is most relevant is that all of them recruited the left PLTC, as portrayed in Fig. 9.2.

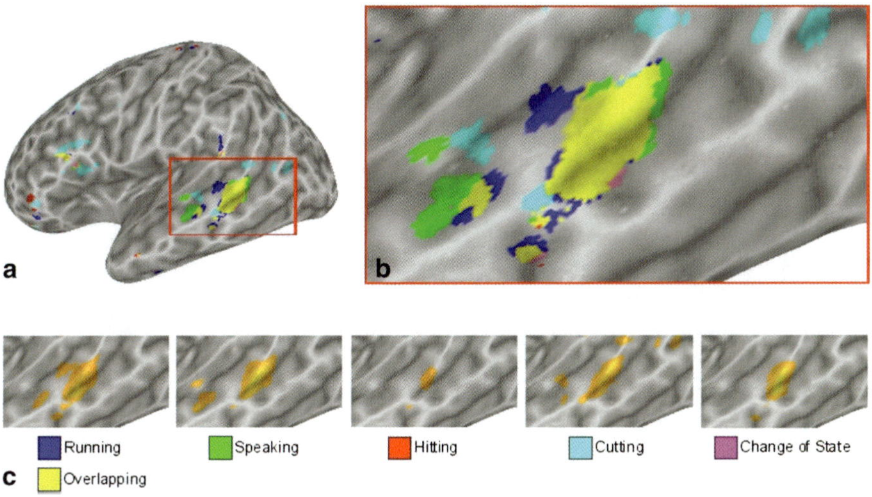

Fig. 9.2 Activation patterns for five classes of verbs—Running, Speaking, Hitting, Cutting, and Change of State—in the left posterolateral temporal cortex (*PLTC*). **a** Activations are rendered on the left hemisphere of an inflated 3-D brain, with the *PLTC* enclosed in a *red box*. **b** Enlargement of the *red box* in (**a**). *Yellow patches* indicate areas of overlapping activation for two or more verb classes, and other colored patches indicate areas of activation unique to particular verb classes, according to the color key in (**c**). **c** Activations for each separate verb class in the territory corresponding to the *red box* in (**a** and **b**). (From Kemmerer et al. 2008, p. 29, with permission from Elsevier)

A recent meta-analysis conducted by Watson et al. (2013) strongly supports the view that the left PLTC plays a major role in the neural representation of action concepts. In particular, by pooling data from 29 different PET and fMRI studies, the investigators were able to demonstrate that the left PLTC is the area of greatest concordance between, first, experiments that focused on the conceptual processing of action *images,* and second, experiments that focused on the conceptual processing of action *verbs*. This finding is clearly consistent with the idea that one of the functions of the left PLTC is to implement the visual–motion features of verb meanings. But, it is important to realize that not all researchers agree with this interpretation.

Most notably, the putative link between the left PLTC and the visual–motion features of verb meanings has been challenged by two fMRI studies by Bedny et al. (2008, 2012). In the first study, Bedny et al. (2008) measured subjects' brain activity while they made semantic similarity judgments about auditorily presented word pairs. The words consisted of three categories of verbs—high motion (e.g., *to kick*), medium motion (e.g., *to bleed*), and low motion (e.g., *to think*)—as well as three corresponding categories of nouns—high motion (e.g., *the tiger*), medium motion (e.g., *the drill*), and low motion (e.g., *the rock*)—with the different degrees of motion for each category deriving from an independent rating experiment. The investigators found that, relative to a baseline condition involving the perception of backward speech, the activation levels in the left PLTC were equally high for the three categories of verbs and equally low for the three categories of nouns. They interpreted these results as evidence that the left PLTC may not represent specifically the visual–motion features of action verbs, but may instead represent more abstract or schematic properties of verb meanings—i.e., properties that are shared by both action and nonaction verbs and that contribute to the content of event structure templates, such as information about agency, causality, transitivity, etc.

In the second study, Bedny et al. (2012) followed up on their first study by administering the same task to both sighted and congenitally blind subjects. Remarkably enough, they obtained very similar results for the two groups of subjects. Multiple analyses indicated that the left PLTC responded significantly more to verbs than nouns, regardless of the amount of motion conveyed by the words, and, even more strikingly, regardless of whether the subjects were sighted or blind. According to Bedny et al. (2012), these findings bolster the view that the left PLTC represents abstract rather than concrete aspects of verb meanings.

Problems still remain, however, because the results of Bedny et al.'s fMRI studies must somehow be reconciled with the results of two other lines of investigation. First, according to several neuropsychological studies that are discussed further below, brain-damaged patients with lesions affecting the left PLTC are often impaired on tasks that require knowledge of the concrete components of verb meanings, including their visual–motion features. And second, contradicting Bedny et al.'s fMRI studies, several other fMRI studies have reported that the left PLTC does in fact respond significantly more to motion-related than nonmotion-related expressions (Tettamanti et al. 2005; Deen and McCarthy 2010; Saygin et al. 2010; Lin et al. 2011; Wallentin et al. 2011; Humphreys et al. 2013). The explanation for these discrepancies is not clear, but it is noteworthy that Bedny et al.'s studies used single words as stimuli, whereas the other studies used sentences, paragraphs, and stories (with the exception of Lin et al. 2011). For example, Wallentin et al. (2011) asked a group of Danish-speaking subjects to simply listen to Hans Christian Andersen's famous fairy tale *The ugly ducking,* and they found that the left PLTC was significantly engaged when all the clauses describing motion events were contrasted against all the clauses describing nonmotion events (while controlling for such nuisance variables as sound intensities, word frequencies, word co-occurrences, emotional content, and physiological noise associated with cardiac pulsation and respiration).

Taking all of these considerations into account, one possibility is that the left PLTC may be innately predisposed to represent several different components of verb meaning, including visual–motion patterns as well as more abstract types of information. Data consistent with this view came from a recent fMRI study which showed that partially segregated sectors of the left PLTC respond to action verbs like *walk* on the one hand and to purely stative verbs like *exist* on the other (Peelen et al. 2012). (It is also notable that other portions of the left PLTC may be sensitive to template-level argument structure and thematic role information, as suggested by Grewe et al. 2007; Shetreet et al. 2007; Wu et al. 2007.) So, even though this large cortical region does not receive normal visual input in congenitally blind individuals, it may nonetheless serve as the default region for storing long-term records of whatever conceptual knowledge can be acquired through other senses about the idiosyncratic motion patterns encoded by action verbs. Further research is needed to explore this issue in greater depth.

9.2.2 Speed and Automaticity of Processing

So far, no studies have employed techniques with high temporal resolution, like event-related brain potentials (ERPs) or magnetoencephalography (MEG), to investigate how rapidly the left PLTC is recruited during the on-line processing of action verbs. However, an important fMRI study by Pirog Revill et al. (2008) does address the question of automaticity. In designing their study, the investigators exploited the well-established fact that during speech perception phonological information is continuously projected to the lexicon, so that an initial consonant cluster like *bl*... automatically activates all the words in the listener's lexicon that begin with those sounds (*black, blanket, bland,* etc.). As the input accumulates, the set of activated words diminishes until only one is still compatible, at which point recognition can be said to occur. Prior to brain scanning, the subjects in the study learned an artificial lexicon with novel multisyllabic words referring to unusual objects and events, the latter including both motion patterns and color/texture changes. Crucially, some of the words differed only in the final syllable, so it was sometimes impossible for a listener to determine whether a word denoted a motion pattern (e.g., *gapitu* "vertical oscillation") or a color/texture change (e.g., *gapito* "whitening") until the final syllable was heard. During the actual fMRI experiment, the subjects heard a series of the newly learned words and judged whether each one described a certain kind of scene. Two major results emerged. First, the left PLTC, extending back into MT+, responded significantly more to motion words than nonmotion words, thereby supporting the hypothesis that the root-level visual–motion features of action verbs are mediated in part by this cortical territory. And second, when the subjects heard nonmotion words, the activation level in the left PLTC/MT+ was greater if the cohort of phonologically similar words contained a motion word than if it did not. This finding is especially intriguing because it suggests that during on-line spoken language comprehension, the visual semantic features of motion words are automatically activated in the left PLTC/MT+, even before the phonological forms of those words have been completely processed and determined to be either compatible or incompatible with the speech input.

9.2.3 Functional Relevance

If, as Hypothesis 1 maintains, the left PLTC subserves the visual–motion features of action verbs, then altering its functional state should affect the processing of those features. Although this prediction could be tested with transcranial magnetic stimulation (TMS), I am not aware of any studies that have done so—at least not yet.

As mentioned above, however, a few neuropsychological studies have provided some initial evidence that the left PLTC does play an essential role in understanding action verbs, a role that may involve representing their visual–motion features (Aggujaro et al. 2006; Tranel et al. 2008; Kalénine et al. 2010; Kemmerer et al. 2012).

In one of the largest neuropsychological investigations of this topic to date, Kemmerer et al. (2012) administered the following battery of six standardized tasks to 226 brain-damaged patients with widely distributed lesions in the left and right hemispheres (a complete list of the items is provided by Kemmerer et al. 2001):

- Naming ($N = 100$ items): For each item, the participant is shown a photograph of an action, and the task is to orally name each one with a specific verb.
- Word–picture matching ($N = 69$ items): For each item, the participant is shown a printed verb together with two photographs of actions, and the task is to determine which action the verb describes.
- Word attribute ($N = 62$ items): For each item, the participant is shown two printed verbs, and the task is to indicate which one designates a type of action that satisfies a certain value for a single attribute (e.g., more tiring; moving the hands in a circle; moving the hands up/down; etc.).
- Word comparison ($N = 44$ items): For each item, the participant is shown three printed verbs, and the task is to determine which one is most different in meaning from the other two.
- Picture attribute ($N = 72$ items): This task is analogous to the word attribute task, but the stimuli are photographs of actions instead of verbs.
- Picture comparison ($N = 24$ items): This task is analogous to the word comparison task, but the stimuli are photographs of actions instead of verbs.

Of the 226 patients who were studied, 61 failed in one or more of the six tasks. Among the few patients who failed the entire battery was a man identified as case 1808. Not only did he average only 62.5 % correct (Table 9.1) but his lesion was highly focal too, being confined to the left PLTC and underlying white matter (Fig. 9.3). In addition, the same region was implicated in a series of group analyses that explored lesion-deficit relationships in the 147 patients for whom brain scans were available. On a task by task basis, brain maps were created that indicated, at each voxel, the likelihood of a lesion being significantly associated with impaired versus unimpaired performance. These analyses yielded significant results in, among other regions, the left PLTC and underlying white matter for five of the six tasks—specifically, naming, word attribute, word comparison, picture attribute, and picture comparison. Overall, these findings from both the single case of 1808 and the larger group of patients constitute strong evidence that the left PLTC is essential for representing the kinds of action concepts that tend to be encoded by English verbs. Moreover, they are consistent with the hypothesis that this cortical region subserves specifically the visual–motion features of those concepts. Further neuropsychological research is needed, however, to determine whether damage to the left PLTC tends to disrupt knowledge of action verbs significantly more than knowledge of nonaction verbs.

Table 9.1 Scores obtained by case 1808 on six tasks probing conceptual knowledge of actions in various verbal and nonverbal ways. Note that z scores of – 2.0 or lower are considered to be significantly below normal. (From Kemmerer et al. 2012, p. 831)

Task	% Correct	z Score
Naming	57	– 5.6
Word–picture matching	78	– 3.1
Word attribute	77	– 4.9
Word comparison	64	– 3.0
Picture attribute	78	– 2.9
Picture comparison	21	– 7.5
Average	*62.5*	*– 4.5*

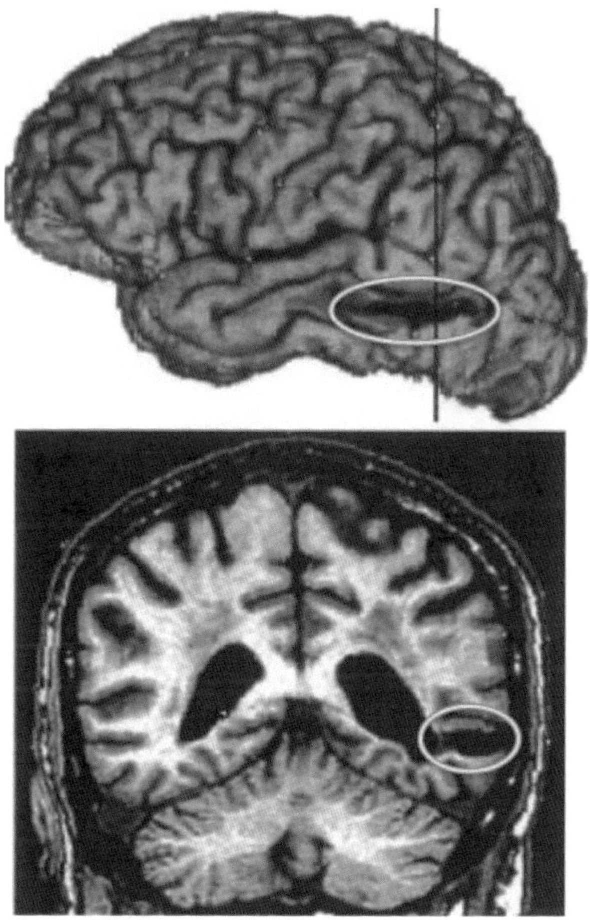

Fig. 9.3 Lesion site of case 1808. The damage, highlighted by the *white circles*, is restricted to the posterior portion of the left middle temporal gyrus and underlying white matter. The *vertical line* in the *upper panel* indicates the plane of the coronal section shown *below*, where the left hemisphere is depicted on the *right side*. (From Tranel et al. 2003, p. 421, with permission from Taylor & Francis, Ltd, www.tandfonline.com)

9.2.4 *Summary*

According to Hypothesis 1, understanding the distinctive types of visual–motion patterns that are encoded by action verbs involves mentally simulating what those patterns look like. Moreover, the hypothesis holds that this simulation process involves activating a large brain region that normally contributes to higher-order motion perception, specifically the left PLTC. As indicated above, this proposal has received preliminary support from several brain-mapping studies. At the same time, however, a number of important questions remain open, some of which are as follows: How exactly are the visual–motion features of verb meanings organized in the left PLTC? Does this region represent both root-level and template-level aspects of event structure? To what extent are verb-induced activations of the left PLTC modulated by such factors as the interpretive demands of the task, the subject's experiential history of perceiving the pertinent kinds of actions, and the linguistic contexts in which the verbs occur? To what extent is the representational organization of the left PLTC influenced by crosslinguistic variation in the verbal encoding of motion? Addressing these questions, and many others, will require much more research.

9.3 Hypothesis 2: The Motor Features of Action Verbs Depend on the Left Premotor and Primary Motor Cortex

The motor cortex resides in the frontal lobes and has a heterogeneous architecture that includes the primary motor cortex and perhaps as many as ten premotor areas—two ventral, two dorsal, and six medial, creating a complex mosaic that Graziano (2009, p. 65) calls "the premotor zoo." The primary motor cortex is traditionally thought of as containing a somatotopically organized map of the body's muscles, with the tongue and lips represented close to the Sylvian fissure, the hand and arm represented at lateral and dorsolateral sites, and the leg and foot represented at the vertex and in the interhemispheric sulcus. It is important to realize, however, that the famous motor homunculus (Penfield and Rasmussen 1950; see also Graziano 2009, p. 8) only captures the general trend. In fact, representations of adjacent body parts overlap a great deal in the primary motor cortex (see the classic paper by Penfield and Boldrey 1937; for a more contemporary view, see Meier et al. 2008). Moreover, recent studies with monkeys suggest that the primary motor cortex as well as many premotor areas are topographically parcellated not only according to the layout of the body but also in terms of different categories of ethologically important behaviors that require the coordination of multiple joints—e.g., climbing/leaping behaviors, reach-to-grasp behaviors, central-space manipulation behaviors, defensive behaviors, and licking/chewing behaviors (Fig. 9.4; for a review, see Graziano and Aflalo 2007).

Since the mid-1990s, some of the most exciting research in cognitive neuroscience has revolved around the discovery that premotor and primary motor areas, as well as several other brain regions, have "mirror" properties such that they are engaged not only when actions are executed by the self but also when they are seen or heard being

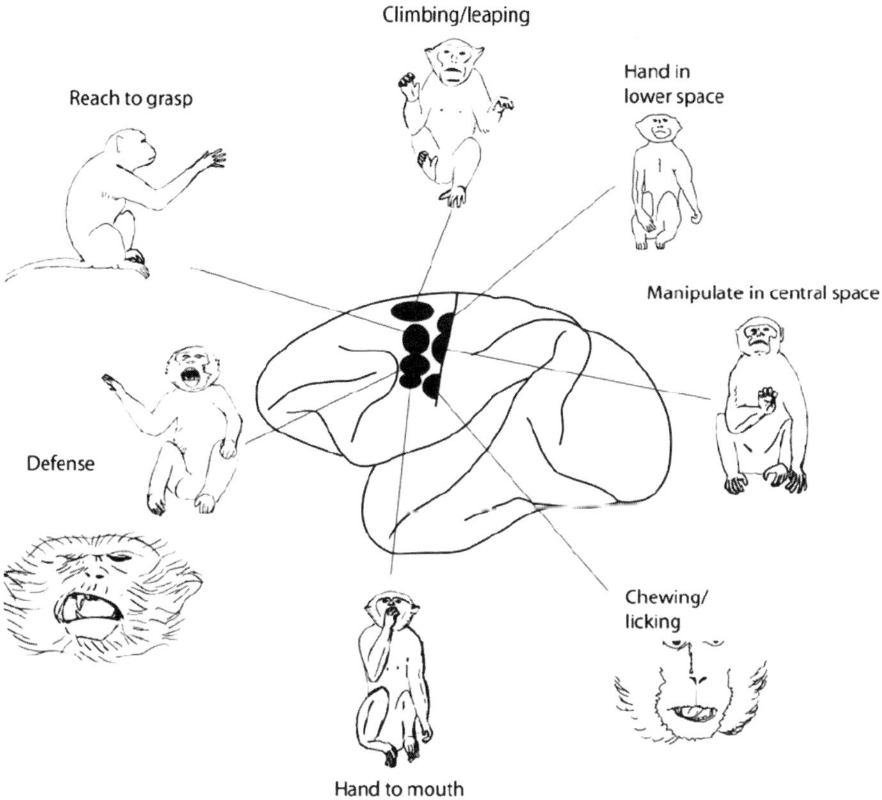

Fig. 9.4 Action zones in the motor cortex of the macaque monkey. These categories of movement were evoked by continuous electrical stimulation of the cortex on the behaviorally relevant timescale of 500 ms. Images traced from video frames. Each image represents the final posture obtained at the end of the stimulation-evoked movement. Within each action zone in the motor cortex, movements of similar behavioral category were evoked. (From Graziano and Aflalo 2007, p. 243, with permission from Elsevier)

performed by other agents (for reviews, see Fadiga et al. 2005; Fogassi and Ferrari 2011; Rizzolatti and Sinigaglia 2010; Oosterhof et al. 2013; Cook et al. 2014; see also the 2008 double-issue of *Social Neuroscience* [vol. 3, issues 3–4] and the 2013 "Forum Discussion" in *Cortex* [vol. 49, issue 10]). These remarkable findings suggest that, in accord with the Embodied Cognition Framework, understanding other people's actions may depend, at least to some degree, on subconsciously simulating them.

Partly because of this line of work, there has been an increasing interest in the provocative idea that when people understand linguistic descriptions of actions, motor regions in their frontal lobes may be recruited (for a previous survey, see Kemmerer and Gonzalez Castillo 2010; see also Pulvermüller 2013). Based on this notion, Hypothesis 2 proposes specifically that the root-level motor features of action verbs—e.g., the different kinds of action programs encoded by *leap, march,* and *skip*—depend on the left premotor and primary motor cortex. Analogous to Hypothesis 1, this hypothesis makes the following predictions: (1) when action verbs are processed, these areas should be engaged; (2) the engagement should be fast and relatively automatic; and (3) it should be functionally relevant to understanding the motor-semantic features of the verbs. Below, I discuss evidence bearing on each of these predictions. (A caveat: The left inferior parietal cortex has also been implicated in some of the motor aspects of action verbs, but this topic is not discussed here due to space limitations. For relevant studies see, e.g., Noppeney et al. 2005; Kemmerer et al. 2008; Liljeström et al. 2008; van Dam et al. 2010; Rueschemeyer et al. 2014)

9.3.1 Activation Patterns

Consistent with the first prediction, many fMRI studies have reported that, compared to various control stimuli, action verbs and sentences activate left premotor and/or primary motor areas in a more or less somatotopic manner—i.e., in such a way that, for the most part, descriptions of leg/foot actions (e.g., *kick*) engage leg/foot areas, descriptions of arm/hand actions (e.g., *pick*) engage arm/hand areas, and descriptions of mouth actions (e.g., *lick*) engage mouth areas (Fig. 9.5). Importantly, a number of these studies included functional localizer scans to verify that some of the premotor and primary motor areas that are ignited when subjects process body-part-specific action verbs and sentences are also ignited when they execute correspondingly body-part-specific movements (e.g., Hauk et al. 2004; Desai et al. 2010; Raposo et al. 2009; Moody and Gennari 2010; but see Postle et al. 2008 and Schuil et al. 2013 for contradictory data).

It is noteworthy, however, that all of the data plotted in Fig. 9.5 came from right-handed individuals whose dominant hand is controlled mainly by the left hemisphere. Why does this matter? Given that nearly 90 % of the people in the world are right handed, it leaves open the possibility that the multiple "hot spots" for arm/hand action verbs and sentences in Fig. 9.5 might not really reflect motor simulations of how the subjects themselves usually execute such actions, but might instead reflect motor simulations of how they usually see other people execute them.

To address this question, Willems et al. (2010) scanned the brain activity of both right- and left-handed subjects while they performed a lexical decision task that included manual action verbs, nonmanual action verbs, and pronounceable pseudowords. Replicating previous studies, the researchers found that in right-handed subjects manual action verbs (compared to nonmanual action verbs) engaged mainly left-sided motor areas for controlling the arm/hand. The fascinating new discovery was that left-handed subjects displayed exactly the opposite hemispheric asymmetry. These results support the view that when people process action verbs, they covertly simulate the way they themselves usually execute the designated types of movements, as opposed to the way they usually see others perform them (for a somewhat different perspective, see Hauk and Pulvermüller 2011). More generally, Willems et al.'s (2010) study suggests that people who use their bodies in systematically different

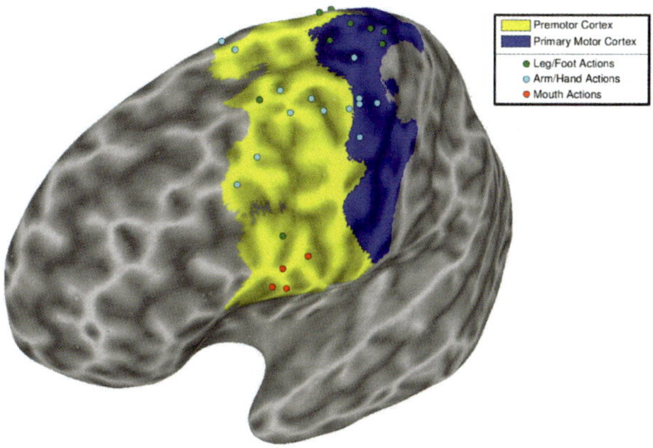

Fig. 9.5 Activation peaks in left premotor and primary motor cortex reported by some of the fMRI studies that have investigated the neural substrates of the motor features of verbs and sentences encoding leg/foot actions (*green circles*), arm/hand actions (*light blue circles*), and mouth actions (*red circles*). Activations are plotted on an inflated brain with boundaries for the premotor cortex (*yellow*) and primary motor cortex (*dark blue*) based on Mayka et al.'s (2006) Human Motor Area Template. The activation peaks are drawn from the following sources: Hauk et al. (2004) (with corrections reported by Kemmerer and Gonzalez Castillo 2010); Tettamanti et al. (2005); Aziz-Zadeh et al. (2006); Rueschemeyer et al. (2007); Kemmerer et al. (2008); Beilock et al. (2008); Boulenger et al. (2009); Raposo et al. (2009); Desai et al. (2010); Pulvermüller et al. (2009); and Willems et al. (2010). (From Kemmerer et al. 2012, p. 843, with permission from Elsevier)

ways exhibit systematically different neural responses to linguistic descriptions of action. Hence, a proviso should be appended to Hypothesis 2 to take handedness into account.

9.3.2 Speed and Automaticity of Processing

If somatotopically mapped premotor and primary motor regions do in fact represent the root-level motor features of action verbs, they should not simply be recruited after such verbs are perceived, but should be recruited quite rapidly (Hauk et al. 2008). In general, ERP studies suggest that lexical-semantic information is accessed as early as 150 ms post-word-onset (Penolazzi et al. 2007; Hoenig et al. 2008), which is well before the roughly 300 ms latency of conscious access (Dehaene 2014). With regard to action verbs, Pulvermüller and colleagues conducted a series of ERP experiments that used source localization techniques to identify the neuronal generators of signals elicited by verbs encoding different body-part-specific categories of actions, and found that within the time window of 150–250 ms, verbs for leg/foot actions engaged dorsal motor-related areas, verbs for arm/hand actions engaged lateral motor-related areas, and verbs for mouth actions engaged ventral motor-related areas (Pulvermüller et al. 2001; Hauk and Pulvermüller 2004). Moreover, similar results were obtained in an MEG study in which subjects passively heard action verbs while their attention was focused on a silent video film, thereby supporting the view that the activation of somatotopically mapped motor regions is a fast and fairly automatic process (Pulvermüller et al. 2005b; see also Shtyrov et al. 2004; Moseley et al. 2013). Additional evidence for this view comes from an ERP study by Boulenger et al. (2008) which found that when verbs for arm/hand actions were presented to subjects subliminally, they not only modulated the "readiness potential" (an index of motor preparation) associated with subsequent reaching movements but also affected the kinematics of those movements. Finally, several behavioral studies bolster the idea that root-level motor aspects of verb meaning are retrieved quickly and more or less automatically (e.g., Boulenger et al. 2006; Zwaan and Taylor 2006; Scorolli and Borghi 2007; Kaschak and Borreggine 2008; Nazir et al. 2008; Taylor and Zwaan 2008).

On the other hand, some of the most interesting and valuable insights in this field of research have come from recent studies showing that the degree to which motor representations are involved in action verb comprehension depends on both the task and the context (for a review, see Tomasino and Rumiati 2013). As an illustration of task effects, in a behavioral study Sato et al. (2008) found that although verbs for arm/hand actions interfered with button presses when subjects performed a semantic task, they did not do so when subjects performed a lexical decision task. As an illustration of linguistic context effects, in an fMRI study Raposo et al. (2009) found that although somatotopically mapped motor areas were engaged when subjects read action verbs in isolation (e.g., *kick*) and in literal sentences (e.g., *kick the ball*), they were not engaged when subjects read such verbs in idiomatic sentences (e.g., *kick the bucket*) (for similar results, see Aziz-Zadeh et al. 2006 and Desai et al. 2013, but for contrary results see Boulenger et al. 2009). Finally, as an illustration of nonlinguistic context effects, in an fMRI study Papeo et al. (2012) found that not only action verbs but also purely stative verbs significantly recruited the motor cortex in certain circumstances—specifically, when they were encountered after subjects first performed a mental rotation task using a motor-oriented rather than a visuospatially oriented strategy. Taken together, these studies, among many others (e.g., Mirabella et al. 2012; Aravena et al. 2012, 2014; Schuil et al. 2013), suggest that verb-induced motor activation is by no means a rigid, inflexible affair, but is instead quite sensitive to attentional and situational factors (for a broader perspective, see Lebois et al. in press).

9.3.3 Functional Relevance

The last prediction of Hypothesis 2 is that altering the operations of left premotor and primary motor areas should affect the processing of the root-level motor features of action verbs. This prediction has been tested in several TMS studies. Although the results of these studies are not entirely consistent, for the most part they confirm the prediction (for discussion see Kemmerer and Gonzalez Castillo 2010). For example, in an influential study by Pulvermüller et al. (2005a), subjects performed a lexical decision task while, 150 ms before the onset of each letter string, either a single TMS pulse or a sham TMS pulse was delivered to either an arm/hand or a leg/foot site in the left or right hemisphere. Stimulation of the left leg/foot region led to significantly faster responses to verbs encoding leg/foot actions than to verbs encoding arm/hand actions, whereas stimulation of the left arm/hand region had the opposite effect. No differences were found, however, when TMS was delivered to the right hemisphere (or when sham TMS was used), which is not surprising since all of the subjects were right handed. In short, this study demonstrates that brief stimulation of body-part-specific motor areas speeds up the retrieval of concordant body-part-specific motor features of verbs.

TMS can also be applied repetitively to disrupt rather than enhance cortical computations, and in a study that adopted this kind of approach, Gerfo et al. (2008) showed that targeting a left arm/hand site significantly slowed down subjects' responses when they changed the inflectional form of arm/hand action verbs, relative to when they changed the inflectional form of purely stative verbs (the inflectional transformations involved shifting between the first-person and second-person forms of Italian verbs). These results provide additional evidence for the idea that, as maintained by Hypothesis 2, somatotopically mapped motor areas are causally involved, as opposed to just incidentally involved, in accessing the root-level motor features of action verbs (see also Repetto et al. 2013; Kuipers et al. 2013).

Several neuropsychological studies have also addressed this topic by investigating brain-damaged patients whose lesions affect frontal motor areas. Two sets of findings that support Hypothesis 2 are as follows. First, patients with motor neuron disease (a.k.a. Lou Gehrig's disease or amyotrophic lateral sclerosis, ALS) have significantly worse knowledge of action verbs than object nouns, due to progressive degeneration of frontal motor areas for controlling muscle groups throughout the body (Bak and Hodges 2004; Hillis et al. 2004, 2006; Grossman et al. 2008). And second, in a study with 21 left-hemisphere stroke patients, Arévalo et al. (2007) found a significant relationship between impaired retrieval of "manipulation" verbs and damage to putative arm/hand motor areas not only in the frontal lobe but also in the parietal lobe.

Furthermore, as mentioned above, Kemmerer et al. (2012) reported an experiment in which 226 patients were administered a battery of six standardized tasks that probed conceptual knowledge of actions in a variety of verbal and nonverbal ways. Although the tasks did not employ a well-controlled set of verbs encoding arm/hand actions, all of the tasks did have a preponderance (roughly 70 %) of arm-/hand-related stimuli. The lesion of one particular patient who failed all six tasks—namely, case 1172—is shown in Fig. 9.6, and his scores are shown in Table 9.2. The lesion affected the midlateral and dorsolateral sectors of the precentral gyrus, including the cortex as well as the underlying white matter—territory that is well established as being crucial for the control of arm/hand actions. At the same time, however, it is apparent that the lesion also affected parts of the inferior frontal gyrus, so it remains possible that the patient's severe verb deficit was due to that damage rather than the damage involving specifically arm-/hand-related tissue. In addition, group-level lesion-deficit analyses involving 147 of the 226 patients revealed that, as exemplified by case 1172, not only

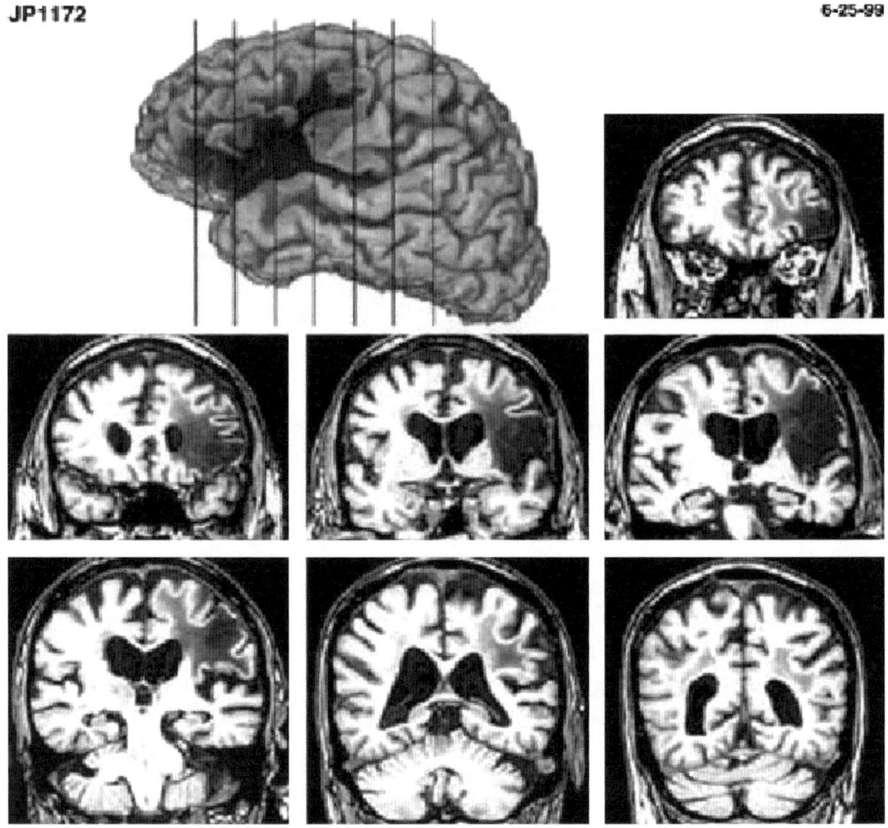

Fig. 9.6 Lesion site of case 1172. The damage is centered in the heart of Broca's area but extends superiorly into the middle part of the premotor cortex, a region known to be important for arm/hand actions. As shown in the *coronal sections,* the damage included white matter underneath the inferior frontal, middle frontal, precentral, postcentral, and supramarginal gyri. (From Kemmerer and Tranel 2003, p. 426, with permission from Taylor & Francis, Ltd, www.tandfonline.com)

arm-/hand-related motor areas but also parts of the left inferior frontal gyrus, were among the regions most reliably linked with impairment on each of the six tasks.

Although the studies just described are consistent with the notion that frontal motor areas contribute to the comprehension of action verbs, other studies suggest that those areas are not absolutely necessary for such comprehension (Papeo et al. 2010; Arévalo et al. 2012; Kemmerer et al. 2013; Maieron et al. 2013). For example, in Arévalo et al.'s (2012) study, 27 left-hemisphere stroke patients were given a task that required judging, on every trial, whether a given word correctly described a picture of an action. The actions involved facial movements, arm/hand movements, and leg/foot movements. Many of the patients had lesions that included frontal motor areas, but contrary to the predictions of Hypothesis 2, significant correlations were not found between impaired performance on body-part-specific action categories

Table 9.2 Scores obtained by case 1172 on six tasks probing conceptual knowledge of actions in various verbal and nonverbal ways. Note that z scores of -2.0 or lower are considered to be significantly below normal. (From Kemmerer et al. 2012, p. 831)

Task	% Correct	z Score
Naming	33	−10.4
Word–picture matching	72	−4.4
Word attribute	79	−4.4
Word comparison	59	−3.7
Picture attribute	82	−2.0
Picture comparison	25	−7.1
Average	*58.3*	*−5.3*

and damage to the corresponding body-part-specific motor areas. Findings like these indicate that the precise function of motor simulation during verb comprehension is still unclear.

9.3.4 Summary

Hypothesis 2 maintains that understanding the types of bodily actions that verbs refer to involves covertly simulating them, using some of the same brain systems that underlie their execution, especially the left premotor and primary motor cortex. The foregoing survey of relevant research indicates that this proposal has already received a fair amount of support. As with Hypothesis 1, however, many issues remain unresolved, some of which are as follows. Do premotor and primary motor regions play different roles in representing the root-level motor features of verb meanings? Are those features organized mainly according to somatotopy (as in Penfield and Rasmussen's 1950 homunculus) or are they mapped more in terms of complex action categories (as in Fig. 9.4)? Why do fMRI studies vary so much in the peak coordinates that are reported for supposedly the same basic kinds of verbs (as in Fig. 9.5)? How exactly is the functional contribution of motor areas to verb comprehension influenced by such variables as the task, the context, and the subject's familiarity with the designated type of action? To what degree is crosslinguistic variation in the lexicalization of action reflected in the organization of the motor cortex? These questions, and many others, will undoubtedly receive more attention in the coming years.

9.4 Conclusion

The purpose of this chapter has been to show how recent advances in cognitive neuroscience have begun to illuminate the representational character of root-level visual and motor features of action verbs. Although this field of inquiry is still in its infancy, much of the available literature suggests that, in accord with the Embodied Cognition Framework, appreciating the idiosyncratic visual and motor features that distinguish between, say, Running verbs (e.g., *stroll, saunter, strut, march, trudge, limp, stagger, tiptoe, sneak, jog, run, sprint,* etc.) requires access to experience-based knowledge stored in modality-specific cortical areas that partially overlap those involved in perceiving and producing the designated types of actions. Regarding visual information, several studies provide evidence that understanding the distinctive kinds of motion patterns encoded by verbs involves mentally simulating how those patterns usually appear, drawing upon the left PLTC, which is independently known to contribute to motion perception. And regarding motor information, a growing number of studies suggest that understanding the distinctive kinds of action programs encoded by verbs involves mentally simulating how those programs are usually implemented, drawing upon the left premotor and primary motor regions, which are well established as playing essential roles in the planning and control of bodily movements. Importantly, there are reasons to believe that during on-line verb processing, these sorts of visual and motor simulations are triggered quite rapidly and make genuine functional contributions to comprehension, at least in some situations. Still, as noted above, many questions about the neural substrates of root-level visual and motor features of verb meanings remain unanswered, leaving plenty of room for further investigation. At the heart of the matter is the long-standing challenge of deciphering how language-specific semantic structures interface with modality-specific systems for perception and action.

References

Aggujaro, S., Crepaldi, D., Pistarini, C., Taricco, M., & Luzzatti, C. (2006). Neuroanatomical correlates of impaired retrieval of verbs and nouns: Interaction of grammatical class, imageability and actionality. *Journal of Neurolinguistics, 19,* 175–194.

Aikhenvald, A. Y., & Dixon, R. M. W. (Eds.) (2006). *Serial verb constructions: A cross-linguistic typology*. Oxford: Oxford University Press.

Aravena, P., Delevoye-Turrell, Y., Deprez, V., Cheylus, A., Paulignan, Y., Frak, V., & Nazir, T. (2012). Grip force reveals the context sensitivity of language-induced motor activity during "action word" processing: Evidence from sentential negation. *PloS ONE, 7,* e50287.

Aravena, P., Courson, M., Frak, V., Cheylus, A., Paulignan, Y., Deprez, V., & Nazir, T. (2014). Action relevance in linguistic context drives word-induced motor activity. *Frontiers in Human Neuroscience*.

Arévalo, A., Perani, D., Cappa, S. F., Butler, A., Bates, E., & Dronkers, N. (2007). Action and object processing in aphasia: From nouns and verbs to the effect of manipulability. *Brain and Language, 100,* 79–94.

Arévalo, A., Baldo, J. V., & Dronkers, N. F. (2012). What do brain lesions tell us about theories of embodied semantics and the human mirror neuron system? *Cortex, 48,* 242–254.

Aziz-Zadeh, L., Wilson, S. M., Rizzolatti, G., & Iacoboni, M. (2006). Congruent embodied representations for visually presented actions and linguistic phrases describing actions. *Current Biology, 16,* 1818–1823.

Bak, T. H., & Hodges, J. R. (2004). The effects of motor neurone disease on language: Further evidence. *Brain and Language, 89,* 354–361.

Barsalou, L. W. (2008). Grounded cognition. *Annual Review of Psychology, 59,* 617–645.

Beauchamp, M. S., & Martin, A. (2007). Grounding object concepts in perception and action. *Cortex, 43,* 461–468.

Bedny, M., Caramazza, A., Grossman, E., Pascual-Leone, A., & Saxe, R. (2008). Concepts are more than percepts: The case of action verbs. *Journal of Neuroscience, 28,* 11347–11353.

Bedny, M., Caramazza, A., Pascual-Leone, A., & Saxe, R. (2012). Typical neural representations of action concepts develop without vision. *Cerebral Cortex, 22,* 286–293.

Beilock, S. L., Lyons, I. M., Mattarella-Micke, A., Nusbaum, H. C., & Small, S. L. (2008). Sports experience changes the neural processing of action language. *Proceedings of the National Academy of Sciences of the United States of America, 105,* 13269–13273.

Binder, J. R., & Desai, R. H. (2011). The neurobiology of semantic memory. *Trends in Cognitive Sciences, 15,* 527–536.

Boulenger, V., Roy, A. C., Paulignan, Y., Deprez, V., Jeannerod, M., Nazir, T. A. (2006). Cross-talk between language processes and overt motor behavior in the first 200 ms of processing. *Journal of Cognitive Neuroscience, 18,* 1607–1615.

Boulenger, V., Silber, B. Y., Roy, A. C., Paulignan, Y., Jeannerod, M., & Nazir, T. A. (2008). Subliminal display of action words interferes with motor planning: A combined EEG and kinematic study. *Journal of Physiology, Paris, 102,* 130–136.

Boulenger, V., Hauk, O., & Pulvermüller, F. (2009). Grasping ideas with the motor system: Semantic somatotopy in idiom comprehension. *Cerebral Cortex, 19,* 1905–1914.

Cook, G., Catmur, C., Press, C., & Heyes, C. (2014). Mirror neurons: From origin to function. *Behavioral and Brain Sciences. 37,* 177–241.

Croft, W. (2012). *Verbs.* Oxford: Oxford University Press.

Damasio, H., Grabowski, T. J., Tranel, D., Ponto, L. L. B., Hichwa, R. D., & Damasio, A. N. (2001). Neural correlates of naming actions and of naming spatial relations. *NeuroImage, 13,* 1053–1064.

Deen, B., & McCarthy, G. (2010). Reading about the actions of others: Biological motion imagery and action congruency influence brain activity. *Neuropsychologia, 48,* 1607–1615.

Dehaene, S. (2014). *Consciousness and the brain: Deciphering how the brain codes our thoughts.* New York: Viking.

Desai, R., Binder, J. R., Conant, L. L., & Seidenberg, M. S. (2010). Activation of sensory-motor areas in sentence comprehension. *Cerebral Cortex, 20,* 468–478.

Desai, R., Conant, L. L., Binder, J. R., Park, H., & Seidenberg, M. S. (2013). A piece of the action: Modulation of sensory-motor regions by action idioms and metaphors. *NeuroImage, 83,* 862–869.

Fadiga, L., Craighero, L., & Olivier, E. (2005). Human motor cortex excitability during the perception of others' action. *Current Opinion in Neurobiology, 15,* 213–218.

Filipovic, L. (2007). *Talking about motion.* Amsterdam: John Benjamins.

Filipovic, L. (2009). Motion events in eyewitness interviews, translation, and memory: Typological and psycholinguistic perspectives. *Language and Linguistics Compass, 3,* 300–313.

Fogassi, L., & Ferrari, P. F. (2011). Mirror systems. *Wiley Interdisciplinary Reviews: Cognitive Science, 2,* 22–38.

Gennari, S. P. (2012). Representing motion in language comprehension: Lessons from neuroimaging. *Language and Linguistics Compass, 6,* 67–84.

Gerfo, E. L., Oliveri, M., Torriero, S., Salerno, S., Koch, G., & Caltagirone, C. (2008). The influence of rTMS over prefrontal and motor areas in a morphological task: Grammatical vs. semantic effects. *Neuropsychologia, 46,* 764–770.
Graziano, M. S. A. (2009). *The intelligent movement machine: An ethological perspective on the primate motor system.* Oxford: Oxford University Press.
Graziano, M. S. A., & Aflalo, T. N. (2007). Mapping behavioral repertoire onto the cortex. *Neuron, 56,* 239–251.
Grewe, T., Bornkessel-Schlesewsky, I., Zysset, S., Wiese, R., von Cramon, D. Y., & Schlesewsky, M. (2007). The role of the posterior superior temporal sulcus in the processing of unmarked transitivity. *NeuroImage, 35,* 343–352.
Grossman, M., Anderson, C., Khan, A., Avants, B., Elman, L., & McCluskey, L. (2008). Impaired action knowledge in amyotrophic lateral sclerosis. *Neurology, 71,* 1396–1401.
Hauk, O., & Pulvermüller, F. (2004). Neurophysiological distinction of action words in the fronto-central cortex. *Human Brain Mapping, 21,* 191–201.
Hauk, O., & Pulvermüller, F. (2011). The lateralization of motor cortex activation to action words. *Frontiers in Human Neuroscience, 5,* Article 149.
Hauk, O., & Tschentscher, N. (2013). The body of evidence: What can neuroscience tell us about embodied semantics? *Frontiers in Psychology, 4,* Article 50.
Hauk, O., Johnsrude, I., & Pulvermüller, F. (2004). Somatotopic representation of action words in human motor and premotor cortex. *Neuron, 41,* 301–307.
Hauk, O., Shtyrov, Y., & Pulvermüller, F. (2008). The time course of action and action-word comprehension in the human brain as revealed by neurophysiology. *Journal of Physiology, Paris, 102,* 50–58.
Hillis, A. E., Oh, S., & Ken, L. (2004). Deterioration of naming nouns versus verbs in primary progressive aphasia. *Annals of Neurology, 55,* 268–275.
Hillis, A. E., Heidler-Gray, J., Newhart, M., Chang, S., Ken, L., & Bak, T. H. (2006). Naming and comprehension in primary progressive aphasia: The influence of grammatical word class. *Aphasiology, 20,* 246–256.
Hoenig, K., Sim, E. J., Bochev, V., Herrnberger, B., & Kiefer, M. (2008). Conceptual flexibility in the human brain: Dynamic recruitment of semantic maps from visual, motor, and motion-related areas. *Journal of Cognitive Neuroscience, 20,* 1799–1814.
Humphreys, G. F., Newling, K., Jennings, C., & Gennari, S. P. (2013). Motion and actions in language: Semantic representations in occipito-temporal cortex. *Brain and Language, 125,* 94–105.
Jastorff, J., & Orban, G. A. (2009). Human functional magnetic resonance imaging reveals separation and integration of shape and motion cues in biological motion processing. *Journal of Neuroscience, 29,* 7315–7329.
Kable, J. W., Lease-Spellmeyer, J., & Chatterjee, A. (2002). Neural substrates of action event knowledge. *Journal of Cognitive Neuroscience, 14,* 795–805.
Kable, J. W., Kan, I. P., Wilson, A., Thompson-Schill, S. L., & Chatterjee, A. (2005). Conceptual representations of action in the lateral temporal cortex. *Journal of Cognitive Neuroscience, 17,* 1855–1870.
Kalénine, S., Buxbaum, L. J., & Coslett, H. B. (2010). Critical brain regions for action recognition: Lesion symptom mapping in left hemisphere stroke. *Brain, 133,* 3269–3280.
Kaschak, M. P., & Borreggine, K. L. (2008). Temporal dynamics of the action-sentence compatibility effect. *Quarterly Journal of Experimental Psychology, 61,* 883–895.
Kemmerer, D., & Gonzalez Castillo, J. (2010). The Two-Level Theory of verb meaning: An attempt to integrate the semantics of action with the mirror neuron system. *Brain and Language, 112,* 54–76.
Kemmerer, D., & Tranel, D. (2003). A double dissociation between the meanings of action verbs and locative prepositions. *Neurocase, 9,* 421–435.
Kemmerer, D., Tranel, D., & Barrash, J. (2001). Patterns of dissociation in the processing of verb meanings in brain-damaged subjects. *Language and Cognitive Processes, 16,* 1–34.

Kemmerer, D., Gonzalez Castillo, J., Talavage, T., Patterson, S., & Wiley, C. (2008). Neuroanatomical distribution of five semantic components of verbs: Evidence from fMRI. *Brain and Language, 107,* 16–43.

Kemmerer, D., Rudrauf, D., Manzel, K., & Tranel, D. (2012). Behavioral patterns and lesion sites associated with impaired processing of lexical and conceptual knowledge of actions. *Cortex, 48,* 826–848.

Kemmerer, D., Miller, L., MacPherson, M. K., Huber, J., & Tranel, D. (2013). An investigation of semantic similarity judgments about action and non-action verbs in Parkinson's disease: Implications for the embodied cognition framework. *Frontiers in Human Neuroscience, 7,* Article 146.

Kiefer, M., & Pulvermüller, F. (2012). Conceptual representations in mind and brain: Theoretical developments, current evidence and future directions. *Cortex, 48,* 805–825.

Kuipers, J. R., van Koningsbruggen, M., & Thierry, G. (2013). Semantic priming in the motor cortex: Evidence from combined repetitive transcranial magnetic stimulation and event-related potentials. *Neuroreport, 24,* 646–651.

Lebois, L. A. M., Wilson-Mendenhall, C. D., & Barsalou, L. W. (in press). Are automatic conceptual cores the gold standard of semantic processing? The context-dependence of spatial meaning in grounded congruency effects. Cognitive Science.

Levin, B. (1993). *English verb classes and alternations: A preliminary investigation.* Chicago: University of Chicago Press.

Levin, B., & Rappaport Hovav, M. (2005). *Argument realization.* Cambridge: Cambridge University Press.

Levin, B., & Rappaport Hovav, M. (2011). Lexical conceptual structure. In K. von Heusinger, C. Maienborn, & P. Portner (Eds.), *Semantics: An international handbook of natural language meaning* (Vol. I, pp. 418–438). Berlin: Mouton de Gruyter.

Levinson, S. C., & Wilkins, D. (Eds.) (2006). *Grammars of space: Explorations in cognitive diversity.* Cambridge: Cambridge University Press.

Liljeström, M., Tarkianen, T., Parviainen, T., Kujala, J., Numminen, J., Hiltunen, J., Laine, M., & Salmelin, R. (2008). Perceiving and naming actions and objects. *NeuroImage, 41,* 1132–1141.

Lin, N., Lu, X., Fang, F., Han, Z., & Bi, Y. (2011). Is the semantic category effect in the lateral temporal cortex due to motion property differences? *NeuroImage, 55,* 1853–1864.

Maieron, M., Fabbro, F. & Skrap, M. (2013). Seeking a bridge between language and motor cortices: A PPI study. *Frontiers in Human Neuroscience, 7,* Article 249.

Majid, A., Boster, J. S., & Bowerman, M. (2008). The cross-linguistic categorization of everyday events: A study of cutting and breaking. *Cognition, 109,* 235–250.

Malikovic, A., Amunts, K., Schleicher, A., Mohlberg, H., Eickhoff, S. B., Wilms, M., Palomero-Gallagher, N., Armstrong, E., & Zilles, K. (2007). Cytoarchitectonic analysis of the human extrastriate cortex in the region of V5/MT+: A probabilistic, stereotaxic map of area hOc5. *Cerebral Cortex, 17,* 562–574.

Malt, B. C., Ameel, E., Imai, M., Gennari, S. P., Saji, N., & Majid, A. (in press). Human locomotion in languages: Constraints on moving and meaning. *Journal of Memory and Language.*

Malt, B. C., Gennari, S., Imai, M., Ameel, E., Tsuda, N., & Majid, A. (2008). Talking about walking: Biomechanics and the language of locomotion. *Psychological Science, 19,* 232–240.

Mayka, M. A., Corcos, D. M., Leurgans, S. E., & Vaillancourt, D. E. (2006). Three-dimensional locations and boundaries of motor and premotor cortices as defined by functional brain imaging: A meta-analysis. *NeuroImage, 31,* 1453–1474.

McGregor, W. (2002). *Verb classification in Australian languages.* Berlin: Mouton de Gruyter.

Meier, J. D., Aflalo, T. N., Kastner, S., & Graziano, M. S. A. (2008). Complex organization of human primary motor cortex: A high-resolution fMRI study. *Journal of Neurophysiology, 100,* 1800–1812.

Meteyard, L., Cuadrado, S. R., Bahrami, B., & Vigliocco, G. (2012). Coming of age: A review of embodiment and the neuroscience of semantics. *Cortex, 48,* 788–804.

Mirabella, G., Iaconelli, S., Spadacenta, S., Federico, P., & Gallese, V. (2012). Processing of hand-related verbs specifically affects the planning and execution of arm reaching movements. *PloS ONE, 7,* e35403.

Moody, C. L., & Gennari, S. P. (2010). Effects of implied physical effort in sensory-motor and prefrontal cortex during language comprehension. *NeuroImage, 49,* 782–793.

Moseley, R. L., Pulvermüller, F., & Shtyrov, Y. (2013). Sensorimotor semantics on the spot: Brain activity dissociates between conceptual categories within 150 ms. *Scientific Reports, 3,* Article 1928.

Nassi, J. J., & Callaway, E. M. (2006). Multiple circuits relaying primate parallel visual pathways to the middle temporal area. *Journal of Neuroscience, 26,* 12789–12798.

Nazir, T. A., Boulenger, V., Roy, A., Silber, B., Jeannerod, M., & Paulignan, Y. (2008). Language-induced motor perturbations during the execution of a reaching movement. *Quarterly Journal of Experimental Psychology, 61,* 933–943.

Noppeney, U., Josephs, O., Kiebel, S., Friston, K. J., & Price, C. J. (2005). Action selectivity in parietal and temporal cortex. *Cognitive Brain Research, 25,* 641–649.

Oosterhof, N. N., Tipper, S. P., & Downing, P. E. (2013). Crossmodal and action-specific: Neuroimaging the human mirror neuron system. *Trends in Cognitive Sciences, 17,* 311–318.

Papeo, L., Negri, G. A. L., Zadini, A., & Rumiati, R. I. (2010). Action performance and action-word understanding: Evidence of double dissociations in left-damaged patients. *Cognitive Neuropsychology, 27,* 428–461.

Papeo, L., Rumiati, R. I., Cecchetto, C., & Tomasino, B. (2012). On-line changing of thinking about words: The effect of cognitive context on neural responses to verb reading. *Journal of Cognitive Neuroscience, 24,* 2348–2362.

Peelen, M. V., & Downing, P. E. (2007). The neural basis of visual body perception. *Nature Reviews Neuroscience, 8,* 636–648.

Peelen, M. V., Romagno, D., & Caramazza, A. (2012). Is verb selectivity in left posterior temporal cortex related to conceptual action knowledge? *Journal of Cognitive Neuroscience, 24,* 2096–2107.

Penfield, W., & Boldrey, E. (1937). Somatic motor and sensory representation in the cerebral cortex of man as studied by electrical stimulation. *Brain, 60,* 389–443.

Penfield, W., & Rasmussen, T. (1950). *The cerebral cortex of man: A clinical study of localization of function*. New York: Mcmillan.

Penolazzi, B., Hauk, O., & Pulvermüller, F. (2007). Early semantic context integration and lexical access as revealed by event-related potentials. *Biological Psychology, 74,* 374–388.

Pinker, S. (1989). *Learnability and cognition: The acquisition of argument structure*. Cambridge: MIT Press.

Pinker, S. (2007). *The stuff of thought: Language as a window into human nature.* New York: Viking.

Pirog Revill, K., Aslin, R. A., Tanenhaus, M. K., Bavelier, D. (2008). Neural correlates of partial lexical activation. *Proceedings of the National Academy of Sciences of the United States of America, 105,* 13111–13115.

Postle, N., McMahon, K. L., Ashton, R., Meredith, M., & de Zubicaray, G. I. (2008). Action word meaning representations in cytoarchitectonically defined primary and premotor.

Pulvermüller, F. (2013). How neurons make meaning: Brain mechanisms for embodied and abstract-symbolic semantics. *Trends in Cognitive Sciences, 17,* 458–470.

Pulvermüller, F., Härle, M., & Hummel, F. (2001). Walking or talking? Behavioral and neurophysiological correlates of action verb processing. *Brain and Language, 78,* 143–168.

Pulvermüller, F., Hauk, O., Nikulin, V., & Ilmoniemi, R. (2005a). Functional links between motor and language systems. *European Journal of Neuroscience, 21,* 793–797.

Pulvermüller, F., Shtyrov, Y., & Ilmoniemi, R. (2005b). Brain signatures of meaning access in action word recognition. *Journal of Cognitive Neuroscience, 17,* 884–892.

Pulvermüller, F., Kherif, F., Hauk, O., Mohr, B., & Nimmo-Smith, I. (2009). Distributed cell assemblies for general lexical and category-specific semantic processing as revealed by fMRI cluster analysis. *Human Brain Mapping, 30,* 3837–3850.

Raposo, A., Moss, H. E., Stamatakis, E. A., & Tyler, L. K. (2009). Modulation of motor, premotor cortices by actions, action words, and action sentences. *Neuropsychologia, 47,* 388–396.

Rappaport Hovav, M., & Levin, B. (1998). Building verb meanings. In M. Butt & W. Geuder (Eds.), *The projection of arguments* (pp. 97–134). Stanford: CSLI.

Repetto, C., Colombo, B., Cipresso, P., & Riva, G. (2013). The effects of rTMS over the primary motor cortex: The link between action and language. *Neuropsychologia, 51,* 8–13.

Rizzolatti, G., & Sinigaglia, C. (2010). The functional role of the parieto-frontal mirror circuit: Interpretations and misinterpretations. *Nature Reviews: Neuroscience, 11,* 264–274.

Rueschemeyer, S. A., Ekman, M., van Ackeren, M., & Kilner, J. (2014). Observing, performing, and understanding actions: Revisiting the role of cortical motor areas in processing action words. *Journal of Cognitive Neuroscience, 26,* 1644–1653.

Rueschemeyer, S. A., Brass, M., & Friederici, A. D. (2007). Comprehending prehending: Neural correlates of processing verbs with motor stems. *Journal of Cognitive Neuroscience, 19,* 855–865.

Sato, M., Mengarelli, M., Riggio, L., Gallese, V., & Buccino, G. (2008). Task related modulation of the motor system during language processing. *Brain and Language, 105,* 83–90.

Saygin, A. P. (2012). Sensory and motor brain areas supporting biological motion perception: Neuropsychological and neuroimaging studies. In K. Johnson & M. Shiffrar (Eds.), *People watching: Social, perceptual, and neurophysiological studies of body perception* (pp. 371–389). Oxford: Oxford University Press.

Saygin, A. P., McCullough, S., Alac, M., & Emmorey, K. (2010). Modulation of the BOLD response in motion sensitive lateral temporal cortex by real and fictive motion sentences. *Journal of Cognitive Neuroscience, 22,* 2480–2490.

Schuil, K. D. I., Smits, M., & Zwaan, R. A. (2013). Sentential context modulates the involvement of the motor cortex in action language processing: An fMRI study. *Frontiers in human neuroscience, 7,* Article 100.

Schultze-Berndt, E. (2006). Sketch of a Jaminjung grammar of space. In S. C. Levinson & D. Wilkins (Eds.), *Grammars of space: Explorations in cognitive diversity* (pp. 63–114). Cambridge: Cambridge University Press.

Scorolli, C., & Borghi, A. M. (2007). Sentence comprehension and action: Effector specific modulation of the motor system. *Brain Research, 1130,* 119–124.

Shetreet, E., Palti, D., Friedmann, N., & Hadar, U. (2007). Cortical representation of verb processing in sentence comprehension: Number of complements, subcategorization, and thematic frames. *Cerebral Cortex, 17,* 1958–1969.

Shtyrov, Y., Hauk, O., & Pulvermüller, F. (2004). Distributed neuronal networks for encoding category-specific semantic information: The mismatch negativity to action words. *European Journal of Neuroscience, 19,* 1083–1092.

Slobin, D. I. (2000). Verbalized events: A dynamic approach to linguistic relativity and determinism. In S. Niemeier & R. Dirven (Eds.), *Evidence for linguistic relativity* (pp. 107–138). Amsterdam: John Benjamins.

Slobin, D. I. (2006). What makes manner of motion salient? Explorations in linguistic typology, discourse, and cognition. In M. Hickmann, & S. Robert (Eds.), Space in languages: Linguistic systems and cognitive categories (pp. 59–82). Amsterdam: John Benjamins.

Snell-Hornby, M. (1983). *Verb descriptivity in German and English.* Heidelberg: Carl Winter.

Taylor, L. J., & Zwaan, R. A. (2008). Motor resonance and linguistic focus. *Quarterly Journal of Experimental Psychology, 61,* 896–904.

Tettamanti, M., Buccino, G., Saccuman, M. C., Gallese, V., Danna, M., Scifo, P., Fazio, F., Rizzolatti, G., Cappa, S. F., & Perani, D. (2005). Listening to action-related sentences activates fronto-parietal motor circuits. *Journal of Cognitive Neuroscience, 17,* 273–281.

Tomasino, B., & Rumiati, R. I. (2013). At the mercy of strategies: The role of motor representations in language understanding. *Frontiers in Psychology, 4,* Article 27.

Tranel, D., Kemmerer, D., Adolphs, R., Damasio, H., & Damasio, A. (2003). Neural correlates of conceptual knowledge for actions. *Cognitive Neuropsychology, 20,* 409–432.

Tranel, D., Martin, C., Damasio, H., Grabowski, T. J., & Hichwa, R. (2005). Effects of noun-verb homonymy on the neural correlates of naming concrete entities and actions. *Brain and Language, 92,* 288–299.

Tranel, D., Manzel, K., Asp, E., & Kemmerer, D. (2008). Naming static and dynamic actions: Neuropsychological evidence. *Journal of Physiology, Paris, 102,* 80–94.

van Dam, W. O., Rueschemeyer, S. A., & Bekkering, H. (2010). How specifically are action verbs represented in the neural motor system? An fMRI study. *NeuroImage, 53,* 1318–1325.

Van Valin, R. D. (2005). *Exploring the syntax-semantic interface*. Cambridge: Cambridge University Press.

Van Valin, R. D. (2006). Some universals of verb semantics. In R. Mairal & J. Gil (Eds.), *Linguistic universals* (pp. 155–178). Cambridge: Cambridge University Press.

Wallentin, M., Nielson, A. H., Vuust, P., Dohn, A., Roepstorff, A., & Lund, T. E. (2011). BOLD response to motion verbs in left posterior middle temporal gyrus during story comprehension. *Brain and Language, 119,* 221–225.

Watson, C. E., Cardillo, E. R., Ianni, G. R., & Chatterjee, A. (2013). Action concepts in the brain: An activation-likelihood estimation meta-analysis. *Journal of Cognitive Neuroscience, 25,* 1191–1205.

Willems, R., Hagoort, P., & Casasanto, D. (2010). Body-specific representations of action verbs: Neural evidence from right- and left-handers. *Psychological Science, 21,* 67–74.

Wilson, S. (1999). *Coverbs and complex predicates in Wagiman*. Stanford: CSLI.

Wu, D., Waller, S., & Chatterjee, A. (2007). The functional neuroanatomy of thematic role and locative relational knowledge. *Journal of Cognitive Neuroscience, 19,* 1542–1555.

Zwaan, R. A., & Taylor, L. J. (2006). Seeing, acting, understanding: Motor resonance in language comprehension. *Journal of Experimental Psychology: General, 135,* 1–11.

Chapter 10
Which Event Properties Matter for Which Cognitive Task?

Jean-Pierre Koenig, Douglas Roland, Hongoak Yun and Gail Mauner

Much of our everyday language use is concerned with describing situations, or what linguists call events and states. Verbs play a critical role in this endeavor, since they describe types or categories of situations (in this chapter, we only discuss events and event types). Speakers and comprehenders know a lot about each event type and much of this information is treated as mutual belief in the sense of Clark (1992). Both Speakers and comprehenders know that in describing a situation where a farmer is loading boxes of tomatoes onto a truck, it is felicitous to say that *the farmer loaded the tomatoes onto the truck,* whether the truck is full or not, but that one can only say *the farmer loaded the truck with tomatoes* if, as a result, the truck is completely full (e.g., Levin and Rappaport Hovav 2005). Thus, if a comprehender hears that *the farmer loaded the truck with tomatoes,* the comprehender understands that the speaker believed that the truck was full.

Comprehenders also have knowledge of what kinds of things are likely to be loaded by different people. Comprehenders are quicker to read *the truck* after *the farmer*

J. -P. Koenig (✉)
Department of Linguistics and Center for Cognitive Science, University at Buffalo, 609 Baldy Hall, Buffalo, NY 14260, USA
e-mail: jpkoenig@buffalo.edu

D. Roland
Graduate School of Arts and Sciences, The University of Tokyo, 3–8-1 Komaba, Meguro-ku, Tokyo 153–8902, Japan
e-mail: doug.roland@gmail.com

H. Yun
Department of Psychology and Center for Cognitive Science, University at Buffalo, 1301–401 Samwhan Apartment, Hyunjung-ro 13, Ilsan-seo-gu, Koyang-si, Kyunggi-do 411–781, South Korea
e-mail: oaktreepark@gmail.com

G. Mauner
Department of Psychology and Center for Cognitive Science, University at Buffalo, 204 Park Hall, Buffalo, NY 14260, USA
e-mail: mauner@buffalo.edu

R. G. de Almeida, C. Manouilidou (eds.), *Cognitive Science Perspectives on Verb Representation and Processing,* DOI 10.1007/978-3-319-10112-5_10

loaded, and *the pistol* after *the assassin loaded* than the other way around (Bicknell et al. 2010). Comprehenders must, at some level, both represent the knowledge that there is a relationship between the syntactic structure used in describing the loading event and whether the action results in the truck being full or not, and the knowledge that assassins are more likely to load pistols, while farmers are more likely to load trucks.

Two important questions in the language sciences have been: What portion of this encyclopedic information is accessed and used during sentence processing? What portion of this information is relevant to the grammar of natural languages? An extensive amount of psycholinguistic research over the past couple of decades has shown that quite a bit of event information is relevant to online sentence processing. In contrast, only a limited amount of information has been found to be relevant to the grammars of natural languages. This contrast leads to the question of why the human parser and the human "grammar maker" seem to rely on different kinds and amounts of information. One possible cause for this divergence is that grammar development and language comprehension are carried out by separate systems that are sensitive to different types of information. This is typically cast in terms of speakers having a separate syntactic subsystem that is only sensitive to a limited set of properties of language (e.g., Pinker 1989). Another possible explanation is that grammar development and language comprehension are carried out by an integrated system, but the task demands of language development and the task demands of language comprehension are different. In this second view, the apparent specificity found in grammar learning is not due to limitations in what the system is sensitive to, but, rather, due to limitations in what information the system finds useful for the task it faces.

In this chapter, we first outline how grammars make limited use of our vast knowledge of events; we show that, grammatical systems that seem more "exotic" from the point of view of more well-known languages still make use of a limited set of properties—even if these properties are not the ones that are typically on the list of properties considered to be grammatically relevant. More importantly for the *why* question, we show that these "exotic" languages still obey the same design constraints as the more well-known systems; we then briefly report on some computational models of online reading experiments which demonstrate quite clearly that a distinct and much larger kind of event knowledge is used by the human parser; and finally, we propose an explanation for the difference in the use of event knowledge. In short, our explanation is that grammars and parsers use different kinds of event knowledge because the tasks listeners and grammar learners must perform are quite distinct.

10.1 The Grammar of Events Is Minimalist

10.1.1 Event Properties that Matter

Several aspects of the grammars of natural languages are sensitive to properties of events and their participants. (For ease of exposition, we sometimes will speak of the *grammar of events* to mean the portion of the grammars of natural languages that is concerned with the description of events.) To name a few: the number of a verb's obligatory syntactic dependants (subject and complements), possibly, the frequency of occurrence of syntactically optional dependants, the grammatical function of these dependants (e.g., what is encoded as the subject or object of the verb), the case of these dependants (e.g., whether the subject, exceptionally, bears a dative case, whether the object, exceptionally, bears a genitive case), valence alternations (i.e., what distinct lists of dependants a verb may have), the mood of the head verb of sentential dependants (whether the head verb of an embedded clause should be in the indicative, optative, or subjunctive). Two patterns have emerged as linguists have investigated an ever-growing number of languages.

First, grammatical processes tend to target *semantically defined classes* of verbs, that is, verbs that share one or more event properties (Pinker 1989; Levin 1993, among others). For example, Pinker points out that the ditransitive construction in English, exemplified in *Mary gave Bill a book,* targets several narrow, semantically defined verb classes (e.g., verbs of future having such as *promise* and *bequeath*). More generally, the monumental work of Levin (1993) demonstrates that there is a close connection between valence alternations and semantic classes of verbs in English. Verbs that belong to the same semantic classes have the same (or very similar) sets of lists of dependants: Verbs that can both be transitive and intransitive, in both their middle and inchoative incarnations, occupy the same region of semantic space; loosely speaking, they are verbs that in their transitive variant describe externally induced changes of state, i.e., typically, changes of state induced by the referent of the subject.[1] When one looks beyond English, verb classes that are very similar semantically to the classes identified by Levin tend to recur as the target of grammatical processes, although not necessarily of the same kind of processes represented by English valence alternations. We illustrate this fact with a look at Hindi ergative case assignment in Sect. 10.1.3.

Second, the range of event properties that determine the encoding of a verb's dependants is very limited. Consider the properties of events that are relevant to linking constraints, i.e., the set of constraints that map semantic arguments onto grammatical functions (leaving aside whether these functions are primitives or derived from phrase–structural relations). The list in Dowty (1991) is fairly limited. The properties that affect the subject and object selection, respectively, are: volition, having a mental representation, causing an event, being in motion, and independent existence; and changing state, being an incremental theme, being causally affected, being stationary, and nonindependent existence. Other researchers might add a few properties, or subtract or rephrase others, but the list would not change much. Similarly, the list of event properties deemed by Pinker (1989) to be relevant to the determination of

[1] McKoon and MacFarland (2000) find corpus examples of verbs normally associated with internal causation appearing in transitive uses, but note of these uses that "if something is said to erode a beach, this cannot be just any something—not a person, not a shovel—it must be something that participates intrinsically in erosion, like wind or water." This finding furthers the notion that there are links between semantic properties and syntactic properties.

(narrow) verb classes targeted by valence patterns (e.g., the ditransitive construction) is quite small: state versus motion; path, direction, and location; causation; manner; properties or categories of the moving entity; temporal distribution; purpose; coreferentiality; and truth value. Again, other researchers might tweak part of this list or add to this list. Note that while some members of this list are potentially open-ended (e.g., manner or properties/categories of the moving object), this is clearly not Pinker's intention. In any case, what is consistent across authors is that the list of grammatically relevant properties is fairly small in comparison to the range of properties of events we know of.

Of course, one rather uninteresting possible explanation for the apparently limited range of grammatically relevant event properties is that linguists have not yet carefully examined languages with a more diverse range of grammatically relevant properties. However, we feel that this is unlikely to be the case. In the rest of this section, we present two case studies of apparently more exotic grammatical systems. Our conclusion will be that these systems are similar in critical respects to more well-known systems, and do not constitute exceptions to the pattern. Furthermore, our description of these less well-known patterns will highlight what we think are critical design properties of the grammar of events and help us understand why grammatically relevant event properties are so limited, as we discuss in Sect. 10.3.

10.1.2 Kin Terms in Oneida (Iroquoian)

The syntax of kin terms in Oneida (a Northern Iroquoian language) is particularly complex so we only focus on what is relevant to our discussion here (see Koenig and Michelson 2010, for details). A few forms will suffice to illustrate our point. Oneida, like other Iroquoian languages, marks its arguments on the verb or noun itself via pronominal prefixes. In the case of most kin terms, intransitive and transitive prefixes are used to mark gender, person, and number of the "subject" and "object" (agent, patient, and transitive prefixes in the Iroquoianist tradition). What is critical for our purposes is the rules that determine which argument of the kin relation denoted by a kin term is the "subject" and which is the "object." To avoid prejudging this issue and because determination of the "subject" and "object" of the kin term is orthogonal to determination of the term's referent or index, we name the kin relation by listing the members of the relation, and underline the member of the relation that corresponds to the kin term's referent. Thus, *mother–<u>child</u>* stands for the kin relation that holds between a mother and a child when the child is the kin term's referent, while *<u>mother</u>–child* stands for the kin relation that holds between a mother and a child when the mother is the kin term's referent. Crucially, in Oneida, whether one uses the root that is chosen to talk about a mother, *nulhá,* as in (1), or the root that is chosen to talk about a child, *yʌha*, as in (2), the "subject" always corresponds to the older-generation kin and the "object" to the younger-generation kin. (Transitive pronominal prefixes encode both the "subject" and "object" of a stem. The gloss *3ZOIC.SG > 3MASC.SG* indicates that the "subject" is third zoic singular and the "object" third masculine singular). The rule in (3) accounts for the "subject" selection of most kin terms in Oneida.

1. lo-nulhá·
   3ZOIC.SG > 3MASC.SG-mother.child
   "his mother"
2. luwa-yʌha
   3FEM.SG > 3MASC.SG-mother.child
   "her son"
3. "Subject"-*selection rule 1* (refers to *generation*): The argument that corresponds to the older generation maps onto the "subject," while the argument that corresponds to the younger generation maps onto the "object."

Rule (3) is inappropriate for a few kin terms, in particular for the root—*ʔkʌha*—"sibling." In this case, the "subject"-selection rule must refer to age, not generation. As (4) and (5) indicate, siblings do not differ generationally, but whoever is the older sibling must be the "subject" and whoever is the younger must be the "object." Rule (6) accounts for the "subject" selection of—*ʔkʌha*—and a couple other stems.

4. lake-ʔkʌha
   3MASC.SG > 1SG-sibling
   "my older brother"
5. khe-ʔkʌha
   1SG > 3FEM.SG-sibling
   "my younger sister"
6. "Subject"-*selection rule 2* (refers to *age,* not generation): The argument that corresponds to the older person in a kin relation maps onto the "subject," while the argument that corresponds to the younger person maps onto the "object."

Rules (3) and (6) are "exotic" and differ markedly from traditional subject-selection rules or linking rules. In that sense, they may challenge Pinker's (1989) claim that linking rules are "quasi innate." But, they share crucial properties with other linking rules. Consider rule (3). It is not unique to Oneida, but seems typical of what Evans (2000) calls *kin verbs,* i.e., it is often found in languages in which kin terms are verbs (or, at least, partially verbs, as in Oneida; see Koenig and Michelson 2010). It is also operative, for example, in Ilgar, an Australian language. Although the content of rule (3) is unknown to nonkin-verb languages, its form is similar to that of the more familiar linking rules. More precisely, rule (3) is based on entailments of sentences that contain the kin term, and therefore applies to all pairs of arguments of the kin relation, as illustrated in (7); that being generationally older is what is relevant is shown by words for uncle and aunt, where generational order and absolute age order do not necessarily coincide; see Koenig and Michelson, op.cit.). Rule (3) applies no matter what properties particular mothers and children have; as long as there is a mothering relation, whoever is the mother will be the "subject," as she is generationally older.

7. For all **X** and *Y*, if MOTHER (**X,** *Y*), then **X** is generationally older than *Y*.

Rule (3) is also formally similar to many other grammar rules that make reference to event properties; namely, it applies to a semantically defined class of roots, kin terms. All stems to which rule (3) applies include kin relations in their meanings.

In contrast to rule (3), rule (6) also differs formally from less "exotic"-linking rules, in that it does not rely on a property *entailed* by the kin term. In other words, that one sibling is older than the other is a property of the individuals that are in a sibling relation, not a property of the relation itself. What is interesting, though, and confirms its oddity, is that the linking rule in (6) is an Oneida innovation. The term *ʔkʌha* used to mean "younger sibling," but came to mean just "sibling." At which point, the property became an incidental property of the fillers of the kin relation's argument positions rather than an entailed property of the relation. Interestingly, the property that determines the subject selection, being older, is still true of all the referents of the kin term's "subjects."

10.1.3 Ergative Case in Hindi

The purpose of linking rules, such as (3) or (6), is to map a word's arguments onto morphosyntactic positions (and, distinguish among arguments in so doing). They apply to all fillers of the argument positions. In some cases, though, the purpose of grammatical rules is to distinguish between different types of fillers. The conditions under which ergative case marking is assigned to subjects in Hindi will illustrate this case with another, apparently "exotic" set of rules (see Shakthi and Koenig 2009; Shakthi 2012).

Ergative case marking in Hindi is sensitive to the verb's aspect, a condition on ergativity that occurs in other languages. Thus, in (8), *Ram* is marked with the ergative case suffix *-ne,* because the main verb is in the perfective, but not in (9), where the verb is in the imperfective. The rule in (10) covers ergative case assignment when the verb is transitive.

8. Ram = ne ghar = ko banaa-yaa
   Ram = Erg house = Dat make-Pfv.M.Sg
   'Ram built the house.'
9. Ram ghar = ko banaa-taa hai
   Ram house = Dat make-Impfv be
   'Ram is building the house.'
10. *Rule 1*: If the verb is *transitive* and *perfective,* the subject is assigned *ergative* case.

Although still sensitive to the verb's aspect, ergative case assignment when the verb is intransitive is subject to additional, more "exotic" conditions illustrated in (11) and (12).

11. Ram khans-aa
    Ram cough-Pfv.M.Sg
    'Ram coughed (without meaning to).'
12. Ram = ne khans-aa
    Ram = Erg cough-Pfv.M.Sg
    'Ram coughed (purposefully).'
13. *khaas* 'cough,' *chiikh* 'sneeze,' *bhauk* 'bark,' *ciik* 'scream,' *cillaa* 'yell,' *muut* 'urinate,' and *thuuk* 'spit'

Some of the verbs denoting bodily functions to which this rule applies are listed in (13) (overall, the rule applies to only about 25 verbs, as many bodily functions are encoded via nominal complements to a light verb). De Hoop and Narashiman (2008) suggest that the subject's referent must have acted volitionally when it bears an ergative case in (11) and (12). It is true that in most attested examples, ergative case marking indicates that the subject's referent performed a bodily function in a nonnatural way, that is, with a *purpose* distinct from the normal coughing, as in (11). But, some examples suggest that the "exotic" additional condition on the assignment of ergative case to the subject of intransitive verbs is somewhat more abstract, and cannot be explained purely as volitionality. Consider the attested example in (14) or the example in (15). The dog cannot, presumably, have the intention required to purposefully not bark in (14). Similarly, there need not be anything unusual about the urination in (15). Rather, it is surprising that everybody urinated at the same time. What seems to be common to all uses of the ergative with intransitive bodily emission verbs is that the action (or, rarely, inaction) was somehow unexpected. One would have expected the dog to bark, and one would not expect everyone in a crowd to simultaneously urinate. Similarly in (12), one would not have expected Ram to cough, given his health. We therefore propose the, for now, informal ergative case assignment rule in (16) to cover intransitive verbs.

14. court mein bahut log moujuud th-ee phir bhii kiisii par bhii *kuttee = ne*
    court in many people present be-Past.3.Pl still any on also *dog = Erg*
    bhauunk-aa tak nahii
    bark-M.Sg even neg
    "Many people were present in court but still the dog did not even bark at anyone"
15. kiisii *ek = ne* nahii *sab = ne* muut-aa
    any *one = Erg* neg *all = Erg* urinate-M.Sg
    "Not just one (person) but everyone urinated."
16. *Rule 2*: If the verb is *intransitive* and *perfective,* it denotes a bodily function, and the action is *unexpected* on the actor's part, then the subject is assigned *ergative* case.

Rule (16) is somewhat "exotic" when it comes to (ergative) case assignment rules; all the more so, since expectations are properties of propositions or situations and case assignment is a formal mark on a dependant of the sentence's head. But, the event property that is marked (being unexpected) is one which is not unknown in other parts of the world. Furthermore, although it is somewhat unusual for a case marking rule to be restricted to a class of verbs as restricted as bodily function verbs (see Malchukov 2008, for a cross-linguistic perspective), the sensitivity of grammars to semantically defined verb classes is well-known. As was the case of Oneida kin term linking rules, Hindi ergative case assignment rules may seem "exotic," but do not invalidate the overarching generalization that grammars are sensitive to a limited number of event properties.

10.2 Sentence Processing Is Promiscuous

In the preceding section, we have seen that grammars make use of a very limited set of event properties. Even in the more "exotic" systems, there seem to be strong constraints on the type of event properties that can influence grammatical systems. Properties that matter to grammar are still part of the meaning of the verb (where meaning is defined, traditionally, in terms of entailments) or part of the semantic contribution of the sentence's syntactic frame (e.g., unexpectedness contributed by the ergative case marking in Hindi) and apply to semantically defined classes of verbs. As mentioned in the introduction, online sentence processing is much more inclusive: Many kinds of event properties seem to matter to sentence processing (see, among many others, Spivey-Knowlton and Sedivy 1995; Tanenhaus et al. 1995; McRae et al. 1997; Altmann and Kamide 1999; Kamide et al. 2003). Kamide et al. (2003), for example, show that the semantic category of the agent affects listeners' looks to picture of potential patients. Thus, upon hearing *the man will ride...* while looking at a picture array that includes pictures of a biker, a girl, a motorcycle, a carrousel, and two other objects, listeners will launch more looks to the motorcycle than when they hear *the girl will ride...* immediately after hearing the verb *ride*. This result, and many other results obtained in the same so-called visual world paradigm, suggests that listeners integrate their knowledge of events (what bikers versus girls are likely to ride), information provided by the linguistic input, and visual information, to predict what the direct object of a verb will be (the object of *ride,* here). Clearly, listeners in Kamide et al.'s Experiment 2 must have used their detailed world knowledge of bikers and events of bikers riding to predict the category of upcoming constituents (i.e., motorcycle).

10.2.1 Semantic Predictability Versus Semantic Similarity

In this section, we want to present data[2] that show that sentence processing is sensitive to (1) the likelihood of a dependant of a verb's semantic category (what we call the *semantic predictability hypothesis*), and (2) the distribution in semantic space of

---

[2] The data presented in this chapter comes from a preliminary version of the work reported in Roland et al. (2012).

the possible categories of a verb's dependant (what we call the *semantic similarity hypothesis*). Before examining each hypothesis in turn, we illustrate the hypotheses on a couple of examples. Consider (17–18).

17. The aboriginal man | *jabbed* | the angry lion | with | a spear | near its prey.
18. The aboriginal man | *attacked* | the angry lion | with | a spear | near its prey.

The previous studies we cited suggest that how fast the region following *with* will be read depends on how likely or semantically predictable a particular instrument is in the scene being described by the sentence up to *with*. Thus, processing will be faster for *spear,* a very likely (predictable) instrument, than for the *hockey stick,* a very unlikely (unpredictable) instrument. As a consequence, if the instrument that occurs is equally likely for the event described by (17) as for the one described by (18) (e.g., if the *spear* is equally likely for both events), we would expect reading times of the underlined region to be equal. But, as we show in this section, processing is sensitive to an even subtler aspect of event knowledge, namely how many other semantically similar instruments could have been the complement of *with* rather than *spear*. More precisely, the semantic similarity hypothesis is that differences in the distribution in semantic space of the likely instruments of *jab* and *attack* might affect processing of the same actual instrument, namely *spear*. Figure 10.1 illustrates this putative difference between the range of instruments for *jab* and *attack:* Intuitively, likely instruments of *jab* (spear, sword, knife, fork, machete, etc.) are more similar to each other than likely instruments of *attack* (spear, sword, knife, gun, rock, stick, etc), because *jab* places requirements on the instrument (e.g., pointy, able to be held in hand) while *attack* does not (e.g., attacking can be done with words, nuclear weapons, etc.). Thus, in processing a sentence with *jab,* listeners and readers would be able to predict more of the properties of the instrument than they would in a sentence with *attack.* These properties, and the categories of instruments that have them, will be more strongly activated, facilitating the processing of the actual instrument, *spear.*

If our hypothesis is correct, reading times of the underlined region might not be equivalent in sentences like (17) and (18) despite the fact that the *spear* is an equally likely instrument of the events being described, because events of jabbing involve instruments that are more semantically similar to the *spear* than events of attacking do. In the rest of this section, we present data that show that semantic predictability and semantic similarity both affect the processing of instrument phrases of the kind underlined in (17) and (18).

To test the distinct contributions of semantic predictability and semantic similarity on sentence processing, we used the reading time data reported by Yun et al. (2006). Yun et al.'s study contained 32 declarative sentences that contained an instrument *with* phrase such as (17) and (18). All sentences had the same syntactic structure and the instruments were carefully normed to be highly plausible in their sentential context, although because we independently model the predictability of the instruments, our results do not depend on the highly plausible instruments being equally plausible. Moreover, there was no correlation between reading times and occurrence of instrument prepositional phrases (PPs) with our verbs in the British National Corpus. Hence, there was no syntactic expectation for an instrument prepositional phrase.

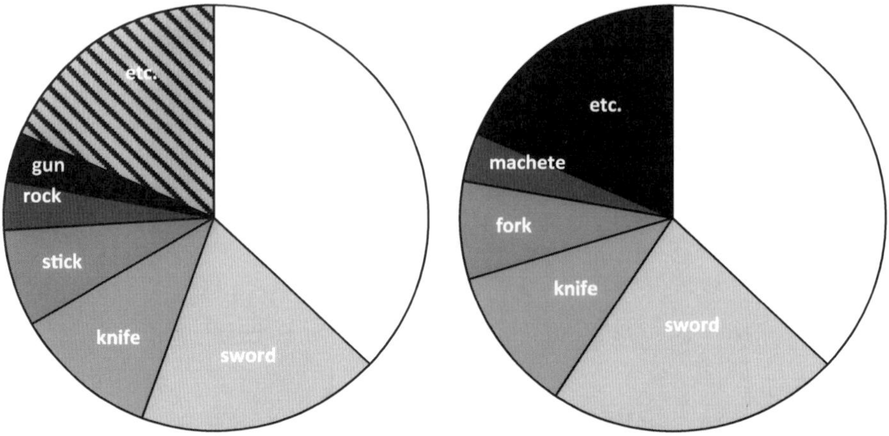

Fig. 10.1 Distribution of possible instruments of *jab* and *attack* in sentences (18) and (17) as determined by completion norming

Yun et al.'s study is particularly appropriate for our purposes because, although the instruments they used were highly plausible, the number and variety of plausible instruments varied across verbs. The semantic similarity hypothesis predicts that reading times for equally plausible instrument phrases will still vary if the distribution of plausible instruments in semantic space differs across verbs. For example, the range of instruments with which one is likely to jab a lion is smaller and more closely related semantically than the range of instrument with which one is likely to attack a lion.

We first measured how semantically predictable the instrument was for each verb. We then measured the distribution in semantic space of likely instruments for sentences such as (17) or (18). We then examined if there was a correlation between these two measures and the reading times for the underlined regions. We also examined whether combining the two measures increased the correlation, as an increase would suggest that semantic similarity has an effect on reading times beyond the effect of semantic predictability.

10.2.2 Testing the Semantic Predictability Hypothesis

The semantic predictability hypothesis holds that syntactic constituents whose meanings are more predictable, given the rest of the sentence will be easier to process, than constituents whose meaning is less predictable. In the case at hand, the syntactic constituents at issue are noun phrase (NP) complements to instrumental *with* and the part of the meaning whose predictability is at issue is the category denoted by the head noun of that NP (i.e., what kind of instrument was used to perform the action). We measured how semantically predictable an instrument was with three types of completion tasks. The first task asked participants to fill in the blank in sentences such as (19). In the second task, another set of participants listed five possible things that could fit in the blank in sentences such as (20). The third task was a variant of Shannon's guessing game (Shannon 1951). Participants saw sentences such as (21) and had to guess the first letter of the word that followed *the*. If they guessed incorrectly, they were asked to guess again until they correctly identified the first letter of the word that followed *the*. Once they correctly guessed the first letter, they were then asked to guess the second letter, and so forth, until all of the letters in the word were correctly identified.

19. The aboriginal man jabbed the angry lion with _______________.
20. The aboriginal man jabbed the angry lion with _____ near its prey.
21. The aboriginal man jabbed the angry lion with the _______________.

We employed the results of these three completion studies to determine whether semantic predictability correlated with the reading times of the underlined regions of sentences like (17) and (18) in Yun et al. (2006). We first correlated reading times with the percentages of times participants completed the sentence in (19) with an NP that contained the instrument used in the online study. We then correlated reading times with the percentages of times the instrument was mentioned first. Finally, we correlated reading times with the percentage of times participants in our Shannon game task guessed correctly the first letter of our online study instrument. If semantic predictability of the filler of an instrument role affects processing of a phrase describing that instrument, we expect reading times to be negatively correlated with our various measures of semantic predictability. All correlations were significant and in the correct direction, that is, there was an inverse correlation between semantic predictability and reading times, as shown in Fig. 10.2. We conclude that how semantically predictable a particular instrument is affects how long readers will take to process a noun phrase that describes that instrument, even when the presence of an instrument is not necessarily expected, as it is rarely expressed.

10.2.3 Testing the Semantic Similarity Hypothesis

The semantic similarity hypothesis holds that the more semantically similar likely fillers of a participant role are (e.g., instruments), the easier it will be to process a constituent whose denotation bears that role (e.g., the NP complement of an instrumental *with*). In the case at hand, the fact that the likely instruments of jabbing in (17) are more similar to each other than are the likely instruments of attacking in (18) means that the underlined phrase in (17) will be easier to process than the underlined phrase of (18). To determine if semantic similarity affected processing, we compared the semantic similarity of the target instrument used in Yun et al.'s online study with sets of instruments listed in the first two completion studies we just mentioned, i.e., the study in which participants finished sentences such as (19) with a single NP and

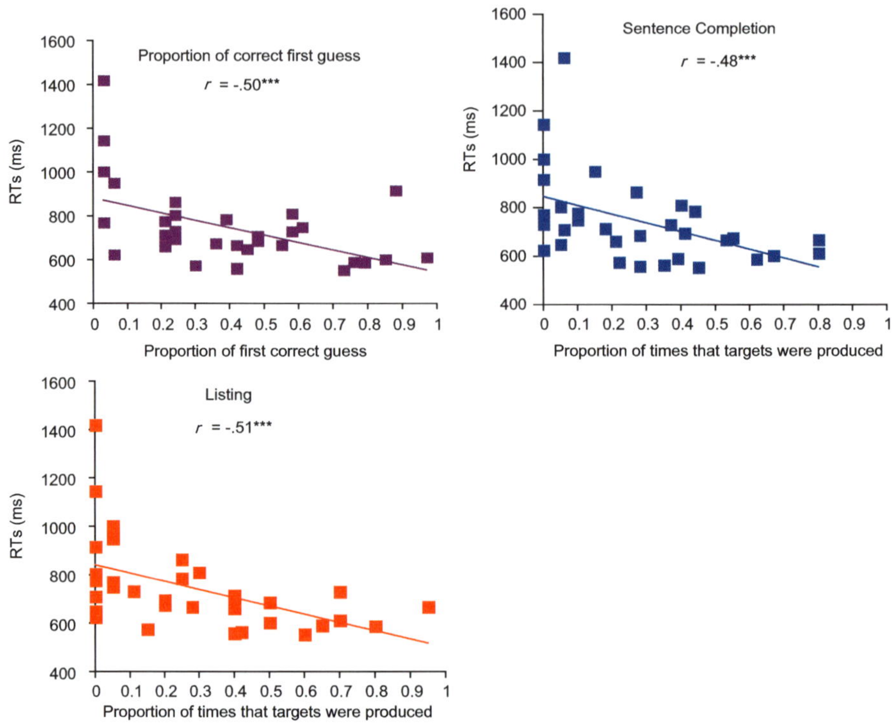

Fig. 10.2 Correlation between Instrument reading times in Yun et al. (2006) and various measures of semantic predictability

the study in which they filled in the blank in sentences such as (20) with up to five instruments. We used two measures of the semantic similarity of our target instruments with the two sets of instruments generated by participants in these two studies. The first measure employed latent semantic analysis (Deerwester et al. 1990), a measure of semantic similarity derived from corpus word co-occurrence information. The second measure computed similarity between the instruments using information contained in WordNet. (We used various measures of WordNet similarity. They all lead to similar results. We report results based on *vector pairs* similarity, Patwardhan and Pederson 2006.) As in the case of semantic predictability, the shared semantic similarity hypothesis predicts that reading times of target instrument NPs will be inversely correlated with the similarity of those instruments with the sets of instruments generated in our two completion tasks. In other words, the more "friends" (i.e., semantically similar) our target instrument NPs have, the easier it is for participants to process them. Figure 10.3 indicates that there was indeed a negative correlation between reading times and our two measures of semantic similarity.

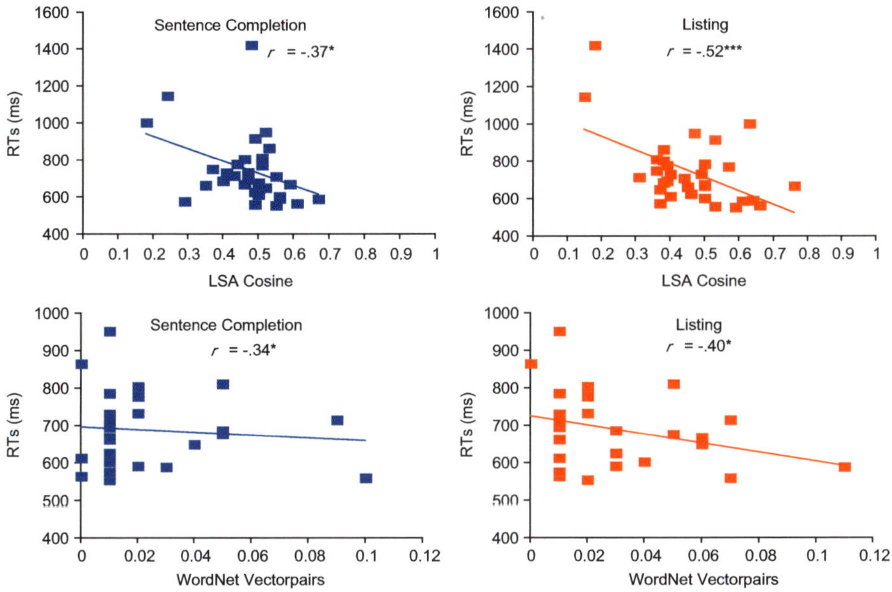

Fig. 10.3 Correlation between Instrument reading times in Yun et al. (2006) and LSA and WordNet measures of Semantic Similarity

10.2.4 Is Semantic Similarity Truly Different from Semantic Predictability?

Like semantic predictability, the semantic similarity of contextually likely instruments eases the processing of an expression that describes that target instrument. But, is semantic similarity different from semantic predictability or are they underlyingly the same? To answer this question, we need to assess whether semantic similarity makes a contribution to the ease of processing instrument denoting expressions that is distinct from that of semantic predictability. We constructed three distinct models: the best model for predicting reading times from semantic predictability, the best model for predicting reading times from semantic similarity of likely instruments to target instruments, and a model combining each of the separate best models. If both semantic predictability and semantic similarity make separate independent contributions to the processing of expressions describing target instruments, then the combined model should be better than models that include only semantic predictability or semantic similarity. The best model of semantic similarity used the number of tries to guess the first letter of the target instrument in the Shannon guessing game study to predict reading times of the instrument NP. Its R^2 value was 0.33. The best model of semantic predictability used the LSA cosine for the similarity to target instrument of the sets of instruments listed by participants in our second completion study to predict reading times of the instrument NP. Its R^2

value was 0.22. A combined model that included both factors had an R^2 value of 0.4. This suggests that both semantic predictability and, crucially, semantic similarity between possible instruments and target instruments play a role in the processing of target instruments denoting NPs.

It should be clear that the event properties that underlie semantic predictability and semantic similarity are not the kinds of properties grammars are sensitive to. We only consider semantic similarity here, for reasons of space. What our computational modeling studies show is that listeners and readers are sensitive to differences in distribution in the semantic space of likely instruments for particular situation types. What matters are differences in the similarity of various instruments with which, for example, one can jab or attack an angry lion. Instruments likely to be used to jab an angry lion share more properties than instruments likely to be used to attack an angry lion. This level of detail in event knowledge is never referenced by grammatical rules, no matter how "exotic" they are. But, why?

10.3 Two Distinct Cognitive Tasks

Pinker (1989) proposes the following explanation for the difference in range of event properties that are relevant to human cognition in general (and the human sentence processor, it seems) and grammar:

> Perhaps there is a set of semantic elements and relations that is much smaller than the set of cognitively available and culturally salient distinctions, and verb meanings are organized around them. Linguistic processes (...) would be sensitive only to parts of semantic representations whose elements are members of this set. (p. 166)

Pinker strongly hints that the human linguistic abilities are innately attuned to this set of semantic elements. If this were the case, it would explain why grammar rules cannot "see" differences among verbs other than the ones we mentioned in Sect. 10.1. Pinker refers to the fact that grammar rules cannot see much of our knowledge of events as its *color blindness,* because properties such as the (typical) color of participants in the described event are among the set of properties that grammatical rules are *not* sensitive to. According to this view, the difference between the semantics of grammar and the semantics of language processing lies in which information is visible to each system. If grammars were truly color-blind, it would also explain Pinker's main concern—children's ability to quickly learn valence alternations, as this limited vocabulary for grammar rules would limit children's hypothesis space when learning valence alternations. We cannot exclude Pinker's hypothesis, particularly in its strongest innateness form, as it is hard to imagine data that could falsify it. However, it is also possible that both grammar and language processing are sensitive to the same diverse range of information, and that the observed differences are due to the different demands of language acquisition (grammar formation) and language comprehension. In this view, the color blindness observed in language is the result of the language acquisition process (or at least the effects of a language going through multiple generations of the acquisition process)—not because the process is inherently insensitive to certain types of semantic factors (e.g., color) and sensitive to others (e.g., volitionality)—but as the result of an acquisition process that looks for correlations between the linguistic input and the situations where that input occurs and then makes conservative generalizations (rules) based on these correlations, such as the system described in Goldberg (2006).

To flesh out this point, we begin with a brief discussion of what the main job of the grammar of events is. Our basic insight is based on an "engineering" perspective—something like a very rough approximation of Marr's (1982) "computational" level of analysis. In other words, the question we are trying to answer is: What event properties should grammar and sentence processing pay attention to, given their function? The most efficient approach is for grammatical rules to only reference a limited amount of our vast knowledge of events. To see why, it is useful to consider, in broad terms, what the purpose of the portion of grammar that pertains to events is.

In broad terms, the function of grammar rules as they are written by linguists is to map the meaning of each of the possible verb lemmas in a language onto a syntactic structure. By syntactic structure, here, we refer to any or all of: structural configurations, ordered list of dependants, and case assignment to dependents, in languages where that is relevant. The number of possible verb lemmas in a language is fairly large. Raters in the English verbal lexicon survey reported in Koenig et al. (2003) and Conklin et al. (2004), knew approximately 4000 verbs. The average polysemy factor, or average number of senses per English verb, is between three (the Collins Cobuild dictionary) and four (WordNet). This means that college-educated American English speakers know between 12,000 and 16,000 lemmas. Alternatively, the number of possible subcategorization frames is fairly small. Using the old list of English subcategorization frames found in Gazdar et al. (1985) as a convenient approximation of the number of such "syntactic structures," this implies that grammar rules must funnel the semantic arguments (and possibly adjuncts) of 12,000–16,000 lemma meanings onto 27 subcategorization frames. Although the use of subcategorization frames of this kind is old-fashioned, and there are some frames missing in this work, nothing crucial hinges on our choice, as what matters is the approximate size of the syntactic distinctions grammar rules must effect. Assumptions about how grammatical rules work, we believe, would not be substantially altered if we had chosen another way of measuring the number of grammatical distinctions that must be made.

The grammar rules we describe in Sect. 10.1 shared two critical properties with more well-known grammatical processes. They were *type general,* that is, they applied to semantically-defined classes of situation types. They were also *token independent,* that is, they applied to all fillers of the argument positions of the relevant lemma's meaning. Token independence and type generality are "rational" properties to include in the design of any mechanism that maps over 12,000 lemmas (and countless tokens of the event type denoted by these lemmas) to a limited set of formal overt distinctions. In order to funnel 12,000–16,000 lemmas into 50–100 formal overt distinctions, verbs must be organized into groups. Consider type generality first. Here, we must map lists of semantic arguments onto subject and object positions or into

NPs bearing particular case marking for a certain set of verbs. There must therefore be a way to select the set of verbs to which a rule applies and reject the set of verbs to which it does not apply. Grammars, we know, use semantically natural event categories to select verbs and reject other verbs. In the case of the Oneida pattern, we discussed in Sect. 10.1.2, the targeted stems were those that denote kin relations. In the case of the Hindi pattern we discussed in Sect. 10.1.3, the targeted stems were those that denote bodily functions.

Let us consider token independence now and imagine that grammars are not token independent. For example, imagine that if one were talking about good food, then the constituent expressing the food would be a direct object, but if the food was not good, then it would be an oblique (or the reverse). Since the quality of what is eaten is not part of the meaning of the verb *eat* (it is not an entailment of the denotation of the filler of its proto-patient argument, to use linguistic jargon), this would require the mapping to subject/object position to differ with the tokens of eating one was describing. Speakers would then be required to pay attention to the properties of participants that may or may not be true of the token of the event type denoted by a verb (and which often would not be known to be true or not). Such token *dependence* would require us to retrieve, evaluate, or guess, information above and beyond the information that is a part of what defines the category of the event being described, information which speakers do not necessarily have at their disposal. Entailments, on the other hand, are guaranteed to hold anytime an event belongs to the category denoted by the lemma. By having grammar rules rely on entailments, speakers can know which rule to apply when using a lemma by virtue of accessing the meaning of the lemma (the event type or event category it denotes).

Not all grammar rules seem to target token-independent entailments of the lemma being considered. Consider the well-known English ditransitive valence alternation:

22. I gave Mary a book.
23. I sent Mary a book.

Whereas the fact that Mary is going to have a book at the end of the event is an entailment of *give* in (22), it is not for *send* in (23) (see Rappaport Hovav and Levin 2008). This is not only because of the vagaries of the post office, say, as we could use the notion of restricted entailment discussed in Koenig and Davis (2001), but because not all tokens of sending result in an (intended) change of possession. If the USA sends men to Mars, Mars will not, as a result, "own" the men. In fact, the use of the ditransitive (at least for those scholars who believe it always encodes an intended change of possession) is partly motivated by the desire to select the subset of tokens of sending that involve intended change of possession. Our proposal that linking rules must be token independent and type general can be extended to model these kinds of cases if we follow Goldberg's (1995) hypothesis that argument structure patterns are the structural equivalent of words and have meanings that combine with the meaning of verbs. Simply put, if the ditransitive pattern (however one chooses to represent it formally) is assigned the meaning of "(intended) transfer of possession,"[3] the semantic side of the ditransitive construction still obeys token independence, i.e., the meaning of the construction must apply to all tokens of lemmas that participate in the construction. A similar analysis can be given to the Hindi ergative case assignment rule we discussed in Sect. 10.1.3. Ergative case assignment to intransitive verbs in Hindi indicated that a particular token of coughing, say, was unexpected; but, of course, not all instances of coughing are unexpected. The use of the ergative case marker distinguishes unexpected tokens of coughing from expected ones in apparent violation of the requirement that the semantic property targeted by a grammatical process be shared by all fillers of an argument position. But, if we analyze ergative case marking on intransitive verbs describing bodily functions as a construction with a particular meaning, we can maintain token independence just like we did for the ditransitive construction: Unexpectedness of the action is a property of all tokens of the verb *in this construction*. There is certainly a "hack" flavor to this resort to constructional meaning. But, we think it does not deter from the general validity of the claim that token independence is a good design principle for the grammar of events, as not all constructions can contribute meaning in the way the Hindi ergative case marking rule does, i.e., some case-marking option must be available that does not add any meaning. If our analysis is on the right track, grammars target meaning of verbs, classes of meaning of verbs, or abstract meanings that are very general meanings of verbs (i.e., constructional meanings) and grammars look the way they do, because they target meanings. Token independence, then, properly understood, reduces to the fact that grammar rules target meanings (although not necessarily individual verbs' meaning).

Type generality and token independence are good design principles, given that event grammars need to map between 12,000 and 16,000 verb senses onto less than a 100 morphosyntactic distinctions. However, even if type generality and token independence constitute the best design for grammars, there must be a mechanism through which this design is implemented. It is beyond this chapter to do more than provide suggestive mechanisms that might be responsible for these observed constraints on grammar rules. A fairly simple assumption about the process of making generalizations during language acquisition could account for many of the observed properties of language, namely, the principle that generalizations are made over the largest coherent grouping. In other words, if a feature is true across several types, and the types form a coherent grouping, then the generalization will be made for the group, rather than at the level of each type. In this manner, if a set of event tokens describe by a verb or a series of verbs share a property, (e.g., the subject is agentive), and the verbs themselves form a group by virtue of having some other properties in common, then the generalization will be made at the level of the event type itself rather than a set of event tokens or at the level of the group of verbs rather than at the

[3] See Goldberg (2006) for the "abstract" meaning of argument-structure constructions and the fact that this meaning corresponds closely to the meaning of "general purpose" verbs like *put*, *give*, and so forth.

level of the individual verb. This tends to result in the features and the verb classes being semantically defined, because semantic features are more likely to be shared by a large set of verbs.

However, the relevant properties for grammar rules do not have to be semantic. Take for example the grammatical patterns observed in conjunction with the feature $\pm$ Latinate. Pinker (1989) and Grimshaw (2005), among others, have observed that most verbs of Latinate origin do not participate in the ditransitive alternation. Thus, despite the semantic similarity of *donate* and *give*, only the latter alternates. Alongside *John donated US$ 5 to the endangered species fund*, we do not have *John donated the endangered species fund US$ 5.* One plausible source of this particular behavior of verbs of Latinate origin is that they do not alternate in their source language (French). As authors have pointed out, what groups those verbs together for English native speakers ignorant of the stock of the verbal lexicon is most likely sound based. More generally, we surmise that semantic factors in grouping verbs that syntactically pattern together are most likely to arise within the development of a single language, but other features such as sound can come into play when two languages with different sound and grammar patterns interact (e.g., the way that Latinate features got into English).

Aside from rather rare sound-based generalizations of the kind found with verbs of Latinate origin (and most likely relevant only in the context of the presence of two lexical stocks within a single language), the fact that generalizations are represented at the level of the group of lexical items for which it holds results in the appearance of *type generality*—the observation that grammar rules apply to semantically defined classes of situation types. These principles operate at all levels of the acquisition processes from the acquisition of lexical meaning to verb selectional restrictions to grammar rules. If the learner was faced with an unusual language where each verb had its own unique mappings between semantic roles and case markings/word order, the learner would learn such a language, but at the expense of not being able to generalize from input to unseen verbs—as the relevant information would be encoded as part of the lexical entry of each verb, rather than at a higher level. However, a different result is more likely in such a situation. If a language started such that an *agent* was mapped to the subject for a random set of verbs and to the object for another random set, and the learner was actively trying to make generalizations across the input, any imbalance in the input (e.g., if a subset of words in early input favored an agent–subject mapping) would result in the learner "regularizing" the language.

We have just suggested that type generality and token independence (with our semantically potent constructions proviso) are good design features for the grammar of events and that the principle that generalizations are made over the largest coherent group of tokens and types makes it possible to "implement" a grammar that obeys these design principles. Is this enough to explain the color blindness of event grammars? Yes and no. Token independence and type generality are enough to explain why nonsemantically potent morphosyntactic constructions (i.e., constructions that do not add semantic information to that present in the verb's meaning) are color-blind. Color, size, odor, and countless other participant properties are excluded from consideration in generalizing over event tokens, because participants' color is not an entailed property for more than a handful of verbs in most languages. So, whereas we have verbs like *redden* and *yellow*, abstracting away a class of event types into something like "changing color" would not be very useful, as it would select only a few verbs. Of course, one *could* imagine a language in which lots of verbs describe colors, changes of colors, and so forth, so that it may make sense to isolate this class of verbs by building in the semantic definition of the class something about color. But, known human languages do not have that many color-oriented verbs. For the same reason, type generality and token independence also account for the kind of semantically potent construction the English ditransitive valence exemplifies, to the extent these constructions' meaning correspond to that of "abstract" verbs. In other words, the types of properties that play a role in grammar rules tend to be semantic, and fairly abstract, high-level types of properties, because most grammar rules apply over large sets of verbs, and properties such as "volitional" are the only kinds of properties likely to be shared by all subjects of a large set of verbs. In contrast, a property like "green" is unlikely to be shared across all fillers of a single role for a single verb, let alone a large set of verbs.

But, our proposal does not as such explain the fact that there is no attested language that is just like Hindi, except that ergative case on the subject (or any other case marker) marks the subject's denotation as being, say, green. In other words, our proposal does not directly explain why the semantic contribution "unexpected action" is attested, but "green" is not. The absence of ergative-marking-green agent languages is not due to good design. It is due, we suggest, to the fact that an agent's color is not part of the causal structure of the world (what our conversations are often about) like volitionality is, or is unlikely to be relevant to a speaker's discourse goal (as signaling the unexpectedness of an action may be). In other words, the absence of ergative-marking-green agent languages is due to the causal or goal stuff the world and our discourses are made out: An agent's color happens to be irrelevant for them.[4]

In summary, ultimately, the explanation for why a larger portion of a language's lexicon is devoted to bodily emissions/functions than color changes is anthropological and is part of the substrate of grammars: What human beings are attuned to and why they develop categories of events they do. But, given that they do have a larger verbal vocabulary for bodily functions than color changes, the fact that grammar rules target classes of verbs that denote bodily functions is a simple engineering decision. You get more bang for your buck when trying to funnel a large set of lemmas into a limited set of formal distinctions.

Let us consider now what the human sentence processor does. Its role is to read the next word or phrase, access the relevant syntactic and semantic information associated with those words or phrases, and integrate this information with the syntactic and semantic representations of the already-encountered expressions. To perform this task, anticipating a part of the syntactic or semantic information of the next expression is quite useful. To that end, then, the processor will predict as many of

[4] However, if, for example, there were a culture where the color green was associated with that culture's supreme being, and all actions performed by green-colored agents were thus considered to be special, then we might expect that a separate case marking for green agents could arise.

the properties of the upcoming expression as possible; in the case of instruments, these properties would include semantic properties, since the usefulness of syntactic information is so limited in this case, as phrases encoding instruments so rarely co-occur with our verbs. Now, the kind of instrument that was likely used in an event does not depend solely on the meaning of the verb. There is no generic instrument we use. Different agents use different instruments on different patients for different tasks. One does not cut a nail with the same kind of instrument used to cut the lawn (one hopes). One does not spear a lion with the same kind of instrument as one debones a lion. So, in semantically predicting the instrument used in a described event, readers must conjure their beliefs about who uses what to do what—the complex set of beliefs that make up our understanding of tools. Of course, the predictions one makes take the form of a probability distribution. Given the event type denoted by the verb (*spear* versus *attack*), the agent involved (an aboriginal man), and the patient involved (an angry lion), there is a range of instruments that are more or less probable to have been used. If, as we suggest, readers activate instruments to the degree they are probable, given the event type, the agent, and the patient they have encountered, clearly, more probable instruments will be integrated faster, since they were more activated. But, more interestingly, the distribution in semantic space of probable instruments will affect reading times. This is because instruments that cluster together in semantic space share many features. High activation of any of these features, because it is borne by a particularly probable instrument, will, in turn, boost the activation of all instruments that share this feature—even the otherwise less probable instruments. The more features are shared across probable instruments, the more each probable instrument will be activated, as these shared features will boost activation of each of them. This explains why semantic similarity has an effect above and beyond semantic predictability; it is a semantic product of the processor trying to predict at every point what the most likely instrument is, semantically.

This chapter has tried to explain a clear difference in the range and kind of information that is relevant to the grammar of events and the online processing of sentences. Rather than rely on an implicit or explicit innateness hypothesis as to the kinds of semantic properties that are "visible" to grammar rules, we suggested that the explanation for the difference lies in task differences between grammar development and utterance processing. Given the charge of the grammar of events, focusing on properties that are type general and token independent is rational. So is focusing on the probability distribution in semantic space of fillers of argument positions of the sentence being read. While we cannot prove that both grammar development and utterance processing are sensitive to the same diverse sets of factors, we argue that there is no need to posit a restriction on which information is available to grammar development. Of course, the story we have told must be fleshed out and some further modeling is needed to show how under standard assumptions, learning mechanisms will zero in on a solution to the mapping problem that is both token independent (either verb-wise or construction-wise) and type general. But what is important for this chapter is that there is a plausible story to tell. There is a plausible cognitive explanation for why grammars are minimalist, and processing is promiscuous that avoids relying on unproven (and possibly unprovable) claims about cognitive architecture.

References

Altmann, G., & Kamide, Y. (1999). Incremental interpretation at verbs: Restricting the domain of subsequent reference. *Cognition, 73,* 247–264.
Bicknell, K., Elman, J. L., Hare, M., McRae, K., & Kutas, M. (2010). Effects of event knowledge in processing verbal arguments. *Journal of Memory and Language, 63*(4), 489–505.
Clark, H. H. (1992). *Arenas of language use*. Chicago: University of Chicago Press.
Conklin, K., Koenig, J.-P., & Mauner, G. (2004). The role of specificity in the lexical encoding of participants. *Brain & Language, 90,* 221–230.
De Hoop, H., & Narashiman, B. (2008). Ergative case-marking in Hindi. In H. deHoop & P. de Swart (Eds.), *Differential subject marking* (pp. 63–78). Dordrecht: Springer.
Deerwester, S., Dumais, S. T., Furnas, G. W., Landauer, T. K., & Harshman, R. (1990). Indexing by latent semantic analysis. *Journal of the American Society for Information Science, 41*(6), 391–407.
Dowty, D. (1991). Thematic proto-roles and argument selection. *Language, 67,* 547–619.
Evans, N. (2000). Kinship verbs. In P. Vogel & B. Comrie (Eds.), *Approaches to the typology of word classes* (pp. 103–172). Dordrecht: Mouton de Gruyter.
Gazdar, G., Klein, E., Pullum, G. K., & Sag, I. A. (1985). *Generalized phrase structure grammar*. Oxford: Blackwell.
Goldberg, A. (1995). *Constructions: A construction grammar approach to argument structure*. Chicago: University of Chicago Press.
Goldberg, A. (2006). *Constructions at work. The nature of generalization in language*. Oxford: Oxford University Press.
Grimshaw, J. (2005). *Words and structures*. Stanford: CSLI.
Kamide, Y., Altmann, G., & Haywood, S. (2003). The time-course of prediction in incremental sentence processing: Evidence from anticipatory eye-movements. *Journal of Memory and Language, 49,* 1133–1156.
Koenig, J.-P., & Davis, A. (2001). Sublexical modality and the structure of lexical semantic representations. *Linguistics and Philosophy, 24,* 71–124.
Koenig, J.-P., Mauner, G., & Bienvenue, B. (2003). Arguments for adjuncts. *Cognition, 89,* 67–103.
Koenig, J.-P., & Michelson, K. (2010). Argument structure of Oneida kin terms. *International Journal of American Linguistics, 76,* 169–205.
Levin, B. (1993). *English verb classes and alternations: A preliminary investigation*. Chicago: University of Chicago Press.
Levin, B., & Rappaport Hovav, M. (2005). *Argument Realization*. Cambridge: Cambridge University Press.
Marr, D. (1982). *Vision*. San Francisco: W.H. Freeman.
McRae, K., Ferretti, T., & Amyote, L. (1997). Thematic roles as verb-specific concepts. *Language and Cognitive Processes, 12,* 137–176.
Malchukov, A. (2008). Animacy and asymmetries in differential case marking. *Lingua, 118,* 203–221.
Patwardhan, S., & Pederson, T. (2006). Using WordNet-based context vectors to estimate the semantic relatedness of concepts. In Proceedings of the EACL 2006 Workshop Making Sense of Sense—Bringing Computational Linguistics and Psycholinguistics (pp. 1–8). Trento, Italy.
Pinker, S. (1989). *Learnability and cognition*. Cambridge: MIT Press.
Rappaport Hovav, M., & Levin, B. (2008). The English dative alternation: The case for verb sensitivity. *Journal of Linguistics, 44,* 129–167.
Roland, D., Yun, H., Koenig, J.-P., & Mauner, G. (2012). Semantic similarity, predictability, and models of sentence processing. *Cognition, 122*(3), 267–279.
Shakthi, P., & Koenig, J.-P. (2009). Hindi aspectual complex predicates. Proceedings of the HPSG '09 Conference (pp. 276–296). Stanford: CSLI.
Shakthi, P. (2012). *Hindi complex predicates at the syntax-semantics interface*. Doctoral Dissertation, University at Buffalo.

Shannon, C. E. (1951). Prediction and entropy of printed English. *Bell System Technical Journal, 30,* 50–64.

Spivey-Knowlton, M., & Sedivy, J. (1995). Resolving attachment ambiguities with multiple constraints. *Cognition, 55,* 227–267.

Tanenhaus, M. K., Spivey-Knowlton, M. J., Eberhard, K. M., & Sedivy, J. E. (1995). Integration of visual and linguistic information in spoken language comprehension. *Science, 268,* 1632–1634.

Yun, H., Mauner, G., & Koenig, J.-P. (2006). Anticipation vs. integration of syntactically infrequent but semantically obligatory arguments. Poster presented at the 19th Annual CUNY Conference on Human Sentence Processing, March 23–25, New York.

Chapter 11
Verb Representation and Thinking-for-Speaking Effects in Spanish–English Bilinguals

Vicky T. Lai and Bhuvana Narasimhan

11.1 Introduction

Does the language we speak influence how we think about the events in our experience? If so, do bilingual speakers construe the same event in different ways, depending on the language they use to verbally encode that event? Or does one of the languages play a more dominant role in influencing event construal? The present study investigates whether bilingual speakers attend to different aspects of a motion event, depending on the language they use to first describe that event. Specifically, we explore whether language-specific verb representations used in encoding motion events influence subsequent performance in a nonlinguistic similarity judgment task in Spanish–English bilinguals.

We will begin by looking at different perspectives on whether language influences thought, including views on linguistic relativity and "thinking-for-speaking." Then we will focus on the domain of motion. We will present linguistic accounts of the semantic representations of motion verbs and discuss the crosslinguistic difference between English and Spanish. Next, we will review empirical studies that examine how verbal encodings influence motion event construal in monolinguals. We will also review empirical studies that explore linguistic relativity versus thinking-for-speaking in bilinguals. We then go on describe the current study. In the final section of the chapter, we discuss our findings in light of thinking-for-speaking effects, how events are conceptualized for language production, and the nature of representations in the bilingual mind.

V. T. Lai (✉)
Department of Psychology, University of South Carolina and Max Planck Institute for Psycholinguistics, 1512 Pendleton Street, Columbia, SC 29208, USA
e-mail: vicky.tzuyin.lai@gmail.com

B. Narasimhan
Department of Linguistics, University of Colorado Boulder,
Hellems 290, 295 UCB, Boulder, CO 80309, USA

R. G. de Almeida, C. Manouilidou (eds.), *Cognitive Science Perspectives on Verb Representation and Processing*, DOI 10.1007/978-3-319-10112-5_11

11.2 Language and Thought

11.2.1 Linguistic Relativity

The issue of how language relates to cognition is a fundamental one that has been the focus of intense debate for decades. Some have argued that language shapes cognition, as proposed by Benjamin Lee Whorf (1956) under the influence of his mentor, Edward Sapir. The central claim of the Whorfian hypothesis is that the grammatical categories of a language can influence how its speakers perceive the world. This position was somewhat ignored in the 1960s, when the universal grammar view prevailed. Since universal grammar presupposes no distinction among languages at the conceptual level, how language relates to cognition became a minor issue.

However, the view that language has an influence on cognition has been revitalized in recent years by "neo-Whorfian" scholars, including Wilkins and Hill (1995); Levinson (1996); Pederson (1995); Lucy (1996); and Boroditsky (2001), among others. A variety of empirical studies have been conducted to examine how speakers' linguistic organization of domains, such as color, space, time, and motion, influence their nonlinguistic conceptualization of these domains (e.g., Berlin and Kay 1969; Heider 1972; Pederson et al. 1998; Boroditsky 2001; Li and Gleitman 2002; Levinson et al. 2002). Findings in support for both positions have been found in these studies, and the directionality and extent of the interaction between language and thought continue to be hotly debated.

In many of these studies, the term "thought" is used interchangeably with "cognition" which is typically measured in tasks that do not involve the use of language at all. These behavioral measures include speakers' manipulation of spatial arrays, reaction time measures, similarity judgment, eye tracking, and recognition memory (Bohnemeyer et al. 2001, 2006; Boroditsky 2001; Levinson et al. 2002; Li and Gleitman 2002; Finkbeiner et al. 2002; Gennari et al. 2002; Papafragou et al. 2002). Speakers' task performance along language-specific lines is taken to demonstrate that cognitive processing is imprinted by language-specific patterns of encoding experience. As the reasoning goes, the patterns that speakers use habitually in order to produce and comprehend language become ingrained and start influencing the ways that speakers construe their experiences from an early age, resulting in language-specific ways of thinking even in situations where language is not used to direct attention or enhance recall in any way.

11.2.2 Thinking-for-Speaking

In addition to focusing on the issue of linguistic relativity, many scholars have also investigated subtler aspects of the relativity hypothesis, whether language plays a role in influencing speakers' performance in nonlinguistic tasks shortly after they have used language to encode a particular aspect of their experience (e.g., Finkbeiner et al.

2002; Papafragou et al. 2008). This line of research hypothesizes that in order to linguistically encode a particular experience, speakers have to first fit their thoughts into a language-particular mold—that is, they engage in "thinking-for-speaking" (Slobin 1987, 1996a). The role of "thinking-for-speaking" appears to be somewhat less controversial among Whorfian and cognitivist scholars, and the role of verbal encoding in subsequent construal of experience across languages dovetails with psychological research, showing how verbal encoding of specific events influences subsequent recall and evaluation of the same events *within a language* (Loftus and Palmer 1974). In Loftus and Palmer's classic study on the interaction of language and memory, subjects viewed films of automobile accidents and were then asked to estimate the speed of the vehicles. Subjects in different groups were presented with the same question with different verbal encodings: "About how fast were the cars going when they smashed/collided/bumped/contacted/hit (into) each other?" The encoding of "smashed" elicited the highest speed estimate while the encoding of "contacted" elicited a slower speed estimate. A week later, subjects returned and answered whether they saw any broken glass in the film. About 32 % of the subjects in the "smash" encoding group answered "yes" even though there was no broken glass in the original film. The study demonstrated that mental representations can be shifted in the direction suggested by the verbal label.

Nevertheless, many open questions remain to be answered about the nature of "thinking-for-speaking" effects. For instance, are thinking-for-speaking effects more likely to arise from differences in lexical semantic representations versus patterns of syntactic encoding? Can a nonlinguistic bias be induced simply by virtue of using a particular language, or does a bias only occur with the use of specific words or constructions? How early in children's development does language begin to shape the construction of semantic categories? Do speakers of multiple languages also have multiple modes of thinking corresponding to each of their languages? Are "thinking-for-speaking" effects more likely to occur in some domains versus others?

11.3 Motion Events

11.3.1 Linguistic Encoding of Motion Events

One domain that is often used to explore the relationship between language and cognition is that of motion. Based on the influential work of Talmy (1985, 2000, 2007), the semantic domain of motion is often discussed in terms of the components of the motion event schema. The "figure" is the object that is moving or located with respect to the reference object, the "ground." The "path" is the movement of the figure with respect to the ground. In addition to the internal components above, a motion event can have an external co-event that is related to the event by "manner" or "cause" (Talmy 2007, pp. 70–71). How speakers of different languages linguistically encode the dynamically changing spatial relationship between figure and ground, the geometric properties of figure and ground, the properties of the path and manner of motion, the causal role of an agent of motion, and other significant aspects of motion events, provide a rich and interesting testing ground for scholars interested in exploring the interaction between language and cognition.

Studies of motion event encoding have taken as their basis, the claim that there exists a typology of languages based on how they encode complex events (Talmy 1985, 2000). According to Talmy, some languages ("satellite-framed" languages), e.g., English, encode the manner of motion in the verb root, e.g., *the bottle floated out of the cave*. Other languages ("verb-framed" languages), e.g., Spanish, typically encode the path of motion (i.e., *float*) in the main verb, expressing manner of motion in a gerundial phrase, e.g., *La botella entró a la cueva flotando "the bottle moved-in to the cave floating"* (Talmy 2007, p. 89). Talmy's account has been extremely influential, although subsequent studies suggest that the crosslinguistic differences in motion event encoding are more restricted than suggested by Talmy. For instance, Aske (1989) proposes that Spanish manner of motion verbs can in fact co-occur with path phrases if they are atelic. For example, *Juan bailó en circulos* "John danced around," *La botella flotó hacia la cueva* "the bottle floated towards the cave," etc. Aske (1989) also pointed out that verb semantics also plays a role in the acceptability of the co-occurrence of manner verb and path phrases in Spanish. For example, the more strongly the motion is implied in the verb, such as "roll" and "run," the more acceptable it is to combine the verb with some path phrases. In a modification of Aske's proposal, Slobin and Hoiting (1994) propose that in verb-framed languages, such as Japanese, Dutch Sign Language, Turkish, the restriction on combining manner of motion verbs with path phrases applies only to those motion events that involve "movement across a boundary" (1994, p. 498).

The locus of these language-specific differences in motion event encoding has been discussed widely in the literature. Talmy's own account suggests that the distinction between "verb-framed" and "satellite-framed" languages is related in part to the semantics of the manner of motion verbs in the two types of languages (Talmy 1985, 2007). The manner of motion verbs in English-type "satellite-framed" languages occur in "lexicalization doublets." For instance, *kick-1* implies an agent's impacting his or her foot into some object (*I kicked-1 the wall with my left foot*), whereas *kick-2* incorporates motion with the sense encoded in *kick-1* (*I kicked-2 the ball across the field with my left foot*): I_A MOVED (by kicking-1) the ball across the field with my left foot (Talmy 2000, p. 31; Talmy 2007, p. 76). The additional specification of directed motion in a variety of manner of motion verbs such as *kick-2* is taken to explain, why these verbs combine so flexibly with directional phrases (e.g., *across the field*), in languages such as English. In contrast, the counterparts of verbs such as "kick-1" in Spanish-type "verb-framed" languages do not occur in lexicalization doublets, and lack an additional sense associated with "kick" that encodes manner of motion combined with a specification of directed motion. So, in these languages, the manner of motion verb is unable to combine with a path phrase and is more typically encoded in a gerundial phrase (for further discussion on the variable behavior of manner of motion verbs, see Levin and Rappaport Hovav 1995).

Subsequent debates have revolved around the issue of whether the crosslinguistic differences have to do with the existence of lexicalization doublets in the verb lexicon of a language as proposed by Talmy, or whether alternative explanations are more valid, e.g., the ability of a manner verb to combine with directed motion constructions (Narasimhan 2003; semi-) productive lexical rules within the lexicon that derive extended verb senses (Levin and Rapoport 1988; Levin and Rappaport Hovav 1999), a principle of semantic composition that allow accomplishment predicates to be constructed out of an activity verb in combination with a goal prepositional phrase (Beck and Snyder 2001), or a syntactic composition operation that combines manner and motion within the verbal constituent (Zubizarreta and Oh 2007).

These accounts are centered around the nature of verb semantic representations: whether the meanings of the verbs are different across languages by virtue of their inherent semantics or by virtue of their ability to acquire more complex meanings by fitting into meaning-bearing clausal templates, undergoing lexical rules in the lexicon, or participating in compositional semantic operations in some languages but not others.

In addition to the nature of the semantic representation of the verb, additional differences between "satellite framed" and "verb-framed" languages have to do with the richness of the manner of motion verb lexicon, and the frequency with which manner of motion verbs are used at all in descriptions of motion events (Slobin 1996b). The frequency of use of manner of motion verbs is linked to the rhetorical style used in narratives in "verb-framed" versus "satellite-framed" languages. As pointed out by Slobin (1996b) in his discussion of motion event descriptions in Spanish and English, Spanish usually sets the static scene in which the motion event takes place first (e.g., "there is a cliff," "there's a water fall," "it's high," etc). When describing the motion event, Spanish speakers describe the path, leaving the manner of motion to be inferred from the scene (e.g., "the agent jumped"). English does not set the stage as frequently, and describes the manner of motion explicitly (e.g., "the agent jumped off of the cliff into the waterfall").

11.3.2 Motion Event Construal

Although theoretical accounts of verb semantic representation differ in how they account for crosslinguistic differences in motion event encoding, they all raise an interesting psycholinguistic question: Do speakers of languages that differ in their linguistic encoding of motion events also differ in how they construe such events? Many researchers have attempted to examine the cognitive consequences of the empirical fact that speakers of "satellite-framed" languages flexibly and frequently use a variety of manner of motion verbs to combine with path phrases whereas speakers of "verb-framed" languages combine manner verbs with path phrases less often and in more restricted contexts of use. Such studies focus on the following question: Are speakers of "satellite-framed" languages, as a consequence of their habitual verbal encoding of manner of motion, also more inclined to attend to manner of motion in comparison to speakers of "verb-framed" languages?

Prior studies have experimentally investigated how speakers of languages with different ways of encoding motion events go on to perform different nonlinguistic tasks. The languages under study include Greek (Papafragou et al. 2002), Japanese (Finkbeiner et al. 2002), Spanish (Gennari et al. 2002), as well as a diverse sampling of 17 genetically and areally distinct languages from around the world (Bohnemeyer et al. 2006). The net finding of these studies is that speakers of verb-framed and satellite-framed languages do not differ in the ways in which they perform in tasks involving recognition, categorization, or similarity judgment. In the Bohnemeyer et al. study, an effect of language was found in a forced-choice similarity judgment task: Speakers had to group together motion events that had the same manner versus the same path (e.g., they were shown a ball rolling up a ramp and asked to match the scene to one of two variant scenes: a ball rolling down a ramp, or a ball bouncing up a ramp). Speakers of some languages had a significantly higher tendency to match events on the basis of same manner rather than same path. However, these language-specific differences did not conform to Talmy's satellite-framed versus verb-framed distinction. The absence of strong Whorfian effects in this domain may have to do with the crosslinguistic validity of the Talmy typology in the first place, given evidence of intra-typological variation in this domain (see Slobin 1996b). It may also have to do with the importance of the main verb in influencing event construal: The assumption that encoding manner of motion in the main verb directs attention to the manner of motion, may simply not be a correct one.

Yet interestingly, the dimensions of motion events that are encoded in the verb do appear to influence nonlinguistic cognitive processing when speakers are required to verbally encode the event first, prior to participating in a nonlinguistic task. Studies by Billman et al. (2000) and Billman and Krych (1998) show differences in event construal that correlate with differences in verbal encoding within a language. English speakers' recognition of events was influenced by whether they heard manner or path verbs when encoding motion events (e.g., *walk* versus *enter*), exhibiting less sensitivity to those aspects of the motion events that they had not verbally encoded. A similar effect of prior verbal encoding is found in studies investigating monolingual speakers of different languages. Gennari et al. (2002) show that speakers of Spanish and English construe motion events differently, when they were first asked to verbally encode the events in their native language. Interestingly, this "thinking-for-speaking" effect is constrained by the type of the task. Speakers' performance differed in the similarity judgment task, but not the recognition task, which is more automatic. And Papafragou et al. (2008) found a similar "preparing-for-speaking" effect in an eye-tracking study. The speakers rapidly fixated on the event components (e.g., manner of motion) that were typically encoded in their native language during the stage when they were preparing to describe the motion events. These findings are compatible with the notion that the language we speak influences how we think when comprehending and producing speech.

11.4 Thinking-for-Speaking in Bilinguals

11.4.1 Prior Studies

A particular intriguing possibility raised by these findings is that bilingual speakers who are fluent in languages that differ in their linguistic encoding of motion events may, in fact, have two different modes of "thinking-for-speaking" depending on the language she/he is hearing and speaking. Studies investigating thinking-for-speaking and relativity effects in bilingual and second-language learner populations have looked at the domains of time (Boroditsky 2001; Lai 2005), physical object (Athanasopolous 2007), and motion (Brown and Gullberg 2008; Gullberg 2011), among others.

In the domain of time, differences in temporal reasoning in different languages were hypothesized to influence how L2 learners reason about time in both their L1 and L2. Boroditsky (2001) investigated Mandarin speakers' descriptions of time, based on the positioning of events along a vertical axis. In Mandarin, time can be reasoned about along a vertical axis. UP represents time in the past and DOWN represents time in the future. Boroditsky hypothesized that L2 Mandarin speakers of English may be influenced by the English way of temporal reasoning, which is based only on the horizontal axis. In a priming experiment, participants first saw two objects arranged on top of each other vertically or beside each other horizontally. Then, participants were presented with English temporal phrases that were either true or false, e.g., "August comes later than/after June." The hypothesis was that viewing a spatial array that conformed to speakers' language-specific conceptualization of time along an axis would facilitate responses in the true–false decision task. It was found that L2 speakers who learned English at a later age in life were still biased to think about time, vertically, in a Mandarin way, even though the experiment was conducted in English. This finding suggests that the L1 conceptualization was still retained by learners although they produced temporal expressions in English.

In Mandarin, time can also be reasoned about along a horizontal axis. Moving a temporal event FORWARD almost always means rescheduling the event to the past, while moving an event BACKWARD means moving it to the future (Lai 2002). In English, though, when a scheduled event is moved FORWARD in time, it is usually ambiguous as to whether the event is moved to the future or to the past (McGlone and Harding 1998), but with a slight preference to move it to the future (Gentner et al. 2002). Lai (2005) hypothesized that the encoding of time in English may have influenced Mandarin-speaking immigrants who were immersed in an English-speaking culture. That is, Mandarin learners of English may be less certain about the direction of FORWARD, unlike monolingual speakers of Mandarin. Three studies were conducted in three speaker groups: Mandarin monolinguals who have been exposed to English education in high school in Taiwan, English monolinguals in four different states in America, and Mandarin–English bilingual speakers who either were immigrants or have stayed in America for more than 4 years. All speakers were asked to move the clock time forward 1 h. The relevant finding here was that the clock time was moved to the past more often by the Mandarin monolinguals than the Mandarin–English bilinguals, even though the bilinguals were instructed in Mandarin during the experiment. In contrast to the Boroditsky study (2001), this finding provides evidence that a shift in time perspective in the direction of the L2 has occurred in the minds of bilinguals.

In the domain of physical object classification, Athanasopolous (2007) examined how language-specific grammatical concepts influence nonlinguistic object categorization preferences in bilinguals. In English, a distinction is made between count and mass nouns: A count noun can be quantified using a numeral ("one apple") whereas a mass noun needs an additional unit of measurement ("one sand" versus "one bucket of sand"). In Japanese, many common nouns are treated as mass nouns, and there is a grammatical category, classifier, used to quantify nouns (e.g., "three apples" in English versus "three pieces of apple" in Japanese). Imai and Gentner (1997) and Cook et al. (2006) found that this crosslinguistic difference influenced monolingual Japanese and English speakers' preferences for object categorization. Japanese speakers tended to categorize objects as being similar on the basis of their material properties, whereas English speakers tended to categorize objects as being similar on the basis of shape. Athanasopolous (2007) looked at this phenomenon and asked whether bilinguals alternate between two language-specific concepts, depending on the language used in the experiment. In a similarity judgment task, participants were presented with a triad of a target (e.g., plastic clip) and two alternates (e.g., a metal clip and plastic pieces). The task was carried out in English monolinguals, Japanese monolinguals, English–Japanese bilinguals instructed in L2 (English), and bilinguals instructed in L1 (Japanese). Results from bilinguals showed that regardless of the language of instruction used in the experiment, a bias for pointing to the same-shaped object in count nouns was found, consistent with the English preference. These findings suggest that English–Japanese bilinguals shifted their categorization behavior towards L2 (English).

Gesture constitutes a valuable tool in furthering our understanding of how L2 and bilingual speakers conceptualize their experience for the purpose of speaking. In the domain of motion events, gesture researchers have examined how co-speech gestures of satellite-framed and verb-framed language speakers reflect their language-specific conceptualization when they describe motion events. When speakers describe motion events, they produce gestures that convey information about aspects of the motion event, e.g., the path of motion (e.g., upwards or downwards) and the manner of motion (e.g., zigzagging, rolling, etc., Brown and Gullberg 2008). Brown and Gullberg investigated how Japanese and English monolinguals and Japanese learners of English verbally encoded and gestured about manner, when describing motion events. In a narrative-retelling task, participants watched a video clip of Tweety Bird cartoon and retold the story to the experimenters.[1] It was found that in both their L1 and L2 productions, Japanese–English learners differed significantly in their encoding of manner in speech from the monolingual English speakers, but not from the

[1] Japanese–English participants told the story once in Japanese and once in English.

monolingual Japanese speakers. A similar pattern was found in learners' encoding of manner in gesture. These findings suggest that rhetorical style is transferred from the L1 to the L2. But in terms of the extent to which gesture backgrounded the manner information, learners appeared to have adopted a rhetorical style more similar to that of English in L1 as well as L2, suggesting an influence on their L1 from the L2.

Our discussion of the studies on second-language learners and bilinguals shows that the findings are somewhat inconsistent. Boroditsky (2001) found that advanced Mandarin–English bilinguals remained L1 like in their temporal reasoning. Lai (2005) found that Mandarin–English bilingual immigrants displayed a shift towards English when reasoning about time. Athanasopolous (2007) found that their Japanese–English bilingual speakers were mainly L2 like and are likely to have acquired the English shape-bias in physical object classification. Brown and Gullberg (2008) found that their Japanese–English participants' co-speech gestures, produced when describing motion events, were gradually shifting towards L2-like behaviors. These studies point to a complex picture of the mental representations in the bilingual mind. In some cases, having acquired two languages resulted in shifts in the semantic representation. A variety of factors modulate whether this shift occurs, including proficiency, age of acquisition, and length of cultural immersion.

11.4.2 The Present Study

The prior studies investigating speakers of multiple languages are primarily concerned with the nature of change in long-term semantic representations in the bilingual mind, suggesting that over the course of learning, bilingual speakers may shift to an L2-specific conceptualization (for the purpose of speaking), retain their L1 preferences in this regard, or construct an intermediate system that is neither "purely" L1 like nor L2 like. But an intriguing question that has not yet been investigated in these studies asks, whether bilingual speakers can be induced to shift flexibly between different modes of "thinking-for-speaking" about situations depending on the language they use to encode the same situations verbally. That is, can the bilingual speaker adapt his or her thinking-for-speaking "on-the-fly" in response to the immediate demands of speech production in a specific language? Or, does she/he rely on a single conceptualization system to construe events regardless of differences in the language used to verbally encode the same situations?

The answer to this question will provide us with interesting insights into the degree to which the process of conceptualization during language production is tailored to particular languages in speakers of multiple languages. In his account of the different stages involved in the production of speech from intention to articulation, Levelt (1996) proposed an initial stage of conceptual preparation during which speakers plan what to say (macroplanning) and how to say the intended message (microplanning). The formulation process of the "preverbal message" is not the same for speakers of different languages (von Stutterheim and Nüse 2003). As discussed earlier, prior experimental work shows that language-specific demands on the formulation of messages play a role in shaping the preparation of utterances as speakers prepare to describe motion events (Papafragou et al. 2008). The present study extends this line of research by investigating the extent to which the conceptual system is capable of supporting more than one mode of conceptualization, each associated with a distinct language. By asking bilingual speakers to describe events in either one of their languages, we can evaluate the extent to which language-particular event construals are evoked during speech production when multiple construals are potentially available. The present study investigates this issue by examining whether the language that bilinguals use, at a given point of time, to describe motion events influences the specific motion event dimension they select as the basis for similarity judgment.

Second, we ask whether nonlinguistic event construal be influenced by language-specific differences in the syntactic packaging of the core components that constitute the semantic representation of motion events. For instance, although both path and manner of motion can be described in both Spanish and English, the grammatical encoding of these concepts differ in the two languages. As discussed in Sect. 11.2.1, English speakers habitually encode the manner of motion in the main verb and the path in "satellite" path phrases (e.g., *the bottle floated out of the cave*), whereas speakers of Spanish typically encode the path of motion (i.e., *float*) in the main verb, expressing manner of motion in a gerundial phrase, e.g., *La botella entró a la cueva flotando "the bottle moved-in to the cave floating*"(Talmy 2007, p. 89). We hypothesize that, since the main verb encodes the most salient information about the event in the clause, thinking-for-speaking effects are most likely to arise from information encoded in the main verb. So in our study, although both the path and the manner of motion are encoded in motion event descriptions by speakers of Spanish and English, we predict that the types of semantic components conflated with motion in the main verb will make different aspects of the motion salient for speakers of the two languages: path in the case of Spanish, and manner in the case of English.

Third, while our experimental procedure involves verbal encoding of the stimulus prior to the nonlinguistic task as in other studies (Billman et al. 2000; Billman and Krych 1998; Gennari et al. 2002; Papafragou et al. 2008), in our study, participants do not spontaneously generate a description of the motion event, but participate in an elicited imitation task in which they hear a description produced by the experimenter which they then repeat. As discussed further in Sect. 11.4, elicited imitation of the same motion event descriptions by all participants ensures consistency in production across participants as well as uniformity in the frequency with which manner and path information is encoded in both languages. Additionally, it allows us to explore whether thinking-for-speaking effects are induced when processing information at (an arguably) a shallower level in an elicited imitation task as compared to the spontaneous generation of linguistic descriptions.

Based on the array of differences in motion event encoding in the two languages discussed earlier, we predict that when interacting in Spanish, bilingual speakers are more likely to attend to the path of motion than when they are interacting in English.

11.5 Description of the Study

In order to evaluate speakers' attention to either the manner or the path of the motion event, we used a similarity judgment task (Kita and Özyürek 2003; Bohnemeyer et al. 2001; Gennari et al. 2002; Papafragou et al. 2002) in which speakers watched a video clip of the target motion event (e.g., ball ROLLS from a cave TO a hut) followed by two alternate clips of motion events that differed from the target event in either the path (ball rolls AWAY FROM a hut to a cave) or the manner of motion (ball SLIDES from a cave to a hut). Prior to the onset of the clip, speakers heard a description of the motion event in either Spanish or English, which they then repeated. Descriptions were constructed based on prior literature describing typical patterns of motion event encoding in the two languages and in consultation with native speakers. In both languages, the description of the motion event included information about the path as well as the manner of motion (see Appendix which lists the stimuli and the descriptions in the two languages). In English, the manner of motion information was provided by the finite verb (*Mr. Red rolled towards the tree*), whereas in Spanish the manner of motion was encoded in an adverbial phrase (*El senor rojo se fue girando hacia el arbol* "Mr. Red went rolling towards the tree"). As discussed earlier, telicity is hypothesized to play an important role in crosslinguistic differences in motion event encoding, so half the descriptions consisted of atelic path phrases ("towards the tree/rock" or *hacia el arbol/la piedra*), whereas the other half consisted of telic, "boundary-crossing" descriptions ("into the hut/cave," *entro en la cabana/la cueva*; Slobin and Hoiting 1994; see also Aske 1989). Since the participants listened to the Spanish and English descriptions provided by the experimenter and then repeated them, there was no interspeaker variation in the motion event descriptions provided by the speakers in the two languages, e.g., in the frequency with which manner or path verbs were used, whether or not an adverbial phrase was provided, etc. Speakers were asked to point to the alternate clip that was more similar to the target clip. Our specific prediction was that bilingual speakers who hear and repeat motion event descriptions in Spanish are more likely to base their judgments of similarity of motion events on shared path, in comparison to bilingual speakers who hear and repeat motion event descriptions in English.

11.6 Method

11.6.1 Participants

A total of 27 bilingual speakers of English and Spanish (14 female and 13 male) participated in this study. When the experiment was completed, the participants were asked to complete a language history questionnaire, created for the multilingualism project at the Max Planck Institute of Psycholinguistics (Gullberg and Indefrey 2003). Data from three participants were discarded because one of them did not interact in fluent English at all and two of them started learning English later than 15 years old. The remaining 24 participants have an average age of 28.5 years (range 19–50). All of the participants learned Spanish since birth. For the participants tested in Spanish, their average age of English acquisition is 6.18 years (range 0–15) and their average duration of living in the USA is 16.8 years (range 5–24). For the participants tested in English, their average age of English acquisition is 3.82 years (range 0–13) and their average duration of living in the USA is 19.2 years (range 5–40). All the participants currently reside in the state of Colorado in the USA. According to self-report, they use both English and Spanish languages on a daily basis. The participants received payment for participation. The study is approved by the local ethics committee.

11.6.2 Design and Materials

We used a subset of a larger set of simple 2D animations depicting motion events created at the Max Planck Institute for Psycholinguistics (Bohnemeyer et al. 2001). These video clips depict an animated round figure moving in various manners between different ground locations. The 16 target clips of motion events used in the current study systematically vary manners of motion (TWIRL, ROLL, JUMP, and SLIDE), pairs of ground locations (a tree and a rock, and a hut and a cave), and direction of motion (leftward and rightward). Each target clip (e.g., *roll to the left*) has a variant clip that uses the same-manner motion (e.g., *roll to the right*) and three variant clips that use the same path but vary in the manner of motion (e.g., *twirl/jump/slide to the left*). The combination of the targets and variant clips resulted in 48 experimental triads. Using a Latin square design (see Bohnemeyer et al. 2001), these triads are distributed into six lists (eight triads per list). Every list displays all its triads in one order of presentation and then the reversed order, resulting in 12 lists. Additionally, 3 practice triads and 16 filler triads are added to each one of the lists.

A given experimental triad displays the target clip for 4 s and then displays the two variant clips side by side on a split screen for another 4 s. On the target scene, two ground objects (e.g., a tree and a rock) appear on the left and the right of a green field with blue sky in the background. An agent that looks like a round "tomato man" moves from one ground object to the other, in one of the four manners (e.g., a tomato man rolls from the tree to the rock). On the split screen, each variant resembles the motion event in the target clip in 1D but differ in another (e.g., a tomato man rolls from the ROCK to the TREE versus a tomato man JUMPS from the tree to the rock). See Fig. 11.1 for an example.

Filler triads are used to prevent participants from settling into a fixed response pattern. In the fillers, two agents (animated entities constituting geometric shapes) interact with each other in four actions that are different from the motion events in the experimental trials. Specifically, two events of change of possession (GIVE and THROW an instrument such as a hammer or a stick) and two events of change of state (BREAK and HIT with an instrument such as a hammer) are employed.

Target Scene (Figure slides from tree to rock)

Two events presented side by side (split-screen)

Fig. 11.1 An example of an experimental trial

A given filler trial begins with the target scene that has the same green field and blue sky as scenes in the experimental trials. Two red and green round agents are in the center of the screen. When the action begins, one agent starts to interact with the other in one of the four actions. For example, one animated entity gives a hammer to another animated entity. This is replaced by a split screen with two action events that resemble the action in the target scene in some way. For example, variant 1 can have the same instrument as in the target scene but using different action, e.g., an animated entity throws a hammer to the other entity. Variant 2 can have the same action as in the target scene but with a different instrument, e.g., an animated entity gives a stick to the other entity. The combination of the color and shape of the agents, their actions, and the instruments used are counterbalanced.

Practice trials are used to familiarize participants with the similarity judgment task. Each trial consists of a red circle and a blue square. For instance, in one of the target clips, the red circle is in a container and then moves out of the container. In the two variant clips, one of them shows the red circle moving into the container and the other shows the blue square moving back and forth inside the container.

11.6.3 Procedure

The participants were randomly assigned to either the English session or the Spanish session, such that half the bilingual speakers participated in the English language session, the other half participated in the Spanish language session (we did not employ a repeated measures design in which the same participant would have been tested in both Spanish and English, since we expected strong carryover effects, such that participation in one language condition would affect subsequent performance in the other language condition). English sessions were conducted in English by a native speaker of English. Spanish sessions were conducted by a native speaker of Spanish. Depending on the session to which they were assigned, participants were first greeted by the experimenter in either English or Spanish. Then the experimenters chatted with the participant for 5 min using the session language. The experimenter then gave the instructions of the task. In each trial, the experimenter first gave the description of the upcoming target scene verbally. Then, the experimenter started the presentation of each triad with a mouse click. During the target clip, the experimenter repeated the description of a given scene. At the end of the target clip, the experimenter paused the video. At this time, the participant repeated the scene description once. The experimenter then continued the video and the split screen with two motion events was shown. The participant had to respond by pointing to either one of the variants, depending on which they found to be most similar to the previous target scene in their opinion. They also had to say "left" or "right" verbally while pointing. To preclude a search for the "correct" answer in their responses, participants were reassured that there was no standard correct answer. The experimenter recorded participants' responses before moving on to the next trial.

11.6.4 Analysis and Results

The participants always repeated the sentences exactly as they were presented to them. For each participant, we calculated the proportion of same-manner choices. The results showed that when spoken to in English, bilinguals selected events that have the same manner of motion as the target scene more often (50 %) than when spoken to in Spanish (21.9 %; Fig. 11.2).

We ran a mixed-effect logistic regression model (Baayen 2008) using contrast coding for the fixed effects, with classification preference ("same path" versus "same manner") as the outcome variable, and language ("Spanish," "English") as the predictor variable. There were two random effect factors, participant and item. We also entered as control variables, type of ground ("hut–cave," "tree–rock"), direction of motion ("left," "right"), manner of motion in the target clip ("twirl," "jump," "slide," "roll"), type of manner contrast shown in the target clip and the variant clip ("twirl–roll," "twirl–jump," "twirl–slide," "jump–slide," "jump–roll," "slide–roll"), and order of presentation of items ("forward," "back").

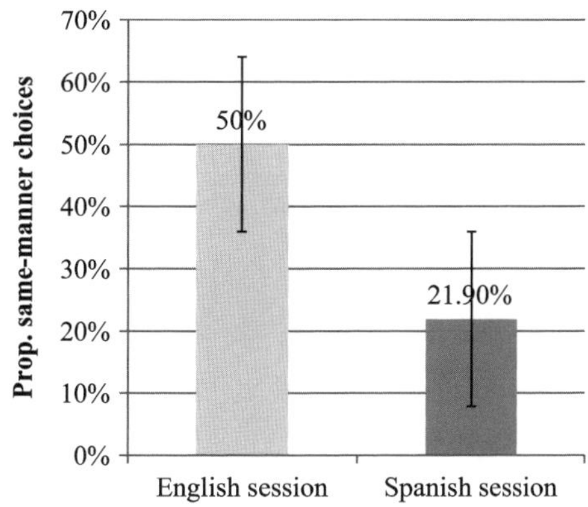

Fig. 11.2 Mean proportion of same-manner choices in English and Spanish sessions (with standard error bars)

Our results (Table 11.1) show that Spanish–English bilinguals describing a motion event in English selected the event that has the same manner of motion as the target scene, significantly more often than bilinguals describing the event in Spanish ($ß = 2.50$, $Z = 2.42$, $p < 0.05$). A likelihood ratio test confirmed the significant effect of language. Although the direction of motion and the "twirl versus roll" manner contrast approach significance, likelihood ratio tests reveal that neither direction nor any of the other variables contribute significantly to the model.

We also inspected the same-manner choices in individual speakers (Table 11.2). In the Spanish sessions, only two bilingual speakers gave same-manner choices more

Table 11.1 Effect of language on "same-path" versus "same-manner" choices in Spanish–English bilinguals

	Estimate	Standard error	Z value	p value
(Intercept)	−0.80	1.28	−0.62	0.53
Language: Spanish	2.50	1.03	2.42	0.02*
Order: forward	−0.72	1.02	−0.70	0.48
Manner contrast: twirl versus roll	2.44	1.30	1.88	0.06
Manner contrast: twirl versus slide	1.81	1.29	1.40	0.16
Manner contrast: jump versus roll	1.93	1.29	1.50	0.13
Manner contrast: jump versus slide	2.04	1.28	1.59	0.11
Manner contrast: roll versus slide	0.24	0.91	0.27	0.79
Direction: right	−0.88	0.46	−1.93	0.05
Ground: tree_rock	−0.47	0.43	−1.11	0.27
Manner_JUMP	−0.43	0.74	−0.58	0.56
Manner_ROLL	0.18	0.78	0.23	0.82
Manner_SLIDE	1.12	0.78	1.44	0.15

$*p < 0.05$

Table 11.2 Mean proportion of same-manner choices in percentage for 24 bilingual speakers in English and Spanish sessions (Italicized numbers indicate greater than or equal to chance level 50%)

English session (%)		Spanish session (%)	
E01	0.0	S01	0.0
E02	12.5	S02	0.0
E03	12.5	S03	0.0
E04	12.5	S04	0.0
E05	25.0	S05	0.0
E06	25.0	S06	25.0
E07	*50.0*	S07	25.0
E08	*75.0*	S08	25.0
E09	*87.5*	S09	25.0
E10	*100.0*	S10	25.0
E11	*100.0*	S11	*50.0*
E12	*100.0*	S12	*87.5*

E English session, *S* Spanish

than 50 % of the trials and five speakers never gave same-manner choices (Table 11.2, column 4). In the English sessions, six bilingual speakers provided same-manner choices more than 50 % of the trials and only one speaker avoided same-manner choices altogether (Table 11.2).

11.7 General Discussion

Our study investigates thinking-for-speaking effects in bilingual speakers of languages that differ in their linguistic packaging of motion event components. Our participants verbally encoded the motion events in either Spanish or English before they performed the nonlinguistic similarity judgment task. We found that the language the bilingual speakers used to describe motion events influenced the dimension of motion they selected as relevant for the purposes of judging similarity between two motion events. Bilingual speakers tested in English tended to select the motion events that had the same manner of motion as the motion events in the target clips. Bilingual speakers tested in Spanish selected motion events that matched the path of motion in the target clips. Further, the type of event description that speakers heard and repeated (atelic descriptions corresponding to the ground locations "tree and rock" versus telic, boundary-crossing descriptions corresponding to "hut and cave") did not play a role in their subsequent judgments of event similarity. There were two interesting trends: the direction of motion and the "twirl versus roll" manner contrast that approached significance. It is likely that left–right directional axis of motion on the horizontal plane is highly salient as it is aligned with the left–right axis in the egocentric frame of reference (Levinson 1996). Changes along this axis may draw more attention than other changes, e.g., the source and goal objects involved ("hut-cave" versus "tree-rock"). The "twirl versus roll" manner contrast is also likely to be prominent, involving rotation of the object around its own axis: either vertical or horizontal (combined with translational motion along the horizontal axis). However, these are post hoc speculations as to the relatively greater prominence of certain geometrical configurations and need to be explored further in a systematic manner.

Our findings demonstrate that speakers of verb-framed versus satellite-framed languages not only differ in how flexibly they combine manner verbs with path phrases, but that, such differences correlate with the dimension of motion to which they attend subsequent to verbal encoding in one language or the other. Our finding that fluent bilinguals shift their event construal depending on the language of verbal encoding suggests that, at least for the purposes of thinking-for-speaking, bilinguals do not rely on a single semantic representation that is associated with either one of their languages or one that constitutes an "intermediate" semantic representation between both languages. Rather, different language-specific event construals come into play during the process of language production, suggesting the availability of multiple "thinking-for-speaking" modes in the bilingual speaker.

Our findings also contribute to the ongoing debate on the interaction between language and thought. Prior research has distinguished between conceptualization as independent of language (the universal view), language based (the Whorfian view), or language oriented (the thinking-for-speaking view; von Stutterheim and Nüse 2003). Our bilingual speakers oriented towards the different aspects of motion events based on the language in use, suggesting that conceptualization for production is oriented by language principles, supporting the intermediate thinking-for-speaking view. In our study, we required our speakers to verbally encode the events; investigation of motion event construal without prior encoding is required to adequately evaluate the universalist and language-based perspectives.

One question that arises in this respect is whether speakers are influenced by the encoding of the motion events in the experimental task or whether they are influenced by overall biases to construe motion events in language-specific ways based on their experience of habitual patterns of encoding in the two languages (see also Gennari et al. 2002). Here, the distinction between telic and atelic event descriptions is relevant. In our study, Spanish speakers heard and repeated descriptions in which the manner information was expressed in a gerundial phrase irrespective of the telicity of the event ("went towards the rock, rolling" or "entered the cave, rolling"). If their overall experience with the patterns in the language played an influential role in their similarity judgments rather than their immediate experience with the linguistic patterns that they produced for the experience, we would expect Spanish speakers to pattern more like English speakers for the "tree-rock" events where they heard atelic event descriptions. This prediction has to do with the fact that Spanish allows manner verbs to combine with path phrases that are atelic ("rolled towards the rock")—a pattern that is similar to the one in English. But we found no differences in similarity judgments based on the telicity of the event description.[2] That is, Spanish speakers were not more likely to make same-manner choices for the scenes accompanied by atelic event descriptions. The fact that we found no differences corresponding to choice of ground location (and event description) suggests that the particular syntactic encoding patterns used in the experiment are involved in influencing event construal rather than an overall bias. However, this conclusion must be a speculative one at the current time, since we do not know whether the combination of manner verbs with atelic path phrases is a frequent, habitual encoding pattern in Spanish in ways that are comparable to English. If it is indeed the case that the specific syntactic patterns used in the experiment influence event construal rather than biases learned over a lifetime of exposure to the language, then it should be possible to induce greater attention to manner (versus path) of motion in Spanish speakers and vice versa in English speakers by manipulating the linguistic constructions employed during the experiment. For instance, since Spanish manner verbs can be combined with directional prepositional phrases (e.g., *La botella flotó hacia la cueva* "the bottle floated towards the cave," Aske 1989) and path verbs can be used in motion expressions in English (e.g., *The bottle approached the cave, floating*), speakers could be asked to produce atelic descriptions in Spanish with manner-of-motion main verbs, or corresponding descriptions in English with path-of-motion main verbs prior to participating in the similarity judgment task. Such a manipulation would allow us to examine how lexicalization of information in the main verb influences thinking-for-speaking independently of typological differences in overall patterns of motion event encoding.[3]

In conjunction with prior work, our study also provides psycholinguistic evidence that the thinking-for-speaking effect is robust, despite differences in verbal encoding methods (Billman and Krych 1998; Gennari et al. 2002; Papafragou et al. 2008). Billman et al. had their participants simply listen to the event descriptions. This can be viewed as a method of comprehension with imitation, a weak imprint of motion verb difference onto the mind. Gennari et al. had their participants use a short phrase with a single verb rather than using several verbs that encoded different subevents of the motion event. This can be viewed as a method using production together with linguistic awareness. Papafragou et al. had their participants freely inspect and describe the events. This can be viewed as a natural reminder of the daily experience to participants, which may be less controlled because different descriptions may give rise to different theoretical reasons why such an effect exists. The present study had the participants repeat predesigned event descriptions, a way to ensure that the verbal encoding is linguistically matched across subjects. The same components of motion (path and manner) were mentioned by all participants, and the types and frequency of the use of manner verbs (as main verbs or gerunds) were the same across participants.

[2] English speakers are not predicted to differ in their similarity judgments based on whether they heard telic versus atelic descriptions because telicity does not influence the ability of manner verbs to combine with path phrases in English to the extent that it does in Spanish.

[3] We thank one of the anonymous reviewers for pointing out this possibility.

Because participants were provided with the linguistically encoded message that they simply had to repeat, the depth of processing involved in the verbal encoding is arguably shallower than when natural generation of motion event descriptions is required, which may require greater focus on the meaning of the message (cf. Erlam 2006). The fact that all four studies found the effect of verbal encoding demonstrates the profound impact of such encoding. A theoretically interesting question is what the nature of such thinking-for-speaking effect is. We suspect that the nature of thinking-for-speaking may have to do with the first phase of short-term memory—the phonological loop (Baddeley and Hitch 1974). The proposed function of the phonological loop is that by providing an articulatory rehearsal component, memory traces that decay rapidly can be slowed down or revived. It is likely that verbal encoding helps the mind hold on to the event information, and depending on how those motion events are linguistically packaged, some components (e.g., manner information in English and path information in Spanish) are made more prominent and therefore are decayed slower. Or, it is likely that information in the verb is more likely to be prominent and retained in the short-term memory than information in an adverbial phrase. Those prominent aspects could then be used and influence the similarity judgment task.

A further contribution of our study lies in its demonstration of the influence of verb-specific representations on thinking-for-speaking. Whereas, both English-speaking and Spanish-speaking participants in our study described both the path and manner of motion in the video clips that they viewed, it was the semantic information encoded in the main verb, rather than in the prepositional or adverbial phrases, that influenced participants' subsequent performance in the similarity judgment task. Further research is required to examine whether aspects of information encoding other than syntactic packaging also influence thinking-for-speaking. For instance, although the experimenters read out the motion event descriptions to the participants using intonation and emphasis that was as consistent as possible, it is possible that subtle variations in their production may have also influenced participants' behavior.[4] Systematic manipulation of these variables will allow us to examine the contribution of prosody to event construal.

A possible future line of research is to examine the variable of language proficiency. In our study, we controlled for proficiency by having our bilingual speakers fill out a detailed language background questionnaire. Though the majority of our speakers self-reported that they started speaking both English and Spanish before the age of seven, they continue to use both languages, and have stayed in the USA for a long period of time; additional information (e.g., amount of interaction speakers have with both language communities) will provide us with more precise measures of proficiency. As discussed earlier, previous studies on language effect in bilingual and advanced second-language learners suggest that proficiency is a factor influencing speakers' mental construal for events, time, and object classification. Investigating

[4] We thank one of the anonymous reviewers for pointing out this possibility.

the interaction between levels of proficiency in multiple languages and the availability of more than mode of event construal will provide us with deeper insights into the nature of the "thinking-for-speaking" effect.

Acknowledgments We thank the editors of the volume and two anonymous reviewers for their valuable comments. We also thank the members of the Language, Cognition and Development Lab at the University of Colorado, Boulder, for their feedback and the research assistants Anwen Fredriksen and Elizabeth Saade for their help in running the experiments.

References

Aske, J. (1989). Path predicates in English and Spanish: A closer look. In K. Hall, M. Meacham, & R. Shapiro (Eds.), *Proceedings of the fifteenth annual meeting of the Berkeley linguistics society* (pp. 1–14). Berkeley: Berkeley Linguistics Society.

Athanasopolous, P. (2007). Interaction between grammatical categories and cognition in bilinguals: The role of proficiency, cultural immersion, and language of instruction. *Language and Cognitive Processes, 22,* 689–699.

Baayen, H. (2008). *Analyzing linguistic data: A practical introduction to statistics using R*. New York: Cambridge University Press.

Baddeley, A. D., & Hitch, G. (1974). Working memory. In G. H. Bower (Ed.), *The psychology of learning and motivation: Advances in research and theory* (Vol. 8) (pp. 47–89). New York: Academic.

Beck, S., & Snyder, W. (2001). Complex predicates and goal PP's: Evidence for a semantic parameter. In A. H.-J. Do, L. Dominguez, & A. Johansen (Eds.), *Proceedings of the 25th Boston University Conference on Language Development* (pp. 114–122). Somerville: Cascadilla Press.

Berlin, B., & Kay, P. (1969). *Basic color terms: Their universality and evolution*. Berkeley: University of California Press.

Billman, D., & Krych, M. (1998). Path and manner verbs in action: Effects of "skipping" or "existing" on event memory. *Proceedings of the Cognitive Science Society, 20,* 156–161.

Billman, D., Swilley, A., & Krych, M. (2000). Path and manner priming: Verb production and event recognition. *Proceedings of the Cognitive Science Society, 22,* 615–620.

Bohnemeyer, J., Eisenbeiss, S., & Narasimhan, B. (2001). Event Triads. In S. C. Levinson & N. Enfield (Eds.), *Manual for the field season 2001* (pp. 100–114). Nijmegen: Max Planck Institute for Psycholinguistics.

Bohnemeyer, J., Eisenbeiss, S., & Narasimhan, B. (2006). Ways to go: Methodological considerations in Whorfian studies on motion events. *Essex Research Reports in Linguistics, 50,* 1–20.

Boroditsky, L. (2001). Does language shape thought? English and Mandarin speakers' conceptions of time. *Cognitive Psychology, 43,* 1–22.

Brown, A., & Gullberg, M. (2008). Bidirectional crosslinguistic influence in L1-L2 encoding of Manner in speech and gesture: A study of Japanese speakers of English. *Studies in Second Language Acquisition, 30,* 225–251.

Cook, V. J., Bassetti, B., Kasai, C., Sasaki, M., & Takahashi, J. A. (2006). Do bilinguals have different concepts? The case of shape and material in Japanese L2 users of English. *International Journal of Bilingualism, 10,* 137–152.

Erlam, R. (2006). Elicited imitation as a measure of L2 implicit knowledge: An empirical validation study. *Applied Linguistics, 27,* 464–491.

Finkbeiner, M., Nicol, J., Greth, D., & Nakamura, K. (2002). The role of language in memory for actions. *Journal of Psycholinguistic Research, 31,* 447–457.

Gennari, S. P., Sloman, S. A., Malt, B. C., & Fitch, W. T. (2002). Motion events in language and cognition. *Cognition, 83,* 49–79.

Gentner, D., Imai, M., & Boroditsky, L. (2002). As time goes by: Evidence for two systems in processing space time metaphors. *Language & Cognitive Processes, 17,* 537–565.

Gullberg, M. (2011). Thinking, speaking, and gesturing about motion in more than one language. In Pavlenko A. (Ed.) *Thinking and speaking in two languages* (pp. 143-169). Multilingual Matters.

Gullberg, M., & Indefrey, P. (2003). *Language background questionnaire. Developed in the dynamics of multilingual processing*. Nijmegen, Max Planck Institute for Psycholinguistics.

Heider, E. R. (1972). Universals in color naming and memory. *Journal of Experimental Psychology, 93,* 10–20.

Imai, M., & Gentner, D. (1997). A crosslinguistic study of early word meaning: Universal ontology and linguistic influence. *Cognition, 62,* 169–200.

Kita, S., & Özyürek, A. (2003). What does cross-linguistic variation in semantic coordination of speech and gesture reveal? Evidence for an interface representation of spatial thinking and speaking. *Journal of Memory and Language, 48,* 16–32.

Lai, V. T. (2002). *The Processing of Temporal Perspectives in Mandarin Chinese*. Unpublished MA thesis.

Lai, V. T. (2005). Language experience influences in the conceptualization of time metaphor. *Proceedings of the II Conference on Metaphor in Language and Thought* (pp. 1–16) Rio de Janeiro: Universidade Federal Fluminense.

Levelt, W. J. M. (1996). Perspective taking and ellipsis in spatial descriptions. In P. Bloom, M. A. Peterson, M. F. Garrett, & L. Nadels (Eds.), *Language and space* (pp. 77–107). Cambridge: MIT Press.

Levin, B., & Rapoport, T. (1988). Lexical subordination. *Papers from the Twenty Fourth Regional Meeting, Chicago Linguistic Society* (pp. 275–289). Chicago: University of Chicago Press.

Levin, B., & Rappaport Hovav, M. (1999). Two structures for compositionally derived events. Proceedings of SALT 9, Cornell Linguistics Circle Publications, Cornell University, Ithaca, NY, pp. 199-223.

Levin, B., & Rappaport Hovav, M. (1995). *Unaccusativity: At the syntax–lexical semantics interface*. Cambridge: MIT Press.

Levinson, S. C. (1996). Relativity and spatial conception and description. In J. J. Gumperz & S. C. Levinson (Eds.), *Rethinking linguistic relativity* (pp. 177–202). Cambridge: Cambridge University Press.

Levinson, S. C., Kita, S., Haun, D. B. M., & Rasch, B. H. (2002). Returning the tables: Language affects spatial reasoning. *Cognition, 84,* 155–188.

Li, P., & Gleitman, L. (2002). Turning the tables: Language and spatial reasoning. *Cognition, 83,* 265–294.

Loftus, E., & Palmer, J. (1974). Reconstruction of automobile destruction: An example of the interaction between language and memory. *Journal of Verbal Learning and Verbal Behavior, 13,* 585–589.

Lucy, J. A. (1996). The scope of linguistic relativity: An analysis and review of empirical research. In J. J. Gumperz & S. C. Levinson (Eds.), *Rethinking linguistic relativity. Studies in the social and cultural foundations of language, No. 17.* (pp. 37–69). Cambridge: Cambridge University Press.

McGlone, M. S., & Harding, J. L. (1998). Back (or forward?) to the future: The role of perspective in temporal language comprehension. *Journal of Experimental Psychology: Learning, Memory, and Cognition, 24,* 1211–1223.

Narasimhan, B. (2003). Motion events and the lexicon: The case of Hindi. *Lingua, 113*(2), 123–160.

Papafragou, A., Massey, C., & Gleitman, L. (2002). Shake, rattle, 'n' roll: The representation of motion in language and cognition. *Cognition, 84,* 189–219.

Papafragou, A., Hulbert, J. C., & Trueswell, J. (2008). Does language guide event perception? Evidence from eye movements. *Cognition, 108,* 155–184.

Pederson, E. (1995). Language as context, language as means: Spatial cognition and habitual language use. *Cognitive Linguistics, 6,* 33–62.

Pederson, E., Danziger, E., Wilkins, D., Levinson, S., Kita, S., & Senft, G. (1998). Semantic typology and spatial conceptualization. *Language, 74,* 557–589.

Slobin, D. I. (1987). Thinking for speaking. Proceedings of the thirteenth annual meeting of the Berkeley linguistics society (pp. 435–444), Berkeley Linguistics Society, Berkeley.

Slobin, D. I. (1996a). From "thought and language" to "thinking to speaking". In J. J. Gumperz & S. C. Levinson (Eds.), *Rethinking linguistic relativity* (pp. 70–96). Cambridge: Cambridge University Press.

Slobin, D. I. (1996b). Two ways to travel: Verbs of motion in English and Spanish. In M. S. Shibatani & S. A. Thompson (Eds.), *Grammatical constructions: Their form and meaning* (pp. 195–220). Oxford: Clarendon Press.

Slobin, D. I., & Hoiting, N. (1994). Reference to movement in spoken and signed languages: Typological considerations. Proceedings of the twentieth annual meeting of the Berkeley linguistic society (pp. 487–505), Berkeley Linguistics Society, Berkeley.

von Stutterheim, C., & Nüse, R. (2003). Processes of conceptualization in language production: Language-specific perspectives and event construal. *Linguistics, 41,* 851–881.

Talmy, L. (1985). Lexicalization patterns: Semantic structure in lexical forms. In T. Shopen (Ed.), *Language typology and syntactic description 3: Grammatical categories and the lexicon* (pp. 57–149). Cambridge: Cambridge University Press.

Talmy, L. (2000). *Toward a cognitive semantics: Typology and process in concept structuring* (pp. 21–146). Cambridge: MIT Press.

Talmy, L. (2007). Lexical typologies. In T. Shopen (Ed.), *Language typology and syntactic description volume 3, grammatical categories and the lexicon* (pp. 166–168). Cambridge: Cambridge University Press.

Whorf, B. L. (1956). *Language, thought, and reality*. Boston: MIT Press.

Wilkins, D. P. & Hill, D. (1995). When GO means COME: Questioning the basicness of basic motion verbs. *Cognitive Linguistics, 6,* 209–259.

Zubizarreta, M. L & Oh, E. (2007). *On the syntactic composition of manner and motion*. Cambridge: MIT Press.

Part V
Acquiring Verbs

Chapter 12
Argument Structure: Relationships Between Theory and Acquisition

Sudha Arunachalam

A central question in language acquisition research is how children come to learn the argument-taking properties of verbs. Sometimes called Baker's paradox, based on Baker (1979), the problem is evident in examples like these: How does the child learn that "bounce" participates in the causative/inchoative alternation, but "laugh" does not (*John laughed the baby)? Or that "give" participates in the dative alternation, but "donate" does not?

Crucial to this question is an understanding of what property of the grammar accounts for these alternations (e.g., whether it is syntax, lexical semantics, or event representations), and argument structure approaches within theoretical linguistics differ on precisely how this kind of knowledge is represented. But despite the close intertwining of the acquisition and representation questions, research on how children learn verbs and how verbs are represented in the grammar has largely proceeded independently of each other. In this chapter, I ask: Can data from language acquisition research inform our understanding of theoretical approaches to argument structure?

I consider several points of contact between the theoretical and acquisition literatures and conclude that the answer to this question is largely *no*. Most existing acquisition data fail to distinguish between current theoretical approaches and argument structure. In part, this may be because of *in-principle* limitations on the kinds of distinctions we can parcel out from acquisition data, but I will nevertheless point out some promising directions for future research.

An early version of this work was presented at the 2008 Verb Concepts: Cognitive Science Perspectives on Verb Representation and Processing workshop. I am grateful to the organizers and audience of that meeting, to David Embick, Alexander Williams, and Jeff Lidz for shaping my thoughts on these issues, and to an anonymous reviewer, Matthew Valleau, Zhenghan Qi, and Matt Goldrick for helpful comments on an earlier version of this chapter. All errors are of course my own.

S. Arunachalam (✉)
Department of Speech, Language, and Hearing Sciences,
Boston University, 635 Commonwealth Ave., Boston, MA 02215, USA
e-mail: sarunach@bu.edu

R. G. de Almeida, C. Manouilidou (eds.), *Cognitive Science Perspectives on Verb Representation and Processing,* DOI 10.1007/978-3-319-10112-5_12

Section 12.1 introduces two main types of grammatical architecture for argument structure that have been proposed in the theoretical literature. Section 12.2 outlines how, in principle, these frameworks relate to the problem of verb acquisition. Section 12.3 considers evidence from the language acquisition literature on how children learn verbs, highlighting major points of contact between the acquisition and theoretical literatures, but concluding that the acquisition data shed little light on the theoretical issues. In Sect. 12.4, I consider why previous attempts at connecting these literatures have fallen short, and outline some promising avenues for future research. Section 12.5 concludes.

12.1 Theoretical Approaches to Argument Structure

Any theory of argument structure must describe how the argument-taking properties of verbs—that is, what structures verbs appear in, as well as what structures they fail to appear in—are encoded. Two main approaches can be distinguished: Lexical Projectionism (e.g., Chomsky 1981; Levin and Rappaport Hovav 1995), and Non-lexical Non-projectionism. Under this latter umbrella, two further distinctions can be made: Compositional Non-projectionist approaches (e.g., Borer 2005; Embick 2004; Embick and Noyer 2006; Hale and Keyser 1993; Marantz 1997) and Construction Grammar approaches (e.g., Goldberg 1995).

All of these approaches posit that grammatical principles somehow play a role in argument-taking behavior. It is, of course, a possibility that syntax–semantics relationships are epiphenomenal and that our knowledge of the argument-taking properties of verbs is based solely on statistical analysis of the input. Children would learn this information by parsing the input and tracking how often verbs appear in particular frames (e.g., Braine and Brooks 1995; Theakston 2004). Children and adults certainly do track the frequency with which verbs appear in various syntactic frames (e.g., Snedeker and Trueswell 2004; Trueswell et al. 1993), and the frequency of experience likely plays a role in acquisition of argument structure knowledge (e.g., Wonnacott and Newport 2005; Wonnacott et al. 2008). However, many linguistic phenomena, such as regularities in argument structure behavior across languages, cast doubt on a purely statistical approach (e.g., Phillips 2006). We therefore assume for current purposes that relationships between form and meaning have at least some representational consequences.

12.1.1 Lexical Projectionism

One possibility is that verbs have highly specified lexico-semantic representations that determine their argument-taking behavior. That is, the lexical representations of verbs specify the number and type of arguments they take. These arguments are mapped onto syntactic positions according to universal linking principles. In this type of approach, the basic syntactic structure in which a verb appears is determined by information in its lexical representation.

Under this scenario, our knowledge about the verb "hit" includes something like the information in (1):

(1) $[[\text{hit}]] = \exists x\ \exists y\ \exists e\ [\text{hit}(e)\ \&\ \text{Agent}(e, y)\ \&\ \text{Theme}(e, x)]$

Such representations *lexicalize* argument-taking behavior, and argument structure *projects* from the lexico-semantic representation; hence, we call this Lexical Projectionism.

In this framework, verbal alternations result from the operation of lexical rules (e.g., Pesetsky 1982; Pinker 1989; Rappaport Hovav and Levin 1998). These rules operate directly on the lexico-semantic representations, and the altered representations project the alternate structure. In this kind of architecture, **John laughed the baby* is ungrammatical because (a) the representation for *laugh* does not project a theme and (b) *laugh* is not a member of a class of verbs that allows the application of a lexical rule adding a theme argument.

12.1.2 Nonlexical Non-projectionism

Another possibility is that the representations of verbs are highly *under*specified for how many arguments they combine with and the semantic roles they assign to them. Instead, verbs appear in syntactic structures which themselves encode arguments and carry meaning. This meaning combines with the core meaning of the verb to yield an interpretation of the sentence. In such models, verbs neither lexicalize nor project argument structure, and so I will call these Non-projectionist models.

Non-projectionist accounts divide into two major types, which I will call Compositional and Construction Grammar approaches. The principal difference between these two kinds of approach lies in how the syntactic structures are derived and represented.

12.1.2.1 Compositional Non-projectionist Approaches

Examples of the Compositional Non-projectionist approach include Distributed Morphology (e.g., Embick 2004; Marantz 1997) and the neo-constructionist work of Borer (2005), as well as others focusing on event syntactic structure as the key to semantic interpretation (e.g., Ramchand 2010). In these approaches, structures are generated according to familiar principles of syntax, and verbs freely appear in them. Because primitives are semantically underspecified for argument-taking behavior, as in (2), rules of semantic composition do more work in generating meaning.[1] These interpretive rules, such as (3), allow interpretation of primitives and argument relations within these structures:

---

[1] There are hybrid positions, as well, under which certain arguments are lexicalized and others are not (e.g., Marantz 1984; see also Kratzer 1996 and Ramchand 1997).

(2) [[hit]] = ∃e [hit(e)]
(3) [[[V' V DP]]] = ∃e [[[DP]] (∃y [[[V]](e) & Theme(e, y)])]

The underspecification of (2) does not of course mean that any verb can appear in any structure; rather, constraints on distribution may stem from the compatibility of the meaning of the primitive and the meaning of the structure, from real-world/conceptual knowledge about the possibility of certain eventualities, and from independent storage in memory of idiosyncratic gaps.

While Lexical Projectionist models assign all participants bearing a thematic role to an argument in a verb's basic argument structure, a Non-projectionist model does not carry such a requirement. In a Non-projectionist model, the number of participants in an event may be specified as part of our knowledge about the verb (e.g., hitting events involve a hitter and a hittee), but because primitives do not determine argument structure, event participants need not map onto verbal arguments in a one-to-one manner (e.g., Williams 2005). Both Lexical Projectionist and Compositional Non-projectionist approaches can capture the same information with respect to verb argument structure, but divide up differently the contributions of event representations, lexico-semantic representations, and syntactic representations to do so.

12.1.2.2 Construction Grammar Approaches

In Construction Grammar, constructions are form–function pairings that are stored in memory. In Goldberg (1995) and subsequent work, the term applies equally to morphemes, words, and entire sentences. The latter may have empty slots for smaller constructions to be integrated into them, or may be completely lexically specified, as in the case of fixed idiomatic expressions.

Construction Grammar shares with other Non-projectionist approaches the fundamental positions that verb argument structure is not projected from the lexical representation of the verb itself, that structures have meaning, and that verbs appear freely in structures "as long as they can be construed as not being in conflict. For example, the specification of the ditransitive construction that requires an animate recipient argument conflicts with the meaning of *storage* resulting in the unacceptability of [*Lisa sent storage a book]" (Goldberg 1995, p. 221).

Nevertheless, these properties are implemented in starkly different ways. In Construction Grammar, the meaning of an expression is the result of integrating the meanings of individual lexical items into the meaning of the entire construction. The ditransitive construction, for example, can be represented as in (4):

(4) Form: Subj [V Obj1 Obj2]
Meaning: transfer (either intended or actual)

Representations under Construction Grammar thus lack functional projections and hierarchical structure; the frame is represented as a whole. In Compositional Non-projectionist theories, on the other hand, the subparts of a structure are interpreted individually, with interpretive rules that determine how the subparts are composed, and further, the meanings associated with syntactic structures are at least to some extent universal, as structures are composed from pieces of functional structure (Marantz 1997; Borer 2005).

12.2 Predictions for Acquisition

These different theoretical approaches make different predictions about how language acquisition might proceed. They require the child to learn different kinds of information and to structure this information in different ways. At first glance, it seems as though it should be easy to distinguish Lexical Projectionist and Non-projectionist architectures using acquisition data; each approach presents a different learning task for the child. Under a Lexical Projectionist architecture, the child needs to learn:

(5) a. individual primitives in such a way that they define the syntactic structures of the language
b. lexical rules for deriving verbal alternations
c. constraints on lexical rules

Under a Non-projectionist architecture, this list looks quite different. The child must learn:

(6) a. the syntactic structures of the language
b. how the structures relate to the semantic interpretation of verbs and arguments
c. which verbs appear in which environments

In practice, however, it is very difficult to find behavioral measures that can distinguish (5) and (6), as we will see. Below, I review several important issues in the acquisition literature to determine whether and how they line up with these theoretical approaches. I begin broadly with some general mechanisms important for language learning (bootstrapping, online sentence processing) and then take a few examples as case studies; these are studies that explicitly claim to test one of the theoretical approaches using data from child language acquisition.

12.3 How Children Learn Verbs

In this section, I consider some general mechanisms that have been attributed to the young verb learner and evaluate their relationship with the argument structure question.

12.3.1 Bootstrapping

The mechanisms by which children learn new verbs have been argued to shed important light on what their verb representations look like (in addition to how these representations are accessed during language comprehension and production). The two main mechanisms proposed in the literature for how children acquire verb knowledge are Semantic Bootstrapping (Grimshaw 1981; Pinker 1984, 1989) and Syntactic Bootstrapping (Landau and Gleitman 1985; Gleitman 1990). These strategies are not mutually exclusive, as is well noted by these scholars; the child is hypothesized to use both, approaching the task from different angles, depending on what knowledge they already possess and the information available in a given learning situation.

Semantic Bootstrapping originated within a Lexical Projectionist framework (Grimshaw 1981), and in its original formulations, Syntactic Bootstrapping asserts Non-projectionist assumptions (Landau and Gleitman 1985; Gleitman 1990). (However, in later work, Lidz and Gleitman discussed Syntactic Bootstrapping within a Lexical Projectionist framework (Lidz et al. 2004); see Sect. 12.3.3.1.) The alignments between Semantic Bootstrapping and Projectionism, and Syntactic Bootstrapping and Non-projectionism, have been taken up elsewhere in the acquisition literature as well (e.g., Borer 2004; Kako 1998; Kako and Wagner 2001).

However, I argue that these alignments are spurious. The evidence demonstrating that children use these two mechanisms could equally be explained within either Lexical Projectionist or Non-projectionist frameworks, and there is nothing in principle about the abilities attributed to the child under either mechanism that precludes either Lexical Projectionist or Non-projectionist assumptions. The following sections describe the two bootstrapping hypotheses and some evidence that children use each of them; we then turn to a discussion of how these mechanisms relate to grammatical architecture.

12.3.1.1 Semantic Bootstrapping

Semantic Bootstrapping centers on the learning of the basic syntactic structures of the language and which verbs can appear in these. It asserts that the child can use verb meaning to determine which frames are licensed. Specifically, on this account, knowledge of verb meaning (a) allows the child to identify what syntactic structures exist in her language and (b) facilitates learning how particular (classes of) verbs relate to syntactic structures.

According to the first point, defended in Pinker (1984, 1989), the child has innate rules that, for example, map the agent of an action onto subject position (but see, e.g., Bowerman (1990), for a learning account of linking rules). Once the child has determined the subject, he or she can construct a compatible tree for the entire sentence according to general, universal, and innate principles of syntax, and in this way, learn about the structural possibilities available in his or her language.

The second point is also a part of Pinker's (1989) proposal, dealing specifically with how children learn which verbs participate in which verbal alternations. When children encounter a new verb, Pinker argues, they place it into "broad-range" and "narrow-range" semantic classes on the basis of its meaning. These classes are defined by the core meaning components shared by all the verbs in the class, and delimit the range of structures in which the verb can appear. For example, all verbs whose lexical semantics involve literal transfer of one object to another can appear in the dative alternation (*John gave the apples to Fred/John gave Fred the apples*). However, there are many verbs that may, but need not, result in a change of possession; only some of these participate in the alternation. Specifically, Pinker points to semantic features such as "instantaneous imparting of force in some manner causing ballistic motion"; these verbs do alternate, such as *throw, toss,* and *kick*. Verbs with the feature of "continuous imparting of force in some manner causing accompanied motion," such as *pull, carry,* and *lower,* do not alternate. These event-specific semantic features place the verbs in different narrow-range semantic classes.

Pinker's hypothesis is that once children have identified the key meaning components of a verb and assigned it to the appropriate class, they can determine what structures that verb can appear in. That is, identification of a verb's meaning results in a representation that determines its syntactic distribution as well. This strategy is particularly useful in determining which verbs appear in which structures of an alternation, and this is the context in which Semantic Bootstrapping is most often discussed in the acquisition literature.

Experimental evidence confirms that children are able to use this mechanism during verb learning. Gropen et al. (1991), for example, demonstrated that children hearing a novel verb while viewing events with a salient manner component assigned the verb manner syntax (specifically, the syntax associated with manner verbs like *pour*), while children viewing events with a salient result component assigned the verb result syntax (associated with verbs like *fill*). More recent evidence has supported the importance of semantic verb classes even while finding that verb class membership alone cannot explain children's and adults' knowledge of verb alternations (Ambridge et al. 2008; 2014; Bidwood et al. 2014).

12.3.1.2 Syntactic Bootstrapping

An important limitation of the Semantic Bootstrapping approach is that it requires that observation of events in the world provides sufficient information to determine the lexical semantics of verbs. As Landau and Gleitman (1985), Gleitman (1990), and subsequent work point out, this is no trivial task. Several kinds of verbs might never be learned if observing the world was the only, or perhaps even principal, strategy, such as verbs describing mental activities like *think*.

Gleitman and colleagues have proposed another mechanism, Syntactic Bootstrapping. Just as with Semantic Bootstrapping, the child exploits relationships between verb meaning and the structures in which they appear. But in Syntactic Bootstrapping, they do so in the reverse direction: the learner uses the syntactic structures in which an unfamiliar verb appears to make some inferences about its meaning.

There is much experimental evidence demonstrating young children's use of this mechanism (e.g., Arunachalam and Waxman 2010; Fisher et al. 1994; Fisher 2002a;

Golinkoff et al. 1996; Hirsh-Pasek and Golinkoff 1996; Naigles 1990; Naigles and Kako 1993; Yuan and Fisher 2009; among many others). For example, in a landmark study, Naigles (1990) familiarized 2-year-olds to scenes in which two actors were simultaneously engaged in two kinds of action: one actor (dressed as a duck) was pushing the other (dressed as a bunny) repeatedly into a squatting position, and both actors were waving one of their own hands in circles. Toddlers heard one of two sentences: a transitive sentence, "The duck is gorping the bunny," or an intransitive sentence, "The duck and the bunny are gorping." In the immediately following test phase, the two actions were pulled apart on two separate screens; one screen depicted just the duck pushing the bunny into a squatting position, and the other screen depicted just the duck and bunny waving. Naigles predicted that toddlers who had heard the transitive sentence would look longer at the scene in which the duck was acting on the bunny because that sentence structure is not compatible with an event in which two actors are acting independently. The results were as predicted: toddlers hearing the transitive sentence preferred the forcing-to-squat scene compared to toddlers hearing the intransitive sentence. These results demonstrate that verb learning need not proceed by observing events in the world and then determining verb syntax on the basis of the inferred meaning; instead, children can also simultaneously coordinate syntactic and observational information to establish new verb representations.

12.3.1.3 Bootstrapping and Architecture

Given empirical evidence that children use both Semantic and Syntactic Bootstrapping mechanisms during verb learning, it cannot logically be the case that Semantic Bootstrapping is only compatible with Projectionism, and Syntactic Bootstrapping only compatible with Non-projectionism. It is therefore important to test the strength of the relationships between these mechanisms and the argument structure theories they were originally aligned with.

Under Semantic Bootstrapping, verb meaning is inferred by observing events, and the structures in which a particular verb can appear are determined on the basis of meaning (e.g., Grimshaw 1981; Pinker 1984, 1989). The Lexical Projectionist explanation of this behavior is that knowledge of verb meaning is used to generate syntactic structures via linking rules, and that alternations result from operations acting on the representation of verb meaning. However, a Non-projectionist explanation is also possible. Suppose that structures carry meanings of a general sort, and that the child has acquired at least some of the basic structure–meaning mappings in his or her language. On observing an event in the world and extracting a candidate meaning (e.g., girl causing ball to bounce), he or she associates it with the structure most compatible with that meaning (e.g., a transitive structure with agent in subject position and patient in object position). This process need not be mediated by the lexico-semantic representation he or she sets down for *bounce* itself.

Under Syntactic Bootstrapping, verb learning is aided by associating events with sentences, including such properties of sentences as the number and type of arguments (Gleitman 1990). The Non-projectionist explanation is that children use their knowledge of the meanings of the syntactic structures in which a verb appears to infer its meaning. That is, structures carry elements of meaning such as causation or transfer, and the child infers that a new verb occurring in one of these structures possesses, or is compatible with, that meaning.[2] Fisher et al. (1994) wrote, "our view is not that there are 'verb classes,' each of which has semantic components and (therefore) licenses certain structures. Rather we suggest that verb frames have semantic implications (truth values), and verbs have meanings" (367).

A Lexical Projectionist, however, could also account for the empirical evidence supporting Syntactic Bootstrapping by arguing that knowledge of linking rules is what allows inference of verb meaning. Syntactic structures themselves need not have meaning if the child is exploiting his or her knowledge of syntax–semantics mappings. Upon hearing a sentence like "The rabbit is gorping the duck," the child can look at his or her linking rules and infer, based on prior knowledge of some verbs, and given the two possible options shown at test—caused-motion or individual motion—that verbs with the caused—motion meaning occur in the observed structure (see also Fisher et al. 1994 for discussion).

Experiments showing the success of Bootstrapping, then, are compatible with both Lexical Projectionist and Non-projectionist representations of syntactic structure. The accounts explain the data differently, but both are able to do so without any apparent loss in generality. Although these mechanisms originated within or are traditionally aligned with particular theoretical frameworks, these alignments are not necessary, and we must look elsewhere for evidence pertaining to links between acquisition data and theoretical approaches to argument structure.

12.3.2 Lexicalism and Sentence Processing

Recently, acquisition researchers have devoted attention to how children access lexical representations during real-time sentence processing (e.g., Fernald and Hurtado 2006; Borovsky et al. 2012; Snedeker and Trueswell 2004; Trueswell and Gleitman 2004; Trueswell et al. 1999), warranting a brief discussion of its consequences for argument structure representations. Much research in sentence comprehension has shown that information about what arguments a verb takes is activated as soon as the verb is activated, for both adults (e.g., Altmann and Kamide 1999; Boland 2005; Trueswell et al. 1994) and young children (Fernald 2004), indicating that this information is part of their earliest verb representations. This kind of research has motivated support for lexicalized grammatical representations within psycholinguistics (e.g., Kim et al. 2002). This raises the obvious question: Is a Non-lexicalist grammar,

[2] Kako and Wagner (2001), for example, argue that Syntactic Bootstrapping provides evidence that syntactic structures carry meaning. However, they outline their claim in *Projectionist* terms: "abstract semantic properties such as causation, motion, transfer, and mental activity 'project' systematically from a verb's lexical–semantic structure into the syntax" (2001, p. 106). If this semantic information is projected from verbs into the syntax, then it must be present in the representation of the verb in the first place. If so, it would be redundant to posit that syntactic structures have meaning independent of lexical items; this meaning is alread specified in verbs themselves.

such as the Compositional Non-projectionist approaches we have considered here, psychologically implausible?

One hindrance to addressing this question is that there are multiple definitions of the word “lexicalist” in the discussion of argument structure (Tomasello 2003). For example, “lexicalist” can refer to having a generative lexicon in which words are derived (e.g., Pustejovsky 1995; Jackendoff 1997); to having large chunks of structure, such as idioms, represented as frozen forms with lexical items; or, as in the psycholinguistic literature, to theories in which processing is influenced by a great deal of knowledge about particular lexical items. The Constraint-based Lexicalist framework in psycholinguistics (MacDonald et al. 1994; Trueswell and Tanenhaus 1994), for example, “assumes a constraint satisfaction approach to ambiguity resolution in which multiple sources of information are used to converge as rapidly as possible on a single interpretation. The central component of this theory is a grammatical processing system that is highly tuned to the structural preferences of individual lexical items, hence ‘lexicalist’” (Trueswell and Gleitman 2004, p. 323). This rules out a system in which accessing grammatical representations is necessarily independent of accessing knowledge about particular lexical items.

However, it does not rule out Nonlexical Non-projectionist architectures. Non-projectionism is compatible with explicit representation of knowledge about particular lexical items, as well as the accessibility of this kind of knowledge within the context of sentence processing. In other words, the claim that language users have available to them abstract structural representations does not entail that they are not *also* tracking frequency information, such as the frequency with which verbs appear in particular frames or co-occur with particular lexical items. (On the contrary, the evidence is overwhelming that we do track such information.) The claim is simply that such statistical information is not *all* we know. Thus, we should avoid basing evaluation of a theory on the use (or nonuse) of the word “lexical” or “lexicalist” or even “lexicon.” Even if children’s sentence processing behavior is highly dependent on their prior experience with particular lexical items, this is not necessarily support for a Lexical (Projectionist) grammatical architecture.

12.3.3 Case Studies

General mechanisms aside, some studies have made explicit claims to link acquisition data and one of the three argument structure approaches. We consider these below.

12.3.3.1 Evidence from Acquisition for Lexical Projectionism

Lidz et al. (2004) present the results of an experiment framed within a Lexical Projectionist approach. Following Naigles et al. (1993), young children (aged 3; 1–3; 10) were asked to use stuffed animals to act out sentences. The sentences were either grammatical (e.g., *The giraffe falls*) or ungrammatical, with the latter class further divided into “near” and “far,” depending on similarity to the structure the verb typically appears in. For the verb *fall*, which usually appears in an intransitive frame with one argument, the near ungrammatical structure was a simple transitive, **The zebra falls the giraffe,* and the far ungrammatical structures both had sentential complements: one tensed (**The zebra falls that the giraffe jumps*) and one untensed (**The zebra falls the giraffe to jump*).

The results of the experiment indicated that children are able to extend verbs to frames in which they do not normally appear. For example, children hearing **The zebra falls the giraffe* acted out a scene in which the zebra pushed the giraffe and caused it to fall. However, the children only extended verbs to those frames that were near. When faced with known verbs in far ungrammatical structures, the children were more likely to be influenced by the lexical semantics of the verb; given a sentence like **The zebra falls that the giraffe jumps*, the children simply made the zebra fall.

Lidz et al. interpreted these data as suggesting that arguments are projected from verbs in children's syntactic representations. They reasoned that if a Non-projectionist architecture, in which structures have meaning and verbs appear freely in them, is correct, children should be equally inclined to use a verb in any structure, and there should be no effect of complement type as instantiated in the near/far distinction. Because they found effects of the near/far classes, it must be the case that the children are able to refine their representation of the verb to project a new structure if that structure is compatible with the verb's meaning, but they cannot do so if the verb's meaning is incompatible with the structure. A Lexical Projectionist explanation of children's willingness to extend **The zebra falls the giraffe* but not **The zebra falls that the giraffe jumps* would then be that the verb "fall" does not project the [V CP] structure, and that there are no lexical rules that can act upon the verb's lexico-semantic representation to allow it to project this structure.

However, I would argue that the acquisition data are also compatible with Non-projectionism. Any architecture must somehow capture constraints on distribution, e.g., that "fall" does not appear with a sentential complement. Within Lexical Projectionism, these constraints are instantiated by lexicalized argument structure representations and rules. The applicability of a rule for any given verb is determined by form-meaning correspondences, for example, that verbs describing manners of motion do not (s–)select propositions (Pesetsky 1982). A Non-projectionist architecture could capture this same generalization but without going on to instantiate it with lexicalizing arguments or linking rules. Instead, the verb "fall" could be incompatible with the sentence complement structure because the meanings of the verb and the structure are incompatible. These data do not, then, require a Lexical Projectionist grammar. These results do convincingly demonstrate that children are unwilling to make arbitrary verb–structure pairings, but this fact can be captured within Projectionist or Non-projectionist architectures.

12.3.3.2 Evidence from Acquisition for Compositional Non-projectionism

Very little work has explicitly discussed links between Compositional Non-projectionist architectures and acquisition. An important exception is Borer (2004), who presents arguments in favor of a Non-projectionist view based on a pattern of acquisition she finds in young Hebrew learners.

Borer observes that Hebrew-learning toddlers make errors in verb production which suggest a disconnect between their knowledge of morphophonology and their knowledge of argument structure. She argues that: "If the child has knowledge of the projection of arguments independently of the properties of the verbs associated with the resulting structure... we predict that it should be possible, in principle, for the child to pass through a stage where the syntactic event structure is fully in place, but vocabulary knowledge is impaired" (2004, p. 297). She argues that the presence of this stage is evidence for a Non-projectionist view of grammar and against a Lexical Projectionist view of grammar because, in the latter approach, verb argument structure is inextricably tied to lexical representations. However, Borer's claim follows if a Lexical projectionist view is committed to representations that specify morphophonological form. But the key point for the Lexical Projectionist program is not specifying the form at the level of phonological instantiation; the precise morphophonological realization need not be spelled out as part of argument structure knowledge. Instead, a Lexical Projectionist grammar could have more abstract representations of form.

For example, Borer notes that toddlers make errors such as

(7) ra'iti "et ha-ciyurim le-"aba
see.1SG OM the-paintings to-daddy
"I showed the paintings to Daddy."

The verb form (binyan) the child produced is correct, but does not match with the ditransitive syntax of the sentence (this verb form only appears with transitive syntax for this root). Borer's interpretation of these data is that the child has projected the correct structures but has not acquired specific knowledge about the relationship between specific vocabulary items and these structures. Her account of these and similar data is that there is a dissociation between vocabulary knowledge and syntactic structures such that the child can project syntactic structures despite impaired knowledge of particular primitives.

We suggest, however, that the Projectionist account need not assume that this process relates morphophonological output forms in detail. Under a Projectionist architecture, the errors that Hebrew learners make could be syntactic structures that are projected from representations that *do not specify* the morphophonological form that is correct for that syntactic structure. With this view of lexicalism, the Lexical Projectionist account can account for children's errors with morphophonology just as a Non-projectionist account can.

12.3.3.3 Evidence from Acquisition for Construction Grammar

Construction Grammar has recently been hailed as a key to understanding language acquisition (see especially Tomasello 2000, 2003). Tomasello has argued that when young children encounter a new verb, they initially construct representations that are very concrete and tied to the particular linguistic context in which the verb was first encountered (e.g., Akhtar and Tomasello 1997; Tomasello 1992, 2003). Early syntactic knowledge, then, would consist of large chunks of structure represented as a whole with specific lexical items embedded. As the child gains more experience with very similar constructions (e.g., *more milk* and *more juice*), he or she begins to generalize across them (e.g., *more X*). (Note that this story is not crucial to any version of Construction Grammar per se, but of a particular instantiation that has been explicitly tied to acquisition data.)

Support for this position comes from evidence that children initially have difficulty extending new verbs (in comprehension and production) to sentences that differ from the sentence in which the verb was introduced. In production, children under 3 years of age are quite conservative in their production of verbs, and they tend to use verbs only in those structures in which they have heard them before (e.g., Tomasello 1992; Lieven et al. 1997). Akhtar and Tomasello (1997) find 2- and 3-year-old children to be conservative in certain comprehension tasks as well; they incorrectly act out instructions with novel verbs they have not heard in that particular structure before.

Although children may be conservative syntactically, this could be because they have not yet discerned the verb's precise meaning or because they are stymied by some of the idiosyncrasies they have observed in syntax–semantics links (Fisher 2002b; Naigles 2002). (Or it could be that the production tasks and act-out comprehension tasks do not reveal the full extent of children's competence, as they require sophisticated overt behavioral responses—a point we return to below.) The simplest explanation for the child pattern seems to be that behavior is conservative as children gather data about the events described by new verbs and try to determine which parts of an event a verb picks out. Soon, children become comfortable enough to use verbs in syntactic structures they have not heard them in before, as long as these structures are compatible with the child's understanding of the relationship between the verb and the event. As they see and hear more verb–event pairings, they refine their understanding of how the verb relates to the event, and they also learn idiosyncrasies about which structures verbs appear in (e.g., when to use the same phonological form for a causative variant, like "break," and when not to, like "rise"; cf. "raise").

While these data are compatible with the Construction Grammar explanation Tomasello and colleagues espouse, both Lexical Projectionist and Compositional Non-projectionist explanations are also possible. On a Lexical Projectionist account, conservative behavior could be due to the child having a particular base argument structure associated with a new verb, and not yet having certain evidence or inclination to apply lexical rules to allow the verb in other structures. On a Compositional Non-projectionist account, conservative behavior could be due to uncertainty as to whether there are constraints on the verb's distribution. Being aware that constraints on distribution exist, but unsure as to what they are for this particular verb, the child may initially rely solely on evidence from the exposure he or she has had with the verb thus far. Alternatively, she could have a limited or incorrect event representation of the event associated with the verb (Fisher 2002b; Naigles 2002); that leaves the child unable to determine which structures the verb is compatible with.

These explanations, too, offer a suggestion for why measures involving lower task demands, such as eye gaze, document earlier abilities to generalize verbs to new structures; the "uncertainty" invoked in these explanations may be overcome in the face of evidence that the verb appears in another structure as well as the availability of candidate events that are compatible with the verb in this other structure. However, another possible explanation for better performance of eye gaze tasks is that children are making a "best guess" on the basis of an incomplete representation for the verb. We return to this issue below.

12.4 Directions

The above discussion has shown that there is a disconnect between what we are able to demonstrate from acquisition data and what we would *like* to demonstrate in order to shed light on the Lexical Projectionism/Non-projectionism debate. Nevertheless, we think that there are some promising avenues for future research.

Recall the lists in (5) and (6) of the learning tasks presented to the child by Projectionist and Non-projectionist architectures. If acquisition data is to shed light on how our linguistic knowledge is organized and represented, then we must identify conditions under which these lists can be distinguished. Perhaps the most promising place to look for such conditions is in acquiring constraints on distribution. In a Lexical Projectionist architecture, constraints on distribution are learned as part of acquiring lexical rules and determining whether a particular lexical rule can apply to a particular verb. This requires sufficient experience with a particular verb, either to have explicit evidence that the lexical rule applies (i.e., hearing the verb in both frames of an alternation) or to know its semantic properties well enough to know in which narrow-range semantic class it belongs. In a Non-projectionist architecture, on the other hand, constraints on distribution are not encoded in rules, but rather result primarily from fit between the lexico-semantic contribution of the primitive and that of the structure, as well as independent storage of more idiosyncratic gaps (e.g., that *donate* does not alternate). In a Non-projectionist architecture, then, we may expect more flexibility or abstractness in being able to generalize newly acquired verbs to new frames.

But while this may be a reasonable prediction based on the properties of the architecture of a Compositional Non-projectionist grammar, some proponents of Construction Grammar take the opposite approach. Specifically, in order to bypass some of the difficult learning problems associated with acquiring primitives and structures (both of which are constructions in a Construction Grammar approach), the learner is hypothesized to begin with concrete lexically specific representations that do not require the child to have grasped abstract syntax–semantics relationships at all (Tomasello 2003). Thus, Construction Grammar approaches differ distinctly from Projectionist and Compositional Non-projectionist approaches in predicting that initially learners have little to no abstract knowledge on encountering a new verb and that only with experience do these concrete representations become more abstract. We therefore can make predictions about how much generalization we expect children to do early on when they have little experience with a new word and therefore little knowledge about its distribution: we would predict that a Construction Grammar approach would allow the least generalizability, followed by a Lexical Projectionist approach, which does assume abstract knowledge, although experience might be required in many cases before children use it, followed by a Compositional Non-projectionist approach, which in principle at least should allow children to "try out" new frames they haven't yet heard a word in.

In order to test hypotheses about whether a learner's initial representation of a new word allows it to be generalized beyond the input to other structures, child language researchers have used an important methodological tool: introducing novel or nonsense words with which children have no prior experience and thus no existing lexical representation. This technique is crucial for understanding the processes children use to create new representations, and the information that those representations contain.

Within the general enterprise of teaching novel words, many studies have looked specifically at how easily young children generalize newly learned verbs to new frames or use their knowledge of the relationships between frames and meanings to identify the meanings of unfamiliar verbs. This rapidly growing body of work is looking for—and finding—abstractness in children's early verb representations (e.g., Arunachalam and Waxman 2010; Bencini and Valian 2008; Bunger 2006; Conwell and Demuth 2007; Fisher 1996, 2002a; Fernandes et al. 2006; Lidz et al. 2003; Naigles 1990; Naigles and Kako 1993; Thothathiri and Snedeker 2008; Viau 2007; Yuan and Fisher 2009). If children's verb representations are indeed abstract from the very beginning, and on the basis of relatively sparse exposure to the verb, then this is evidence that they may not need significant experience to identify a verb's narrow semantic class and thus what lexical rules apply, as predicted by Lexical Projectionist accounts, nor do they need to begin with concrete, unanalyzed representations, as predicted by Construction Grammar accounts (Naigles 2002; Fisher 2002b).

For example, several studies have built on Naigles's classic (1990) study (see Sect. 12.3.1.2) in which 2-year-olds used the sentential context in which a verb appeared to isolate the correct component of a scene. However, this study is limited in its insight into the question of abstractness. Toddlers heard a sentence describing a subpart of the complex scene in front of them, and they had to isolate the correct component. They were thus faced with observable evidence about the novel verb's meaning as they were hearing the syntactic information. Yuan and Fisher (2009) developed a clever paradigm to determine whether toddlers can extract abstract information about verb meaning from syntactic information *alone*, and use these abstract syntax–semantics correspondences to identify a visual referent afterward. This study and several following (e.g., Arunachalam 2013; Arunachalam and Waxman 2010; Arunachalam et al. 2013; Dautriche et al. in press; Gertner and Fisher 2012; Scott and

Fisher 2009; Yuan et al. 2012) present novel verbs in informative syntactic contexts *before* presenting the child with the visual information from the events. In order to succeed in such a task, toddlers are likely extracting an abstract verb representation from the sentence alone (e.g., X acts on Y, causing Y to do something) and later filling in the details (e.g., X pushes Y, causing Y to bend at the knees).

This line of work reveals remarkably sophisticated abilities, at young ages, to use syntactic information to make broad inferences about an unfamiliar verb's meaning, and as such, are evidence that toddlers do not require extensive experience with a novel verb and its referent eventuality in order to determine its argument structure. But still, such findings are controversial; as we have seen, others have found that children's behavior with newly learned verbs is conservative and does not evince abstractness (e.g., Abbot-Smith et al. 2001; Akhtar 1999; Akhtar and Tomasello 1997; see Tomasello 2000, 2003 for discussion). This difference is largely due to methodological differences. Studies finding abstract knowledge have largely used production tasks, in which the child was required to produce a novel verb in a structure in which it had not been modeled, or act-out comprehension tasks, in which the child was required to carry out the action described by a sentence. Those finding evidence of abstract representations, on the other hand, have used preferential looking (Golinkoff et al. 1987; Hirsh-Pasek and Golinkoff 1996) or forced-choice pointing tasks (e.g., Arunachalam and Waxman 2010) to study comprehension, in which children need not execute such complex behaviors to demonstrate their knowledge.

Detractors of these latter studies rightly argue that at least some evidence of abstractness, especially in preferential looking tasks, is due to incomplete or partial semantic representations, or that children may be simply looking to the best referent among the available choices (e.g., Chang et al. 2006; Tomasello 2000). This is a valid criticism. After all, 2- and 3-year-olds will preferentially look at a red airplane if asked to find a strawberry in a visual display containing a red airplane and a yellow airplane (Johnson and Huettig 2011; Johnson et al. 2011). Surely, toddlers do not think that the airplane is a true match for the heard word; rather, their looking behavior reflects their abilities to make a partial match between the features of the depicted referents and those of the named concept.

This presents a strong call to those of us seeking evidence of abstract knowledge in children's early linguistic representations. We must devise tests that are more rigorous but that still have low enough task demands and appropriate pragmatic contexts to encourage children to display abstract knowledge. Tests of incremental sentence processing are well suited for this task. If children can be shown to predict upcoming referents in such a way that they reflect verb general knowledge, then it is unlikely that they are doing so on the basis of partial representations or task strategies, as the processing must be done online. Research using event-related potentials (ERPs) to study children's moment-to-moment sentence processing, too, is likely to shed light on these issues. ERPs may in some cases reveal deeper integration than eye tracking. ERP research has shown that toddlers as young as 14 months show an N400 response, which is generally construed to reflect semantic processing (e.g., Friederici et al. 1993; Holcomb and Neville 1991; Kutas and Hillyard 1984), when they hear a word that does not match the picture they are viewing (e.g., hearing "shoe" while looking at a picture of a ball; Friederich and Friederici 2004, 2005a). Further, by 19 months, they show N400 effects when hearing sentences with semantic anomalies (e.g., The cat drinks the ball; Friederich and Friederici 2005b). By 6 years of age, they show P600 effects—a marker of syntactic processing—to syntactic violations (Hahne et al. 2004).

This technique might be useful to assess whether the representations of newly learned verbs encode their arguments. For example, a novel verb describing a transfer event that has been introduced in a prepositional dative frame might elicit a P600 and/or N400 if it is subsequently heard in a double object frame—if children cannot generalize the verb to this other structure. This would be evidence against abstraction and, potentially, evidence for a lexicalized Construction Grammar framework. In support of this general research direction is a recent study by Ye et al. (2007) finding that "construction-based semantic violations" elicit an N400 effect in adults, traditionally associated with semantic and not syntactic violations, and they construed this as evidence for a Construction Grammar approach because the constructions themselves carry meaning. (Note that these particular N400 effects could also be structure-based violations framed within a Compositional Non-projectionist approach, as the only critical component is that structures themselves are associated with meanings.)

12.5 Conclusions

The goal of this chapter has been (a) to clarify some of the distinctions between argument structure frameworks as they relate to the acquisition question, (b) to demonstrate the limitations of recent attempts to link acquisition data to particular argument structure approaches, and (c) to review promising data revealing abstract knowledge in children's early verb representations and suggest some further avenues for research in this area.

Although Lexical Projectionist and Nonlexical Non-projectionist approaches each apparently pose different learning tasks to children, the acquisition literature we reviewed does not convincingly support one of these approaches to the exclusion of the other. Even evidence relating to Semantic and Syntactic Bootstrapping, each of which was developed with reference to one of these two frameworks, fails to discriminate the two argument structure approaches.

A major difficulty in connecting the theoretical and acquisition literatures is that evidence in the theoretical linguistics literature generally comes from phenomena that are difficult to assess in young children, such as adjectival passives, resultatives, and nominalization. One lesson to extract is that instead of looking for acquisition data to confirm or disconfirm a particular approach, we should take linguistic theory seriously as a basis for interpreting children's behavior (see, e.g., Marantz 2005; Poeppel and Embick 2005 for a similar exhortation at the interface between theoretical linguistics and neuroscience).

From the point of view of acquisition, further identifying two kinds of Non-projectionist approaches, which I termed Compositional Non-projectionist and Construction Grammar approaches, may be useful. I described a research program demonstrating abstractness in children's early verb representations and suggested that such abstractness is evidence against strongly lexicalized versions of Construction Grammar, and in support of either Lexical Projectionist or Compositional Non-projectionist approaches. If children's representations are abstract from the very beginning, as these results strongly suggest, the appeal of Construction Grammar approaches is mitigated.

Further, this work suggests a program for future research. To examine the nature of children's representations, a crucial part of the program will continue to be to teach them novel verbs. This offers insight into the processes children use to create new representations and the information that those representations contain. Moreover, rigorously manipulating the information they receive when they are introduced to the novel verb (e.g., only giving them syntactic input) will allow us to determine the basis on which they form these representations. This kind of approach may very well lead us to distinguishing between Lexical Projectionist and Compositional Non-projectionist approaches as well.

References

Abbot-Smith, K., Lieven, E., & Tomasello, M. (2001). What pre-school children do and do not do with ungrammatical word orders. *Cognitive Development, 16,* 1–14.

Akhtar, N. (1999). Acquiring basic word order: Evidence for data-driven learning of syntactic structure. *Journal of Child Language, 26,* 339–356.

Akhtar, N., & Tomasello, M. (1997). Young children's productivity with word order and verb morphology. *Developmental Psychology, 33,* 952–965.

Altmann, G. T. M., & Kamide, Y. (1999). Incremental interpretation at verbs: Restricting the domain of subsequent reference. *Cognition, 73,* 247–264.

Ambridge, B., Pine, J. M., Rowland, C. F., & Young, C. R. (2008). The effect of verb semantic class and verb frequency (entrenchment) on children's and adults' graded judgements of argument-structure overgeneralization errors. *Cognition, 106,* 87–129.

Ambridge, B., Pine, J. M., & Rowland, C. F. (2014). The retreat from locative overgeneralization errors: A novel verb grammaticality judgment study. *PLoS ONE, 9*(5), e97634.

Arunachalam, S. (2013). Two-year-olds can begin to acquire verb meanings in socially impoverished contexts. *Cognition, 129,* 569–573.

Arunachalam, S., & Waxman, S. R. (2010). Meaning from syntax: Evidence from 2-year-olds. *Cognition, 114,* 442–446.

Arunachalam, S., Escovar, E., Hansen, M. A., & Waxman, S. R. (2013). Out of sight, but not out of mind: 21-month-olds use syntactic information to learn verbs even in the absence of a corresponding event. *Language and Cognitive Processes, 28,* 417–425.

Baker, C. L. (1979). Syntactic theory and the projection problem. *Linguistic Inquiry, 10,* 533–581.

Bencini, G. M. L. & Valian, V. (2008). Abstract sentence representation in 3-year-olds: Evidence from comprehension and production. *Journal of Memory and Language, 59,* 97–113.

Bidgood, A., Ambridge, B., Pine, J. M., & Rowland, C. F. (2014). The retreat from locative overgeneralization errors: A novel verb grammaticality judgment study. *PLoS ONE, 9*(5), e97634.

Boland, J. (2005). Visual arguments. *Cognition, 95,* 237–274.

Borer, H. (2004). The grammar machine. In A. Alexiadou, et al. (Eds.), *The unaccusativity puzzle: Explorations of the syntax-lexicon interface*. Oxford: Oxford University Press.

Borer, H. (2005). *Structuring sense II: The normal course of events*. Oxford: Oxford University Press.

Borovsky, A., Elman, J. L., & Fernald, A. (2012). Knowing a lot for one's age: Vocabulary skill and not age is associated with anticipatory incremental sentence interpretation in children and adults. *Journal of Experimental Child Psychology, 112,* 417–436.

Bowerman, M. (1990). Mapping thematic roles onto syntactic functions: Are children helped by innate linking rules? *Linguistics, 28,* 1253–1289.

Braine, M. D. S., & Brooks, P. J. (1995). Verb argument structure and the problem of an overgeneral grammar. In M. Tomasello & W. E. Merriman (Eds.), *Beyond names for things: Children's acquisition of verbs* (pp. 353–376). Hillsdale: Lawrence Erlbaum.

Bunger, A. (2006). How we learn to talk about events: Linguistic and conceptual constraints on verb learning. Doctoral dissertation, Northwestern University.

Chang, F., Dell, G. S., & Bock, K. (2006). Becoming syntactic. *Psychological Review, 113,* 234–272.

Chomsky, N. (1981). *Lectures on government and binding*. Dordrecht: Foris.

Conwell, E., & Demuth, K. (2007). Early syntactic productivity: Evidence from dative shift. *Cognition, 103,* 163 179.

Dautriche, I., Cristia, A., Brusini, P., Yuan, S., Fisher, C., & Christophe, A. (In press). Toddlers default to canonical surface-to-meaning mapping when learning verbs. *Child Development*. doi:10.1111/cdev.12164.

Embick, D. (2004). On the structure of resultative participles in English. *Linguistics Inquiry, 35,* 355–392.

Embick, D., & Noyer, R. (2006). Distributed morphology and the syntax/morphology interface. In G. Ramchand & C. Reiss (Eds.), *The Oxford handbook of linguistic interfaces*. Oxford: Oxford University Press.

Fernald, A. (2004). The search for the object begins at the verb. Presented at the 29th Annual Boston University Conference on Language Development. Boston, Nov. 4–7.

Fernald, A., & Hurtado, N. (2006). Names in frames: Infants interpret words in sentence frames faster than words in isolation. *Developmental Science, 9,* F33–F40.

Fernandes, K. J., Marcus, G. F., Di Nubila, J. A., Vouloumanos, A. (2006). From semantics to syntax and back again: Argument structure in the third year of life. *Cognition, 100,* B10–B20.

Fisher, C. (1996). Structural limits on verb mapping: The role of analogy in children's interpretations of sentences. *Cognitive Psychology, 31*(1), 41–81

Fisher, C. (2002a). Structural limits on verb mapping: The role of abstract structure in 2.5-year-olds' interpretations of novel verbs. *Developmental Science, 5,* 56–65.

Fisher, C. (2002b). The role of abstract syntactic knowledge in language acquisition: A reply to Tomasello (2000). *Cognition, 83,* B13–B23.

Fisher, C., Hall, D. G., Rakowitz, S., & Gleitman, L. R. (1994). When it is better to receive than to give: Syntactic and conceptual constraints on vocabulary growth. *Lingua, 92,* 333–375.

Friederich, M., & Friederici, A. D. (2004). N400-like semantic incongruity effect in 19-month-olds: Processing known words in picture contexts. *Journal of Cognitive Neuroscience, 16,* 1465–1477.

Friederich, M., & Friederici, A. D. (2005a). Phonotactic knowledge and lexical-semantic processing in one-year-olds: Brain responses to words and nonsense words in picture contexts. *Journal of Cognitive Neuroscience, 17,* 1785–1802.

Friederich, M., & Friederici, A. D. (2005b). Semantic sentence processing reflected in the event-related potentials of one- and two-year-old children. *Neuroreport, 16,* 1801–1804.

Friederici, A. D., Pfeifer, E., & Hahne, A. (1993). Event-related brain potentials during natural speech processing: Effects of semantic, morphological and syntactic violations. *Cognitive Brain Research, 1,* 183–192.

Gleitman, L. (1990). Structural sources of verb meanings. *Language Acquisition, 1,* 1–63.

Goldberg, A. E. (1995). *Constructions: A construction grammar approach to argument structure*. Chicago: University of Chicago Press.
Golinkoff, R. M., Hirsh-Pasek, K., Cauley, K. M., & Gordon, L. (1987). The eyes have it: Lexical and syntactic comprehension in a new paradigm. *Journal of Child Language, 14,* 23–45.
Golinkoff, R. M., Jacquet, R., Hirsh-Pasek, K., & Nandakumar, R. (1996). Lexical principles may underlie the learning of verbs. *Child Development, 67,* 3101–3119.
Grimshaw, J. (1981). Form, function, and the language acquisition device. In C. L. Baker & J. J. McCarthy (Eds.), *The logical problem of language acquisition*. Cambridge: MIT Press.
Gropen, J., Pinker, S., Hollander, M., & Goldberg, R. (1991). Affectedness and direct objects: The role of lexical semantics in the acquisition of argument structure. *Cognition, 41,* 153–195.
Hahne, A., Eckstein, K., & Friederici, A. D. (2004). Signatures of syntactic and semantic processes during children's language development. *Journal of Cognitive Neuroscience, 16,* 1302–1318.
Hale, K., & Keyser, S. J. (1993). On argument structure and the lexical expression of syntactic relations. In K. Hale & S. J. Keyser (Eds.), *The view from building 20* (pp. 111–176). Cambridge: MIT Press.
Hirsh-Pasek, K., & Golinkoff, R. M. (1996). *The origins of grammar: Evidence from early language comprehension*. Cambridge: MIT Press.
Holcomb, P. J., & Neville, H. J. (1991). Natural speech processing: An analysis using event-related brain potentials. *Psychobiology, 19*(4), 286–300.
Jackendoff, R. (1997). *The architecture of the language faculty*. Cambridge: MIT Press.
Johnson, E. K., & Huettig, F. (2011). Eye movements during language-mediated visual search reveal a strong link between overt visual attention and lexical processing in 36-month-olds. *Psychological Research, 75,* 35–42.
Johnson, E. K., McQueen, J. M., & Huettig, F. (2011). Toddlers' language-mediated visual search: They need not have the words for it. *The Quarterly Journal of Experimental Psychology, 64*(9), 1672–1682.
Kako, E. T. (1998). The event semantics of syntactic structures. Doctoral dissertation, University of Pennsylvania.
Kako, E. T., & Wagner, L. (2001). The semantics of syntactic structures. *Trends in Cognitive Sciences, 5,* 102–108.
Kim, A., Srinivas, B., & Trueswell, J. C. (2002). The convergence of lexicalist perspectives in psycholinguistics and computational linguistics. In P. Merlo & S. Stevenson (Eds.), *Sentence processing and the lexicon: Formal, computational, and experimental perspectives*. Philadelphia: John Benjamins.
Kratzer, A. (1996). Severing the external argument from its verb. In J. Rooryck & L. Zaring (Eds.), *Phrase structure and the lexicon*. Dordrecht: Kluwer.
Kutas, M., & Hillyard, S. A. (1984). Brain potentials during reading reflect word expectancy and semantic association. *Nature, 307,* 161–163.
Landau, B., & Gleitman, L. (1985). *Language and experience*. Cambridge: Harvard University Press.
Levin, B., & Rappaport Hovav, M. (1995). *Unaccusativity: At the syntax-lexical semantics interface*. Cambridge: MIT Press.
Lieven, E. V. M., Pine, J. M., & Baldwin, G. (1997). Lexically-based learning and grammatical development. *Journal of Child Language, 24,* 187–219.
Lidz, J., Gleitman, H., & Gleitman, L. (2003). Understanding how input matters: The footprint of Universal Grammar on verb learning. *Cognition, 87,* 151–178.
Lidz, J., Gleitman, H., & Gleitman, L. (2004). Kidz in the 'hood: Syntactic bootstrapping and the mental lexicon. In D. G. Hall & S. R. Waxman (Eds.), *Weaving a lexicon*. Cambridge: MIT Press.
MacDonald, M. C., Pearlmutter, N. J., & Seidenberg, M. S. (1994). The lexical nature of syntactic ambiguity resolution. *Psychological Review, 101,* 676–703.
Marantz, A. (1984). *On the nature of grammatical relations*. Cambridge: MIT Press.

Marantz, A. (1997). No escape from syntax: Don't try morphological analysis in the privacy of your own lexicon. In A. Dimiatridis et al. (Eds.), *Proceedings of the 21st Annual Penn Linguistics Colloquium*. Philadelphia: Penn Working Papers in Linguistics.
Marantz, A. (2005). Generative linguistics within the cognitive neuroscience of language. *The Linguistic Review, 22,* 429–445.
Naigles, L. (1990). Children use syntax to learn verb meanings. *Journal of Child Language, 17,* 357–374.
Naigles, L. (2002). Form is easy, meaning is hard: Resolving a paradox in early child language. *Cognition, 86,* 157–199.
Naigles, L., & Kako, E. (1993). First contact in verb acquisition: Defining a role for syntax. *Child Development, 64*(6), 1665–1687.
Naigles, L., Gleitman, H., & Gleitman, L. (1993). Syntactic bootstrapping and verb acquisition. In E. Dromi (Ed.), *Language and cognition: A developmental perspective*. Norwood: Ablex.
Pesetsky, D. (1982). Paths and categories. Doctoral dissertation, MIT Press. Cambridge, MA.
Phillips, C. (2006). Three benchmarks for distributional models of syntax. In R. Zanuttini, H. Campos, E. Herburger, & P. Portner (Eds.), *Negation, tense, and clausal architecture: Cross-linguistic investigations*. Washington, DC: Georgetown University Press.
Pinker, S. (1984). *Language learnability and language development*. Cambridge: Harvard University Press.
Pinker, S. (1989). *Learnability and cognition*. Cambridge: MIT Press.
Poeppel, D., & Embick, D. (2005). Defining the relation between linguistics and neuroscience. In A. Cutler (Ed.), *Twenty-first psycholinguistics: Four cornerstones*. Hillsdale: Erlbaum.
Pustejovsky, J. (1995). *The generative lexicon*. Cambridge: MIT Press.
Ramchand, G. (1997). *Aspect and predication: The semantics of argument structure*. Oxford: Clarendon.
Ramchand, G. (2010). *Verb meaning and the lexicon*. Cambridge: Cambridge University Press.
Rappaport Hovav, M., & Levin, B. (1998). Building verb meanings. In M. Butt & W. Geuder (Eds.), *The projection of arguments: Lexical and compositional factors*. Stanford: CSLI.
Snedeker, J., & Trueswell, J. (2004). The developing constraints on parsing decisions: The role of lexical-biases and referential scenes in child and adult sentence processing. *Cognitive Psychology, 49,* 239–299.
Theakston, A. L. (2004). The role of entrenchment in children's and adults' performance on grammaticality judgment tasks. *Cognitive Development, 19*(1), 15–34.
Thothathiri, M., & Snedeker, J. (2008). Syntactic priming during language comprehension in three- and four-year-olds. *Journal of Memory and Language, 58,* 188–213.
Tomasello, M. (1992). *First verbs: A case study of early grammatical development*. Cambridge: Cambridge University Press.
Tomasello, M. (2000). Do young children have adult syntactic competence? *Cognition, 74,* 209–253.
Tomasello, M. (2003). *Constructing a language: A usage-based theory of language acquisition*. Cambridge: Harvard University Press.
Trueswell, J. C., & Gleitman, L. R. (2004). Children's eye movements during listening: Evidence for a constraint-based theory of parsing and word learning. In J. M. Henderson & F. Ferreira (Eds.), *Interface of language, vision, and action: Eye movements in the visual world*. New York: Psychology.
Trueswell, J. C., & Tanenhaus, M. K. (1994). Toward a lexicalist framework of constraint-based syntactic ambiguity resolution. In C. Clifton & L. Frazier (Eds.), *Perspectives on sentence processing*. Hillsdale: Erlbaum.
Trueswell, J. C., Tanenhaus, M. K., & Kello, C. (1993). Verb-specific constraints in sentence processing: Separating effects of lexical preference from garden-paths. *Journal of Experimental Psychology: Learning, Memory, and Cognition, 19,* 528–553.
Trueswell, J. C., Tanenhaus, M. K., & Garnsey, S. M. (1994). Semantic influences on parsing: Use of thematic role information in syntactic ambiguity resolution. *Journal of Memory and Language, 33,* 285–318.

Trueswell, J. C., Sekerina, I., Hill, N. M., & Logrip, M. L. (1999). The kindergarten-path effect: Studying on-line sentence processing in young children. *Cognition, 73,* 89–134.

Viau, J. (2007). Possession and spatial motion in the acquisition of ditransitives. Doctoral dissertation, Northwestern University.

Williams, A. (2005). Complex causatives and verbal valence. Doctoral dissertation, University of Pennsylvania.

Wonnacott, E., & Newport, E. L. (2005). Novelty and regularization: The effect of novel instances on rule formation. In A. Brugos, M. R. Clark-Cotton, & S. Ha (Eds.), *Proceedings of the 29th Annual Boston University Conference on Language Development* (Vol. 29, pp. 663–673). Somerville: Cascadilla.

Wonnacott, E., Newport, E. L., & Tanenhaus, M. K. (2008). Acquiring and processing verb argument structure: Distributional learning in a miniature language. *Cognitive Psychology, 56,* 165–209.

Ye, Z., Zhan, W., & Zhou, X. (2007). The semantic processing of syntactic structure in sentence comprehension: An ERP study. *Brain Research, 1142,* 135–145.

Yuan, S., & Fisher, C. (2009). "Really? She blicked the baby?": Two-year-olds learn combinatorial facts about verbs by listening. *Psychological Science, 20,* 619–626.

Yuan, S., Fisher, C., & Snedeker, J. (2012). Counting the nouns: Simple structural cues to verb meaning. *Child Development, 83,* 1382–1399. doi:10.1111/j.1467-8624.2012.01783.x.

Chapter 13
The Beginning of Morphological Learning: Evidence from Verb Morpheme Processing in Preverbal Infants

Alexandra Marquis and Rushen Shi

Abbreviations

1SG	First person singular
2PL	Second person plural
2SG	Second person singular
3SG	Third person singular
AUX	Auxiliary verb
CVC	Consonant–vowel–consonant
CCVC	Consonant–consonant–vowel–consonant
CCVCV	Consonant–consonant–vowel–consonant–vowel
IMP	Imperative
INF	Infinitive
PP	Past participle
s	Seconds

13.1 Introduction

Verbs in many languages are believed to be harder to acquire than nouns. This is in part because in these languages, verbs bear a high load of information via morphological affixes entailing tense, aspect, number, etc. Affixes are bound morphemes

A. Marquis (✉)
École d'orthophonie et d'audiologie, Faculté de médecine, Université de Montréal,
C.P. 6128, Succursale Centre-Ville, Montréal, QC H3C 3J7, Canada
e-mail: alexandra.marquis@umontreal.ca

R. Shi
Département de Psychologie, Faculté des Sciences Humaines,
Université du Québec à Montréal, C.P. 8888, Succursale Centre-Ville,
Montréal, QC H3C 2E9, Canada
e-mail: shi.rushen@uqam.ca

R. G. de Almeida, C. Manouilidou (eds.), *Cognitive Science Perspectives on Verb Representation and Processing,* DOI 10.1007/978-3-319-10112-5_13

that never surface alone without a root or stem. Young children who have not yet acquired their language must thus cope with a multitude of forms for the same verb (e.g., *call, calls, called, calling*). Moreover, morphological affixations often modify the syllable structures of verbs (e.g., the sound /l/ is the coda in the monosyllabic root *call* /kʰɔːl/ but becomes the onset of the second syllable in *calling* /kʰɔː.lɪŋ/), rendering the recognition of stems and affixes more difficult. Two competing views may be considered regarding verb representations. Decomposed roots and suffixes (e.g., *call, -s, -ed, -ing*) may be stored as separate pieces. Stored pieces can be later linked together at the level of morphological paradigms. Or, in a nondecomposition view, variable forms of a verb may be encoded in the lexicon as single units (e.g., *call, calls, called, calling*) without segmentation of the suffixes. The difference of these two views has important implications for the understanding of initial morphological development in infants. The nondecomposition view is consistent with the assumption that infants may not be capable of parsing affixes and stems, nor does the grammar require them to do the parsing. On the contrary, the decomposition view predicts that infants can segment stems and affixes. Furthermore, on this view, a verb stem in the unaffixed bare form and in affixed forms may be treated as variants of the same unit despite considerable differences due to phonological operations.

Obviously, the full knowledge of morphological paradigms involves the acquisition of meaning for stems and affixes. A crucial question concerns the initial state of morphological learning: do infants rely on semantics to break into the learning of morphological paradigms? That is, do infants first have to learn word meaning in order to understand the relatedness of bare/unaffixed and affixed forms of a verb? Or do they instead begin learning morphological relatedness of word forms by distributional analyses without relying on semantics? We suggest that the latter is more plausible, especially for verb learning, given the evidence that infants have more difficulty learning verb meaning at the early stage of acquisition (e.g., Gentner 1982). Although infants' very first encoded pieces (e.g., stems and suffixes, or nonsegmented affixed verbs, depending on the theory) may be achieved through successful analyses of the input without semantics in both decomposition and nondecomposition theories, different predictions can be made about subsequent morphological learning.

In this chapter, we argue for the decomposition view for infants' morphological development. In particular, we suggest that infants at the initial learning stage parse verb stems and affixes without relying on semantics, but on the basis of high token frequency of affixes and high type frequency of stems (i.e., regular morphological operations). The encoding of affixes is likely, given the crosslinguistic evidence that preverbal infants from 6 months of age begin to recognize highly frequent function words such as determiners (e.g., Hallé et al. 2008; Höhle and Weissenborn 2003; Shafer et al. 1998; Shi et al. 2006a, c; Shi and Lepage 2008), long before they understand the meaning of these words. In addition, we suggest that infants can perceive a stem in different morphological contexts as variants of the same abstract form. Moreover, this knowledge can bootstrap infants into the learning of verb meaning, in the sense that the learning system expects the bare stem and segmented stems from affixed contexts to have the same core meaning, as it would when acquiring the meaning of a nonaffixed word from different utterance locations. Thus, in our view, infants' earliest stem-affix segmentation already represents rudimentary morphological knowledge, which may facilitate the subsequent learning of verb meaning, which in turn may lead to the generalization and abstraction of full morphological paradigms. This position contrasts widely with theories that require infants to first learn verb meaning in order to learn the morphological relations shared among unaffixed and various affixed forms of a verb. The nondecomposition account, for example, would need semantics for infants to understand the link between the wholly encoded unit *calling* and the unaffixed form *call*.

There is evidence from perceptual studies that by 6 months of age, infants already begin to recognize highly frequent word forms without knowing their meaning (e.g., Bortfeld et al. 2005; Shi et al. 2006b). Previous research showed infants' capacities to perform statistical analysis of speech (e.g., Johnson and Jusczyk 2001; Saffran et al. 1996). In these studies, infants were capable of segmenting word-like units based on the statistical distribution of syllables of a novel language. What remains to be tested is whether infants can perform such distributional analysis on subsyllabic units that corresponds to bound morphemes such as suffixes. In our research, we seek to answer this question, and further, to determine whether the outcome of suffix parsing represents rudimentary understanding of morphology.

The question of morphological parsing is tied to the question of embedded words. In the case of morphologically varying forms such as *call–calling* /kʰɔ:l/-/kʰɔ:.lɪŋ/, the first part of *calling* (i.e., *call-*) and the bare form *call* should be interpreted as the same abstract unit. Nonmorphological words involving embedded forms cannot be interpreted in the same way. For example, *sir–circle* share no internal relation. Natural languages must require mechanisms that balance between these different linguistic functions: relatedness for morphosyntactic needs and unrelatedness for increasing vocabulary contrasts (e.g., adding phonemes or a syllable to create new words, *sir–circle*, *man–mandate*). We suggest that frequency is the mechanism that makes this balance work: high token frequency of the suffix co-occurring with high type frequency of the stem leads to stem-suffix segmentation. For example, the *-ing* suffix in English occurs highly frequently with a large number of different stems (e.g., *calling, walking, eating, drinking, dozing, leaving,* etc.), allowing *-ing* and stems to be segmented. Furthermore, infants operate with a bias that allows them to expect the segmented stems (e.g., *call-* from *calling, called*) and the bare stem (e.g., *call*) of a verb to be the same unit and to later receive the same meaning. This bias is governed by frequency conditions and is inhibited in the case of *sir–circle* since the frequency requirement is not met.

Existing research on infants' interpretation of embedded words provide some indirect supporting evidence. Jusczyk and colleagues (Jusczyk et al. 1999) found that English-acquiring infants did not accept *ham* as a variant of *hamlet,* suggesting that they encoded the disyllabic noun as a whole, and perceived the words *ham* and *hamlet* as unrelated. Although infants at 8 months of age sometimes mis-segment embedded forms such as *tar* in *guitar* due to prosodic factors, by 10–11 months they can overcome the weak prosody of *gui-*, perceive the distributional integrity of the two syllables, and treat the disyllabic word *guitar* and the monosyllabic word *tar* as different forms. Yet, perception of embedded forms in morphological cases presents a different story. A study on verb segmentation (Mintz 2013) showed that English-learning infants seem to understand the link between a novel bare stem and the stem plus the present progressive *-ing* (e.g., *lérjov* and *lérjoving*). However, because infants in Mintz' study (2013) were familiarized with suffixed forms and tested with the bare stem versus a novel word, results can also be interpreted as indicating that infants only mapped the form of the test word with the beginning of the familiarized words without necessarily processing the forms in a morphological fashion. A study in which infants are directly tested on their stem representation is thus needed. Such a study will need to present a suffixed form and test infants on both possible parses (i.e., stem vs. partial stem). In this chapter, we present our recent empirical data that demonstrate that infants not only associate suffixed and unsuffixed verb forms but they also encode detailed representations for word-internal morphemes in a way that reflects rudimentary knowledge of morphological alternations.

13.2 The Beginning of Verb Parsing: Experiment 1

We used French, a language with substantial verb morphology, as the test case for examining early morphological learning. Before investigating infants' processing of verb form alternations, we first needed to determine the age at which they begin recognizing unaffixed verb forms in sentential contexts. One verb segmentation study with infants (Nazzi et al. 2005) has been previously published, showing that English-learning infants can segment bisyllabic verbs (e.g., *discount, permit*) starting from 13.5 months of age. We hypothesized that infants may segment monosyllabic verbs at a younger age. We thus tested 8- and 11-month-old French-learning[1] infants (16 for each age group) using a visual preferential procedure (Cooper and Aslin 1990), a procedure that has been used in previous infant segmentation studies (e.g., Curtin et al. 2005; Shi and Lepage 2008). In this Experiment (Marquis and Shi 2008), infants were familiarized with trials presenting repeatedly a consonant–vowel–consonant (CVC) unaffixed verb form, either /bif/ or /tar/, until they reached 30 s of total looking time while listening to the target. Following the familiarization phase, infants were tested with trials of sentences containing the bare verb form /bif/ versus trials of sentences containing the bare verb form /tar/. Note that such bare forms appear in most present tense conjugations, except for first and second person plural, and they appear in the singular imperative. We expected that if infants recognized the target verb, they should discriminate trials containing the target and those not containing the target during the test phase.

During the Experiment, the infant sat on his or her parent's lap in front of a monitor and a loudspeaker in an acoustic chamber. The parent heard masking music through headphones. An observer in the adjacent room observed the infant and

[1] Note that all infants in this chapter were exposed to French for a minimum of 70 % of their awake time.

pressed down a computer key whenever the infant looked towards the monitor. A computer program (Cohen et al. 2000) presented audiovisual stimuli, calculated the infant's looking times online, and controlled the progression of the Experiment from the familiarization to the test phases. Each trial was initiated by the infant's look towards the monitor. Between trials, a flashing light appeared on the monitor to attract the infant's attention.

Infants were randomly assigned to either /bif/ or /tar/ familiarization condition. The first test trial presented either the /bif/ or /tar/ sentences, counterbalanced across infants, and the two types of test trials were presented in alternation for a total of ten test trials (see Appendix 1). The target verbs /bif/ "to cross out" and /tar/ "to weigh" are infrequent in spoken French (Beauchemin et al. 1992); thus, probably unknown to young children. Their low frequency will allow us to assess not only infants' ability to segment unaffixed verb roots but also their ability to segment novel verbs. Stimuli were recorded by a native Quebec-French female speaker in infant-directed speech style.

Our prediction was that if infants were capable of recognizing the unaffixed verb roots, they should pay greater attention to the sentences containing the familiarized target than to those containing the other novel verb during the test phase. We expected a looking preference for the target sentences given that previous published infant segmentation studies using natural speech stimuli typically showed looking preferences for targets (e.g. Jusczyk and Aslin 1995; Jusczyk et al. 1999; Polka and Sundara 2003; Shi and Lepage 2008). We also predicted that the 11-month-olds, having had more experience with their native language, should perform better than the 8-month-olds in segmenting unaffixed verbs from continuous speech.

Results confirmed our predictions. Average looking time per trial for each trial type during the test phase was calculated, and the data were analyzed in a 2 × 2 mixed analysis of variance (ANOVA), with sentence type (sentences containing the familiarized target verb vs. sentences containing the nonfamiliarized verb) as the within-subject factor, and age (8 vs. 11 months) as the between-subject factor. We obtained a sentence type × age interaction, $F(1, 30) = 6.41, p = 0.017$, and no main effect of sentence type nor age, indicating that the two age groups differed. In follow-up t-tests, the group of 11-month-olds looked significantly longer during target trials than during nontarget trials, paired $t(15) = 2.284, p = 0.037$ (see Fig. 13.1), revealing that the 11-month-olds recognized the target verbs. In contrast, the 8-month-olds listened equivalently to both types of test trials, paired $t(15) = 1.147, p = 0.269$, thus showing no evidence of recognizing unaffixed verb forms in sentences. All t-tests reported in this chapter are two-tailed.

These results (Marquis and Shi 2008) show that by 11 months of age, infants are capable of segmenting unaffixed verb roots from continuous speech. Moreover, the fact that they were capable of segmenting these infrequent verbs suggests that our assumption is correct about semantics being unnecessary for infants' early speech processing. The next step was to test whether infants are capable of segmenting verb stems from suffixed forms and associating suffixed verb forms with their corresponding unsuffixed stem.

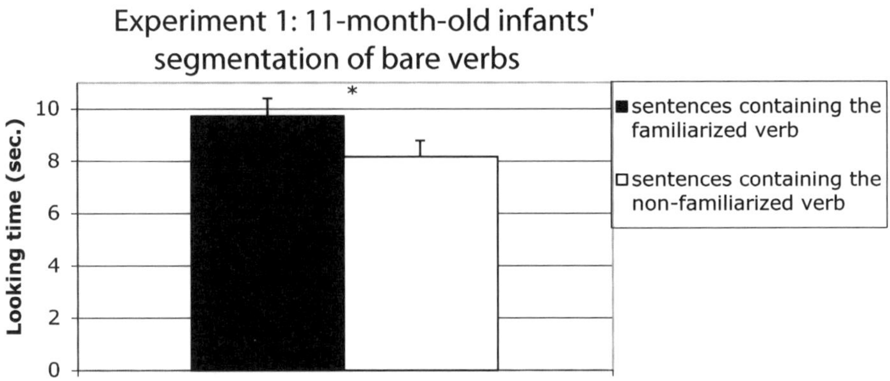

Fig. 13.1 Eleven-month-old infants' orientation times (means with standard errors of the means) to the test sentences containing the familiarized target verb and nonfamiliarized verb (Experiment 1)

13.3 Infants' Parsing of Verb Stems and Suffix: Experiment 2

We established in Experiment 1 that French-learning infants are capable of segmenting unaffixed verb stems from continuous speech stream as early as 11 months of age. We then set out to examine whether infants at this age can segment affixed verbs into smaller morphemic pieces, stem and suffix, as predicted in the decomposition theory. In Experiment 2 (see Marquis and Shi 2012), French-learning 11-month-old infants were familiarized to a nonce consonant–consonant–vowel–consonant (CCVC) form, either /glYt/ (phonetically realized as /glYt/) or /trid/ (phonetically realized as /trId/), until they reached 30 s of total looking time. Following familiarization, infants were tested with the same procedure as in Experiment 1. One group of 16 infants were tested with condition 1: sentences containing /gly.te/ versus sentences containing /tri.de/ (i.e., the CCVC forms conjugated with a French verb morpheme, the suffix /e/, which is the form of an infinitive, a past participle, or the second person plural present/imperative forms; see Appendix 2). These morphemes (i.e., the infinitive, the past participle and second person plural present/imperative) are homophonous, such as the infinitive *manger* /mɑ̃.ʒe/ "to eat," the past participle *mangé* /mɑ̃.ʒe/ "eaten," and second person plural present/imperative *mangez* /mɑ̃.ʒe/ "eat/youPL eat." The other group of 16 infants was tested with condition 2: sentences containing /gly.tu/ versus sentences containing /tri.du/, both of which are disyllabic monomorphemic nonsense words (see Appendix 3). This /u/ vowel exists in the French language and appears in word final position but is not used as a verb morpheme, thus we could use this nonsense suffix as a control to directly test infants' morphological knowledge of /e/.

Nonsense verbs were used in our Experiments for two reasons: (1) to ensure that all forms were novel to infants so that we could reliably assess their generalized ability to perform morphological parsing when encountering unknown verbs; (2) the first vowel of these forms is subject to vowel alternations, /glYt/–/gly.te/–/gly.tu/,

/trɪd/–/tri.de/–/tri.du/; according to Quebec-French phonological rules, tense high vowels are laxed when they appear in closed syllables (e.g., Walker 1984); we were therefore able to test whether infants have the ability to segment underlying stems and the suffix /e/ despite such surface changes, an ability beyond simply mapping the bare root form of the familiarization phase with part of the suffixed forms of the test phase. The same native French speaker as in Experiment 1 recorded the new stimuli.

Infants were randomly assigned to either /glYt/ or /trɪd/ familiarization condition. After familiarization, the first group of infants was tested with /gly.te/ sentences versus /tri.de/ sentences in alternating trials, and the second group was tested with /gly.tu/ sentences versus /tri.du/ sentences, for a total of 14 trials for each test group. The counterbalancing followed the same steps as in Experiment 1.

We predicted that during the test phase, if the first group succeeded in processing the vowel alternation rule in a morphological fashion, that is, recognizing the frequent /e/ verb suffix, segmenting the underlying form of the stem and matching it with the familiarization target, they should then prefer to listen to the test sentences containing the suffixed target verb. On the other hand, a failure in this Experiment could support the nondecomposition view according to which bare stem forms and affixed forms are separately stored as whole units without affix segmentation.

Furthermore, if infants are truly able to segment word-internal stems and suffixes, the group of infants tested with sentences containing /gly.tu/ versus those containing /tri.du/ monomorphemic forms (i.e., the targets ending with the nonmorpheme /u/) should demonstrate no preference. That is, the /gly.tu/ and /tri.du/ forms present in the test sentences should be treated as unrelated to the familiarized /glYt/ and /trɪd/ forms. This interpretation would apply only if the first group (the /gly.te/–/tri.de/ test condition) showed a significant looking preference for the targets.

Infants' average looking time for each trial type during the test phase was calculated, and the data were analyzed in a 2 × 2 mixed ANOVA with sentence type (sentences containing the familiarized target root versus sentences containing the nonfamiliarized root) as the within-subject factor and condition (condition 1 real verb suffix /e/ vs. condition 2 nonmorphemic /u/) as the between-subject factor. This comparison yielded a significant interaction of sentence type × condition, $F(1, 30) = 4.836$, $p = 0.036$, while no other main effect was obtained (see Fig. 13.2). As predicted, infants tested with the targets ending with the real suffix /e/ looked significantly longer while listening to the sentences containing the affixed form of the familiarized root in comparison to the sentences containing the novel suffixed verb, paired $t(15) = 3.113$, $p = 0.007$. This could be interpreted as evidence for the decomposition of the targets as /glyt + e/ and /trid + e/. Importantly, for the group tested with the targets ending with the nonce suffix /u/, looking times for the sentences containing the familiarized root and those containing the nonfamiliarized root were not significantly different, paired $t(15) = 0.945$, $p = 0.359$, demonstrating that infants in condition 2 did not decompose the dissyllabic forms into /glyt + u/ and /trid + u/. Had infants just been parsing any initial overlapping sound sequences in this Experiment without attending to the morphological status of word endings, a preference for target sentences obtained with the first group of infants should have also been found with the second group.

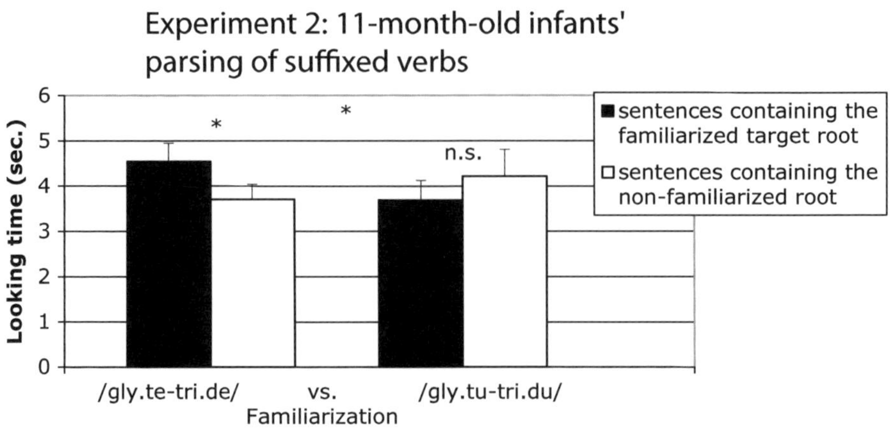

Fig. 13.2 Eleven-month-old infants' orientation times (means with standard errors of the means) to the test sentences containing the familiarized target verb and nonfamiliarized verb (Experiment 2). The two *left* columns are the real suffix /e/ condition 1, and the two *right* columns are the nonsense suffix /u/ condition 2

Taken together, these results (Marquis and Shi 2012) favor the decomposition view of early morphological learning. The novel verbs were parsed only in cases involving a real suffix, as /glYt/, /trid/, and /e/. The monomorphemic forms /gly.tu/ and /tri.du/, which have a phonologically comparable structure as /gly.te/ and /tri.de/, were not parsed into smaller units. Infants in condition 2 treated the word forms /glYt/ and /gly.tu/ (and /trid/, /tri.du/) as unrelated lexical items.

13.4 Rudimentary Knowledge of Morphological Paradigms: Experiments 3 and 4

The empirical data from Experiment 2 suggest the decomposition of stems and affixes by French-learning infants at 11 months of age. One question remains concerning whether our results unambiguously indicate the knowledge of a morphological paradigm, i.e., whether infants represent /glYt/–/gly.te/ as alternations of the same abstract verb. It is possible that infants in Experiment 2 segmented the suffixed verb, but only represented the initial phonemes of the stem with phonetic details, leaving the final phoneme or phonemes unspecified. This interpretation is possible because the test comparison trials involved a completely novel suffixed verb bearing no phonological resemblance to the familiarized stem (i.e., /tri.de/ after /glYt/, or /gly.te/ after /trid/). The same results could have been obtained if infants only recognized the frequent suffix /e/ and the initial part of the affixed verbs. It would then be uncertain whether infants can fully represent verb form alternations. We therefore decided to conduct a more direct test of the knowledge of morphological paradigms.

In the new Experiments, we familiarized infants with a suffixed nonsense verb /gla.te/ and tested them with the stem form /glat/ versus a partial stem /gla/. This design forced infants to make a decision about which form during the test phase was related to the familiarized form /gla.te/. If infants had rudimentary knowledge about the verb paradigm involving the highly frequent and regular /e/ suffix, they should then segment /e/ and recognize /glat/ as the variant related to /gla.te/. The partial stem /gla/ should be regarded as having no relation with /gla.te/ although the two forms overlap at the initial portion. A control condition was designed, which included a nonsense form /gla.tu/ as the familiarization form (note again that /u/ is not a suffix in French), and also /glat/ versus /gla/ trials during the test phase. In this control condition, we predicted that infants should perceive /gla.tu/ as a disyllabic monomorphemic word and regard the two forms presented in the test phase as equally unrelated to /gla.tu/. This interpretation of null results would hold only if the results of the Experimental condition involving /gla.te/ turned out to be the pattern that we predicted. Thus, for the current Experiment, we created new nonwords, different from those used in Experiment 2. The vowel alternation was removed here (/a/ does not alternate). That is, the vowel of the target and test words remained constant. Because detailed knowledge of verb paradigms is more advanced than suffix parsing, we deliberately avoided vowel changes so as to better determine whether infants understood verb alternations when both forms in the test phase have exact partial phonetic overlap with the target form in the familiarization phase. The same speaker recorded the new stimuli as in the previous Experiments.

In Experiment 3, 11-month-old Quebec-French-learning infants (16 per condition) were familiarized with a nonsense word, either /gla.te/ (affixed) or /gla.tu/ (monomorphemic), during 33 s. All infants were tested with /glat/ versus /gla/ trial types. Multiple exemplars of each word were presented. The first test trial was either /glat/ or /gla/, counterbalanced across infants. If at this age, infants understand morphological alternations, then a looking preference for the stem form /glat/ should be observed only for the infants familiarized with the suffixed form /gla.te/. For the infants familiarized with the monomorphemic form /gla.tu/ (ending with the nonsuffix /u/), two outcomes were possible: (1) no preference for either test words (see the discussion of the interpretation of this control condition in the above paragraph) or (2) a preference for /gla/, suggesting a syllabic bias for parsing that was unrelated to morphological processing.

Infants' looking times during the test phase were analyzed in a 2 $\times$ 2 mixed ANOVA, with parse type (/glat/ vs. /gla/) as the within-subject factor and familiarization (/gla.te/ vs. /gla.tu/) as the between-subject factor. Results revealed a significant interaction of parse type $\times$ familiarization, $F(1, 30) = 4.194$, $p = 0.049$, while no other main effect was obtained. For the group of infants familiarized with the real French suffix /e/ (i.e., /gla.te/), even though their looking times appear to be longer for the /glat/ stem trial ($M = 7.438$ s, SE = 1.242 s) than for the /gla/ partial stem trial ($M = 5.225$ s, SE = 0.650 s), this difference did not reach significance, paired $t(15) = 1.522$, $p = 0.149$. As for the group of infants familiarized with the nonce suffix /u/, looking times for the /glat/ ($M = 5.113$ s, SE = 1.128 s) and the syllabic /gla/ trial types ($M = 6.663$ s, SE = 1.070 s) were also not significant, paired $t(15) = 1.379$,

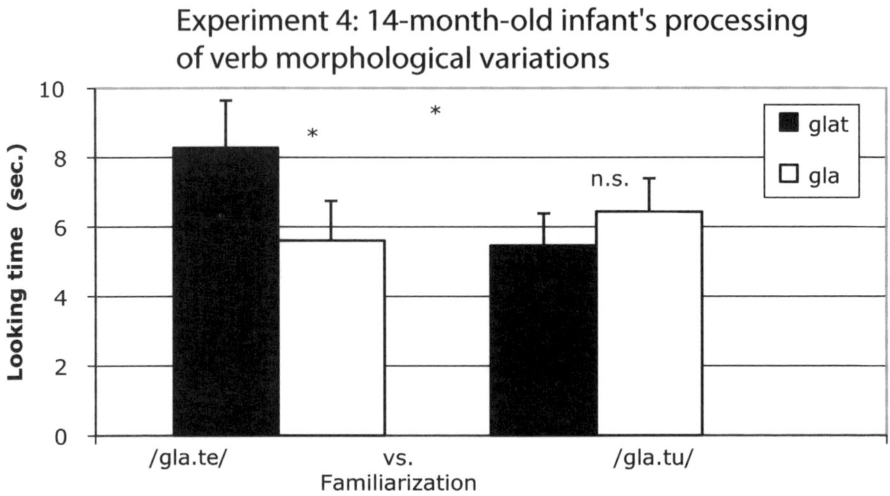

Fig. 13.3 Fourteen-month-old infants' orientation times (means with standard errors of the means) to /glat/ and /gla/ (Experiment 4). The *left* two columns are the results for the group familiarized with the verb affixed with the real suffix /e/, and the *right* two columns are the results for the group familiarized with the disyllabic monomorphemic form

$p = 0.188$. Therefore, there was no clear evidence that the 11-month-old infants in this Experiment stored affixed verb forms as decomposed stem and affixes, although they begin to show a tendency towards this direction. We thus pursued our investigation with 14-month-olds, an older group of French-learning infants.

In Experiment 4 (Shi and Marquis 2009), 14-month-old infants (16 per condition) were familiarized and tested in exactly the same way as those in Experiment 3. It was predicted that by 14 months of age infants should demonstrate more robust morphological knowledge. The looking responses during the test phase of the /gla.te/ (affixed) and the /gla.tu/ (nonaffixed) familiarization groups were analyzed in the same 2×2 mixed ANOVA as in Experiment 3, with parse type (/glat/ vs. /gla/) as the within-subject factor and familiarization (/gla.te/ vs. /gla.tu/) as the between-subject factor. Results revealed a significant interaction of parse type $\times$ familiarization, $F(1, 30) = 6.297$, $p = 0.018$, while no other main effect was obtained (see Fig. 13.3). As predicted, the group of infants familiarized with the affixed form /gla.te/ looked significantly longer while listening to the /glat/ stem trial than to the /gla/ partial stem trial, paired $t(15) = 2.724$, $p = 0.016$. For the group of infants familiarized with the unaffixed /gla.tu/ form, looking times for /gla/ and /glat/ were not significantly different, paired $t(15) = 0.913$, $p = 0.376$. Results are shown in Fig. 13.3.

Results of Experiment 4 (Shi and Marquis 2009) demonstrate that by 14 months of age, French-learning infants understand morphological alternations. They can link forms that are related in highly regular verb paradigms. These findings follow the decomposition theory of morphological learning that we argue for. The high token frequency of the suffix /e/ co-occurring with the high type frequency of verb stems that infants have encountered during their prior linguistic experience should have enabled them to parse the stem /glat/ and /e/ suffix, and to interpret the form /glat/ as being associated with the affixed form /gla.te/. Infants who were familiarized with the unaffixed /gla.tu/ responded quite differently. They correctly treated /gla.tu/, /glat/, and /gla/ as separate, unrelated lexical items.

13.5 General Discussion

This chapter concerns the most fundamental aspects of verb learning. We attempt to understand how infants might resolve the problems of segmentation and morphological variations of verb forms. At the beginning stage of language learning, infants must determine the possible lexical units of their language from continuous speech, that is, to find the units that will allow them to build a vocabulary. Segmentation of even the most stable word forms, those that do not go through morphological changes, is itself a challenging task for infants, as the acoustic and phonetic realizations of words naturally vary due to many factors such as sentential position, coarticulation, phonological operations, focus, speech rate, affect, speaker differences, etc. The story is particularly complex for languages that involve a high degree of morphological variations. Inflectional morphology, for example, may pose a major challenge for early verb learning. Infants not only have to segment verb forms, but must also learn that certain variable forms are associated in verb paradigms (e.g., *call, calls, calling, called*).

In the first section of this chapter, we discussed two theoretical positions for verb representations, the nondecomposition theory and the decomposition theory. According to the nondecomposition theory, morphological alternations of the same verb (e.g., *call, calls, calling, called*) are encoded in the lexicon as nonseparable single units. Therefore, the implication for acquisition is that infants would most likely need semantic evidence to learn that these forms are related in a verb paradigm. We believe that the reliance on semantics for early verb learning is unlikely since there is evidence that verb meanings are harder to induce than noun meanings (e.g., Gillette et al. 1999). Instead, we propose a theory of decomposition without semantics for initial verb acquisition. Specifically, we suggest that verbs that contain internal regular morphology (such as *calling*) are decomposed into stems and suffixes at the earliest stage of learning. The mechanism of this learning involves the frequency of stem and suffix distributions: high token frequency of the suffixes co-occurring with high type frequency of the stems leads to word internal decomposition. We also suggest that infants should interpret the segmented stem (from affixed forms) and the bare stem of the same verb as variants of the same abstract unit, and that the learning system would subsequently expect these variants to receive the same core meaning. Moreover, we assume continuity of the decomposed morphological representations from infancy to adulthood. That is, the decomposed units are represented in the lexicon during initial learning by infants and remain so in the mature lexicon.

The empirical evidence that we discussed in this chapter supports the decomposition theory. In our Experiments, we examined the knowledge of verb morphology in infants at the age when they hardly have any meaningful vocabulary. We showed that infants who acquire French, a language with considerable verb morphological variations, are capable of parsing stems from affixed forms and associate such forms with the bare stem. They are able to do so on the basis of distributional analyses, without relying on verb meaning. Infants' responses in our Experiments reflected their prior learning. Thus, when hearing a novel verb ending with the regular verb suffix /e/ during our Experiments, infants parsed them into the stem and suffix because they had months of accumulated exposure to this highly frequent suffix co-occurring with many different stems. In other words, the different stems that infants had encountered in their natural environment were of high type frequency, which lowered the transitional probability between any stem and the suffix, therefore supporting the decomposition of affixed verbs into smaller morphological units. Furthermore, we obtained direct evidence supporting the knowledge of the French verb paradigm involving /e/ in 14-month-old infants. After hearing a suffixed novel verb form and then being given the choices of a related bare stem versus an unrelated but partially matching word form, infants correctly linked the bare stem and the affixed form of the verb. Overall, the combined results from our perceptual Experiments provide robust evidence for the decomposition account of infants' initial learning of verb form alternations.

In addition to evidence from perceptual learning, children's speech production also supports the decomposition theory. Infants' early speech is known to be telegraphic, lacking functional morphemes (e.g., Brown 1973). Infants' spontaneous speech typically omits inflections in cases where the usage is obligatory. In controlled speech production Experiments, toddlers only dropped real grammatical suffixes but not nonsense suffixes (e.g., Gerken et al. 1990), suggesting that they had separate representations of the stem and suffix, but they had a whole representation for disyllabic monomorphemic forms. These observations with older infants are consistent with the findings from our perceptual Experiments with preverbal and early verbal infants, indicating that children decompose stems and suffixes from the beginning of verb learning. Furthermore, omission patterns in production show that children associate stems from various contexts with the same word meaning. Additional evidence for the decomposition theory can be seen in children's overgeneralization productions. For example, English children often produce incorrect forms such as *finded* as the past tense of *find* (e.g., Pinker 1995), indicating that regular suffixes are separately represented and serve as the default paradigms for English children. Forms of this kind are not produced by adults, and must therefore be the result of a rule-generating process in children. Similarly, French children often wrongly conjugate irregular verbs with regular endings as in **je l'ai batté* instead of *je l'ai battu* "I've beaten him/her" (e.g., Royle 2007). These errors also indicate that verbs are decomposed into stem and affixes.

Decomposition appears to be a basic element for most morphological theories in linguistics and psycholinguistics. Many existing theories focus on adult representations, and therefore are not concerned with the question of decomposition versus nondecomposition of inflected words during acquisition. Nevertheless, they usually contain inflection rules, at least for regular verbs, and the rules would presumably operate with some kind of representations corresponding to suffixes and stems. This would be the case for the morphological framework under generative phonology (Chomsky and Halle 1968), which posits rules for regular as well as irregular verbs. Stems and inflectional morphemes would logically be the representational units for the rule constructs, knowledge that is most likely acquired during childhood. Unlike models under generative phonology, Bybee's theory (1985, 1988) considers word frequency as a crucial factor that influences adult lexical representations. Low-frequency words with regular inflections are derived in the lexicon, but high-frequency words with regular inflections are stored as a whole. Logically, the high-frequency regulars were once infrequent at the early stage of acquisition. Thus, they may have been represented in decomposed units before the frequency of these affixed words reached the high-frequency threshold. But regardless of whether Bybee's theory would regard these regulars as decomposed or not at the initial stage of acquisition, decomposition must have occurred at some stage that enabled the learning of morphological rules, which her theory uses for the derivation of low-frequency words with regular inflections. The rule-rote model by Pinker (e.g., Pinker 1991; Pinker and Prince 1994) can apply directly to early morphological acquisition: irregular verbs are rote memorized, and nondecomposed, in the lexicon, whereas regular verbs are generated by a rule process that must involve decomposed suffixes and stems. Our work explains an acquisition stage prior to the knowledge in Pinker's model, we propose a learning mechanism by which infants establish these representational units and reach a basic understanding of morphological alternations.

In sum, our goal here is to develop a model of early morphological learning. We approach this acquisition problem by studying infants' initial segmentation of verb forms and their interpretation of verb morphological alternations. Our empirical findings demonstrate that infants begin to parse verbs into decomposed stems and suffixes by 11 months of age, and they have rudimentary knowledge of regular verb paradigms by 14 months of age. We further show that this learning is entirely based on distributional analyses of the input without the need for semantics. The fact that infants in our studies treated morphologically alternating forms of a verb as variants of the same verb suggests that they should then expect these forms to have the same core meaning, an outcome which we predict for future Experiments that specifically test infants' interpretation of word meaning. The remarkable morphological knowledge that we demonstrated in young infants constitutes an essential part of the early grammatical representations, which may directly impact the subsequent acquisition of semantics as well as more refined morphosyntactic structures.

Acknowledgment This research was supported by grants from NSERC, SSHRC, FQRSC, and CFI to the second author, and by an FQRSC scholarship to the first author. We wish to thank Andréane Melançon, Elena Kulagina, Bruno Gauthier, and Mélanie Lepage for research assistance.

13.6 Appendix 1. Experiment 1 Stimuli

Familiarization: either /bif/ or /tar/

Test: *La jolie maman* /bif/ *les poèmes.*
"The pretty mommy cross out—3SG the poems."
Elle /bif/ *les verbes.*
"She cross out—3SG the verbs."
Évidemment on /bif/ *la virgule.*
"Obviously we cross out—3SG the comma."
C'est le juron qu'il /bif/.
"It's the curse that he cross out—3SG."
Le petit bébé /bif/ *le graffiti.*
"The little baby cross out—3SG the graffiti."
La syllabe je /bif/.
"The syllable I cross out—1SG."
/bif/-*tu le calembour?*
"Do you cross out—2SG the pun?"
and
La nouvelle maman /tar/ *le magot.*
"The new mommy weigh—3SG the pile."
Il /tar/ *les grappes.*
"He weigh—3SG the clusters."
Visiblement je /tar/ *la parcelle.*
"Visibly I weigh—1SG the fragment."
C'est le lot qu'elle /tar/.
"It's the share that she weigh—3SG."
Le gentil bébé /tar/ *les gondoles.*
"The nice baby weigh—3SG the gondolas."
Les rouets je /tar/.
"The spinning wheels I weigh—1SG."
/tar/-*tu la coquille?*
"Do you weigh—2SG the shell?"

13.7 Appendix 2. Experiment 2 Condition 1 Stimuli

Familiarization: either /glYt/ or /trid/

Test: *Maman a* /gly.te/ *le magot.*
"Mommy AUX /gly.te/$_{\text{NONCE.PP}}$ the pile."
/gly.te/ *les grappes est amusant.*
"/gly.te/$_{\text{NONCE.INF}}$ the grapes is amusing."
T'as /gly.te/ *la parcelle.*
"You AUX /gly.te /$_{\text{NONCE.PP}}$ the parcel."

C'est le lot qu'on a /gly.te/.
"It's the lot that we AUX /gly.te/$_{\text{NONCE.PP}}$."
Bébé va /gly.te/ *les gondoles.*
"Baby AUX /gly.te/$_{\text{NONCE.INF}}$ the gondolas."
Les roues, j'ai /gly.te/.
"The wheels, I AUX /gly.te/$_{\text{NONCE.PP}}$."
/gly.te/ *la coquille!*
"/gly.te/$_{\text{NONCE.2PL.IMP}}$ the shell!"
and
Maman a /tri.de/ *les poèmes.*
"Mommy AUX /tri.de/$_{\text{NONCE.PP}}$ the poems."
/tri.de/ *les verbes est amusant.*
"/tri.de/$_{\text{NONCE.INF}}$ the verbs is amusing."
T'as /tri.de/ *la virgule.*
"You AUX /tri.de/$_{\text{NONCE.PP}}$ the comma."
C'est le jeu qu'on a /tri.de/.
"It's the game that we AUX /tri.de/$_{\text{NONCE.PP}}$."
Bébé va /tri.de/ *le graphème.*
"Baby AUX /tri.de/$_{\text{NONCE.INF}}$ the grapheme."
La phrase, j'ai /tri.de/.
"The sentence I AUX /tri.de/$_{\text{NONCE.PP}}$."
/tri.de/ *le juron!*
"/tri.de/$_{\text{NONCE.2PL.IMP}}$ the curse word!"

13.8 Appendix 3. Experiment 2 Condition 2 Stimuli

Familiarization: either /glYt/ or /trid/

Test: **Maman a* /gly.tu/ *le magot.*
"*Mommy AUX /gly.tu/$_{\text{NONCE.CCVCV}}$ the pile."
*/gly.tu/ *les grappes est amusant.*
"*/gly.tu/$_{\text{NONCE.CCVCV}}$ the grapes is amusing."
**T'as* /gly.tu/ *la parcelle.*
"*You AUX /gly.tu/$_{\text{NONCE.CCVCV}}$ the parcel."
**C'est le lot qu'on a* /gly.tu/.
"*It's the lot that we AUX /gly.tu/$_{\text{NONCE.CCVCV}}$."
**Bébé va* /gly.tu/ *les gondoles.*
"*Baby AUX /gly.tu/$_{\text{NONCE.CCVCV}}$ the gondolas."
**Les roues, j'ai* /gly.tu/.
"*The wheels, I AUX /gly.tu/$_{\text{NONCE.CCVCV}}$."
*/gly.tu/ *la coquille!*
"*/gly.tu/$_{\text{NONCE.CCVCV}}$ the shell!"
and
**Maman a* /tri.du/ *les poèmes.*

"*Mommy AUX /tri.du/$_{\text{NONCE.CCVCV}}$ the poems."
*/tri.du/ *les verbes est amusant.*
"*/tri.du/$_{\text{NONCE.CCVCV}}$ the verbs is amusing."
**T'as* /tri.du/ *la virgule.*
"*You AUX /tri.du/$_{\text{NONCE.CCVCV}}$ the comma."
**C'est le jeu qu'on a* /tri.du/.
"*It's the game that we AUX /tri.du/$_{\text{NONCE.CCVCV}}$."
**Bébé va* /tri.du/ *le graphème.*
"*Baby AUX /tri.du/$_{\text{NONCE.CCVCV}}$ the grapheme."
**La phrase, j'ai* /tri.du/.
"*The sentence, I AUX /tri.du/$_{\text{NONCE.CCVCV}}$."
*/tri.du/ *le juron!*
"*/tri.du/$_{\text{NONCE.CCVCV}}$ the curse word!"

References

Beauchemin, N., Martel, P., & Théoret, M. (1992). *Dictionnaire de fréquence des mots du Francais parlé au Québec: Fréquence, dispersion, usage, écart réduit (Dictionary of word frequency in Québec spoken French: Frequency, scattering, usage, reduced deviation).* New York: P. Lang.

Bortfeld, H., Morgan, J. L., Golinkoff, R. M., & Rathbun, K. (2005). Mommy and me: Familiar names help launch babies into speech stream segmentation. *Psychological Science, 16,* 298–304.

Brown, R. (1973). *A first language: The early stages.* Cambridge: Harvard University Press.

Bybee, J. L. (1985). *Morphology: A study of the relation between meaning and form.* Philadelphia: John Benjamins.

Bybee, J. L. (1988). Morphology as lexical organization. In M. Hammond & M. Noonan (Eds.), *Theoretical morphology* (pp. 119–141). San Diego: Academic.

Chomsky, N., & Halle, M. (1968). *The sound pattern of English.* New York: Harper and Row.

Cohen, L. B., Atkinson, D. J., & Chaput, H. H. (2000). *Habit 2000: A new program for testing infant perception and cognition,* (Version 2.2.5c) [Computer software]. Austin: University of Texas.

Cooper, R. P., & Aslin, R. N. (1990). Preference for infant-directed speech in the first month after birth. *Child Development, 61*(5), 1584–1595.

Curtin, S., Mintz, T. H., & Christiansen, M. H. (2005). Stress changes the representational landscape: Evidence from word segmentation. *Cognition, 96,* 233–262.

Gentner, D. (1982). Why nouns are learned before verbs: Linguistic relativity versus natural partitioning. In S. Kuczaj II (Ed.), *Language development, Volume 2: Language, thought and culture,* (pp. 301–334). Hillsdale: Lawrence Erlbaum.

Gerken, L., Landau, B., & Remez, R. E. (1990). Function morphemes in young children's speech perception and production. *Developmental Psychology, 26*(2), 204–216.

Gillette, J., Gleitman, H., Gleitman, L., & Lederer, A. (1999). Human simulation of vocabulary learning. *Cognition, 73*(2), 135–176.

Hallé, P., Durand, C., & de Boysson-Bardies, B. (2008). Do 11-month-old French infants process articles? *Language and Speech, 51,* 23–44.

Höhle, B., & Weissenborn, J. (2003). German-learning infants' ability to detect unstressed closed class elements in continuous speech. *Developmental Science, 6*(2), 122–127.

Johnson, E. K., & Jusczyk, P. W. (2001). Word segmentation by 8-month-olds: When speech cues count more than statistics. *Journal of Memory and Language, 44*(4), 548–567.

Jusczyk, P. W., & Aslin, R. N. (1995). Infants' detection of the sound patterns of words in fluent speech. *Cognitive Psychology, 29*(1), 1–23.

Jusczyk, P. W., Houston, D. M., & Newsome, M. (1999). The beginnings of word segmentation in English-learning infants. *Cognitive Psychology, 39,* 159–207.
Marquis, A., & Shi, R. (2008). Segmentation of verb forms in preverbal infants. *Journal of the Acoustical Society of America, 123*(4), EL105–EL110. doi:10.1121/1.2884082.
Marquis, A., & Shi, R. (2012). Initial morphological learning in preverbal infants. *Cognition, 122,* 61–66. doi:10.1016/j.cognition.2011.07.004.
Mintz, T. H. (2013). The segmentation of sub-lexical morphemes in English-learning 15-month-olds. *Frontiers in Psychology, 4*(24). doi:10.3389/fpsyg.2013.00024.
Nazzi, T., Dilley, L. C., Jusczyk, A. M., Shattuck-Hunagel, S., & Jusczyk, P. W. (2005). English-learning infants' segmentation of verbs from fluent speech. *Language and Cognitive Processes, 19,* 97–136.
Pinker, S. (1991). Rules of language. *Science, 253*, 530–535.
Pinker, S. (1995). Why the child holded the baby rabbits: A case study in language acquisition. In L. Gleitman & M. Liberman (Eds.), *Invitation to cognitive science: Language* (Vol. 1, 2nd ed, pp. 107–133). Cambridge: MIT Press.
Pinker, S., & Prince, A. (1994). Regular and irregular morphology and the psychological status of rules of grammar. In S. D. Lima, R. L. Corrigan, & G. K. Iverson (Eds.), *The reality of linguistic rules* (pp. 321–352). Philadelphia: John Benjamins doi:10.1075/slcs.26.
Polka, L., & Sundara, M. (2003). Word segmentation in monolingual and bilingual infant learners of English and French. In M. J. Solé, D. Recasens & J. Romero (Eds.), *Proceedings of the 15th International Congress of Phonetic Sciences*, (pp. 1021–1024). Barcelona: Futurgraphic.
Royle, P. (2007). Variable effects of morphology and frequency on inflection patterns in French preschoolers. *The Mental Lexicon Journal, 2*(1), 103–125.
Saffran, J. R., Aslin, R. N., & Newport, E. L. (1996). Statistical learning by 8-month-old infants. *Science, 274,* 1926–1928.
Shafer, V. L., Shucard, D. W., Shucard, J. L., & Gerken, L. (1998). An electrophysiological study of infants' sensitivity to the sound patterns of English speech. *Journal of Speech, Language, and Hearing Research, 41*(4), 874–886.
Shi, R., & Lepage, M. (2008). The effect of functional morphemes on word segmentation in preverbal infants. *Developmental Science, 11*(3), 407–413. doi:10.1111/j.1467-7687.2008.00685.x.
Shi, R., & Marquis, A. (2009). Mechanisms of segmentation and morphological learning in infants. In J. Chandlee, M. Franchini, S. Lord, & G.-M. Rheiner (Eds.), *Proceedings of the 33rd Boston University Conference on Language Development* (pp. 480–491). Somerville: Cascadilla.
Shi, R., Cutler, A., Werker, J., & Cruickshank, M. (2006a). Frequency and form as determinants of functor sensitivity in English-acquiring infants. *Journal of the Acoustical Society of America, 119*(6), EL61–EL67. doi:10.1121/1.2198947.
Shi, R., Marquis, A., & Gauthier, B. (2006b). Segmentation and representation of function words in preverbal French-learning infants. In D. Bamman, T. Magnitskaia, & C. Zaller (Eds.), *Proceedings of the 30th Annual Boston University Conference on Language Development* (pp. 549–560). Somerville: Cascadilla.
Shi, R., Werker, J. F., & Cutler, A. (2006c). Recognition and representation of function words in English-learning infants. *Infancy, 10*(2), 187–198. doi:10.1207/s15327078in1002_5.
Walker, D. C. (1984). *The pronunciation of Canadian French*. Ottawa: University of Ottawa Press.

Author Index

R. G. de Almeida, C. Manouilidou (eds.), *Cognitive Science Perspectives on Verb Representation and Processing*, DOI 10.1007/978-3-319-10112-5

Subject Index

R. G. de Almeida, C. Manouilidou (eds.), *Cognitive Science Perspectives on Verb Representation and Processing,* DOI 10.1007/978-3-319-10112-5

Printed by Printforce, the Netherlands